TM 9-818

10-TON 6 x 4 MACK TRUCK

TECHNICAL MANUAL

BY WAR DEPARTMENT

©2013 Periscope Film LLC
All Rights Reserved
ISBN#978-1-940453-21-7
www.PeriscopeFilm.com

DISCLAIMER:

This manual is sold for historic research purposes only, as an entertainment. It contains obsolete information and is not intended to be used as part of an actual operation or maintenance training program. No book can substitute for proper training by an authorized instructor.

©2013 Periscope Film LLC
All Rights Reserved
ISBN#978-1-940453-21-7
www.PeriscopeFilm.com

TM 9-818
1

PART ONE—VEHICLE OPERATING INSTRUCTIONS

Section I

INTRODUCTION

	Paragraph
Scope	1

1. SCOPE.

   a. This technical manual* is published for the information and guidance of the using arms personnel charged with the operation, maintenance, and minor repair of this materiel.

   b. In addition to a description of the 10-ton 6 x 4 Truck (Mack), this manual contains technical information required for the identification, use and care of the materiel. The manual is divided into two parts. Part One, section I through section VII, contains vehicle operating instructions. Part Two, section VIII through section XXXV contains vehicle maintenance instructions for using arms personnel charged with the responsibility of doing maintenance work within their jurisdiction. Part Two is followed by a list of references and an index.

   c. In all cases where the nature of the repair, modifications or adjustment is beyond the scope or facilities of the unit, the responsible ordnance service should be informed, so that trained personnel with suitable tools and equipment may be provided, or proper instructions issued.

   d. This manual contains operating and organization maintenance instructions from the following Quartermaster Corps 10-series technical manuals, and, together with TM 9-1818A and TM 9-1818B, it supersedes them:

   (1) TM 10-1197, 19 September 1941.

   (2) TM 10-1421, 19 September 1941.

   (3) TM 10-1545, December 1942.

---

*To provide operating instructions with the materiel, this technical manual has been published in advance of complete technical review. Any errors or omissions will be corrected by changes or, if extensive, by an early revision.

TM 9-818
2

10-TON 6 x 4 TRUCK (MACK MODEL NR)

Figure 1—Left Front View of 10-ton, 6 x 4 Truck

Section II

DESCRIPTION AND TABULATED DATA

	Paragraph
Description	2
Data	3

2. DESCRIPTION.

a. The cargo truck described in this manual consists of a six-wheel, Diesel-powered chassis, with drive through the four rear wheels, a cab having a convertible type of canvas top, and a wooden cargo body with tail gate, tarpaulin and bows, and troop seats (figs. 1, 2, 3 and 4).

b. Chassis. Units of the chassis are in conventional automotive arrangement, in assembly with the customary frame structure, having two channel members as fundamentals.

(1) POWER PLANT (fig. 5). The vehicle is powered by a six-cylinder, four-cycle, valve-in-head, Diesel engine (figs. 6 and 7) which is three-point mounted at the front of the chassis. Fuel injection is by an American Bosch pump, a multiple-unit type, and the built-in drive has incorporated an automatic-advance coupling known as a Synchrovance. Cooling is by a frontal-type radiator having a vertical tube-and-fin core. The transmission (fig. 8) is mounted on the rear end of the engine, and is driven through a single-disk friction clutch. The transmission has ten speeds forward and two reverse—that is, five fast ratios and five slow ratios forward. Fifth of the fast group is an overdrive, and fifth of the slow group is direct drive. There are two shift levers—a five-position and a two-position lever.

(2) DRIVE. A propeller shaft transmits the power back to the two rear axles. There is a power divider, not a free differential, in the front rear axle from which the in-line drive to the rear axle is through a short propeller shaft (fig. 10). The front axle is not a driving axle.

(3) AXLES. The front axle (fig. 9) is of the non-driving type of conventional, automotive, I-beam design with steering knuckles. Steering control is by the steering wheel through the steering gear and linkage and the front wheels. Rear axles (fig. 10) are of the full-floating type with flat-banjo housings, which have spindle tubes and center housings unified as one piece.

(4) SUSPENSION. At the front, springs are semielliptic with the axle clamped at the center. There are hydraulic shock absorbers at the front (fig. 11). The rear axles are attached through rubber shock insulators to the ends of two longitudinal, beam-type springs which are mounted at their centers on trunnions, on which they can pivot. Axle torque is taken by two tubular rods with ball end connections, which are attached to the tops of the gear carrier housings, and are

10-TON 6 x 4 TRUCK (MACK MODEL NR)

Figure 2 — Right Rear View of 10-ton, 6 x 4 Truck

TM 9-818
2

DESCRIPTION AND TABULATED DATA

Figure 3 — Front View of 10-ton, 6 x 4 Truck

TM 9-818
2

10-TON 6 x 4 TRUCK (MACK MODEL NR)

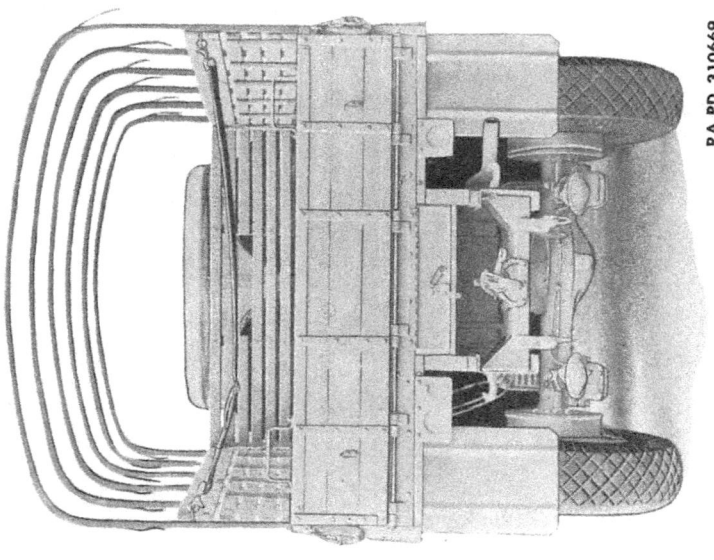

Figure 4 — Rear View of 10-ton, 6 x 4 Truck

DESCRIPTION AND TABULATED DATA

TM 9-818
2

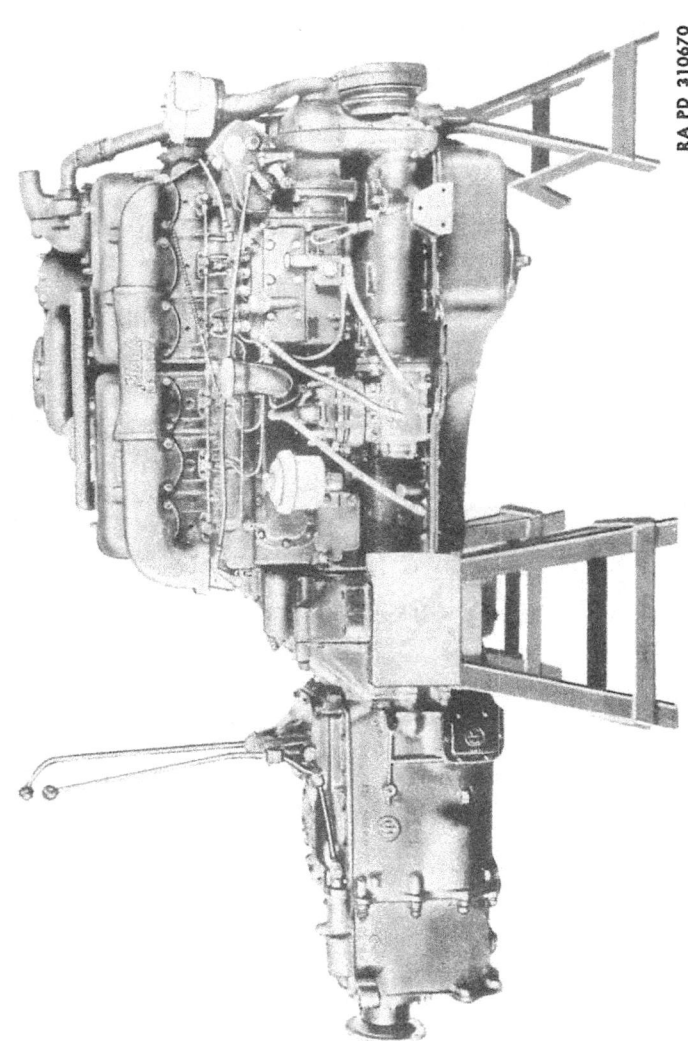

Figure 5 — Right Side of Complete Power Plant

10-TON 6 x 4 TRUCK (MACK MODEL NR)

anchored on the frame cross-member located midway between the rear axles (fig. 13). The two rear axles have single tires, and these track with the front tires.

(5) BRAKES.

(a) The service brakes act on all six wheels (figs. 14 and 15). These are actuated by air through individual brake chambers

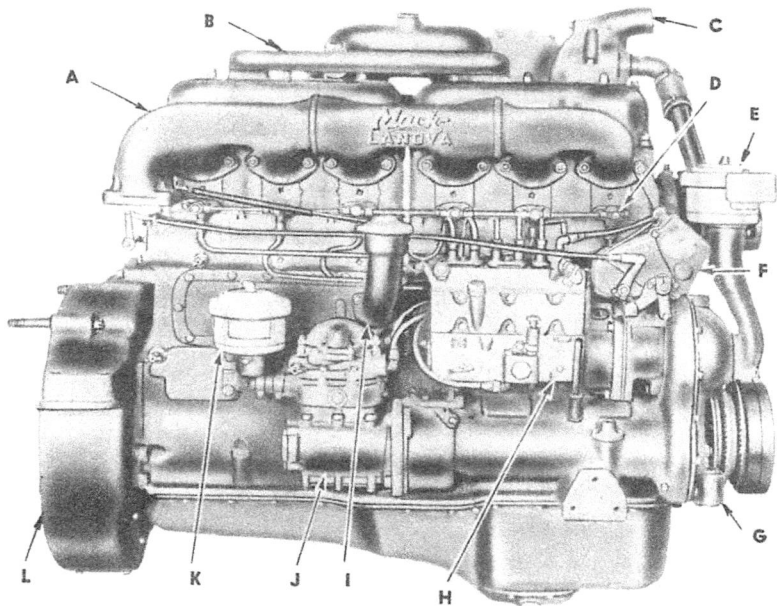

A	EXHAUST MANIFOLD	H	FUEL INJECTION PUMP
B	BREATHER CONNECTION	I	CRANKCASE FILLER
C	WATER OUTLET	J	AIR COMPRESSOR
D	INJECTION NOZZLE	K	COMPRESSOR AIR CLEANER
E	BREATHER AIR CLEANER	L	ENGINE REAR SUPPORT
F	GOVERNOR		(CLUTCH HOUSING)
G	ENGINE FRONT SUPPORT		

RA PD 310502

Figure 6—Right Side of Engine

mounted at the wheels. The air compressor is flange-mounted on the engine with enclosed gear drive, and it is lubricated and water-cooled from the engine.

(b) The parking brake, operated by a hand lever in the cab through linkage, is a disk brake having four shoes. It is on the propeller shaft approximately midway between the transmission and the

TM 9-818

DESCRIPTION AND TABULATED DATA

front rear axle, and is carried by a frame cross-member. The parking brake is entirely independent of the service brakes (fig. 16).

(6) ELECTRICAL SYSTEM. The source of electrical current is a generator and four 6-volt batteries. Through a relay type switch, the motor for starting the engine is operated at 24 volts. Otherwise the electrical system is 12 volts. There are individual grid heaters at the

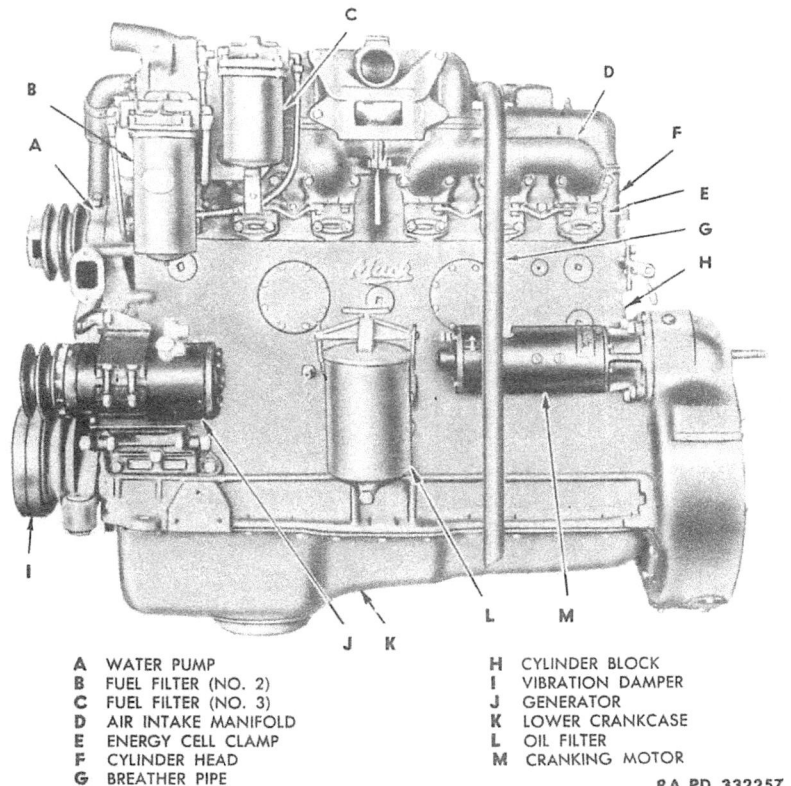

A WATER PUMP
B FUEL FILTER (NO. 2)
C FUEL FILTER (NO. 3)
D AIR INTAKE MANIFOLD
E ENERGY CELL CLAMP
F CYLINDER HEAD
G BREATHER PIPE
H CYLINDER BLOCK
I VIBRATION DAMPER
J GENERATOR
K LOWER CRANKCASE
L OIL FILTER
M CRANKING MOTOR

RA PD 332257

Figure 7—Left Side of Engine

junctions of the inlet manifold branches and the cylinder heads for cold-weather starting. These are in two groups of three each in series, and they are operated by a manual relay switch. Lighting is blackout type throughout, with a single right headlight, two fender parking lights and two tail and stop lights. There are also the necessary instrument panel lights and a low-pressure indicator buzzer for the air brake system. Bonds, condensers and shielding achieve radio sup-

10-TON 6 x 4 TRUCK (MACK MODEL NR)

pression. The electrical system has fuse and circuit-breaker protection.

(7) Brush Guard, Bumper, and Towing Connections. A grille-type brush guard protects the frontal radiator. There is a channel bumper at the front, with a post-type driver's guide at its right-hand end. There are tow hooks at the front and rear, and a pintle hook at the rear of the chassis frame at the center, between the ends

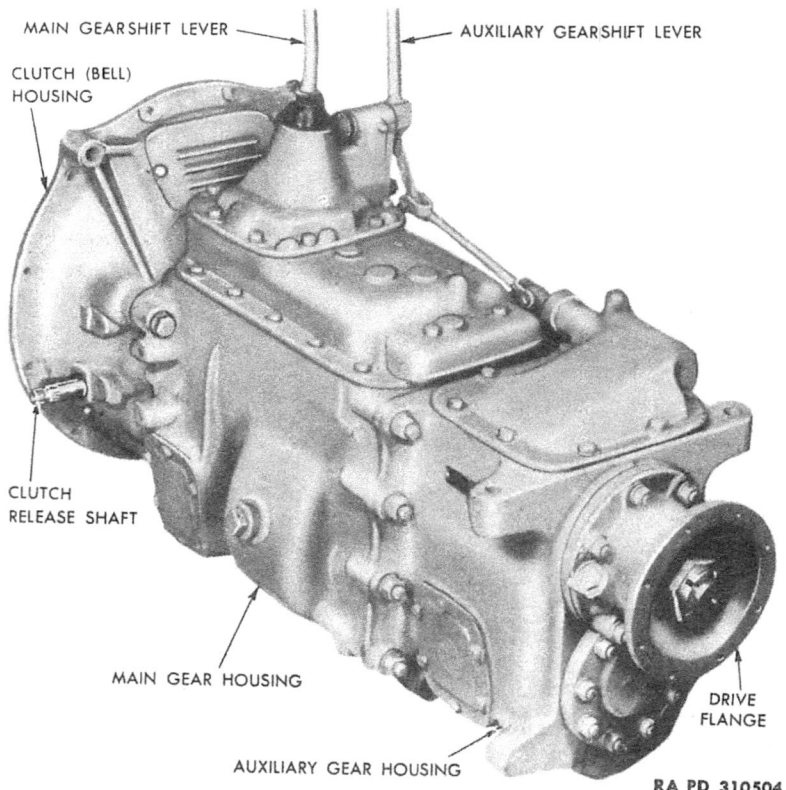

Figure 8—Left Rear View of Transmission

of the frame rails. For trailer attachment, there are electrical and air brake connections at the rear (figs. 3 and 4).

(8) Spare Tires. Two spare tires, one for the front and one for the rear wheels, are mounted in carrier attached directly to the chassis frame close behind the cab (fig. 17).

c. **Cab.** Seating is provided for two, with driver's seat separate; but three can be accommodated. The seat compartment is of steel

TM 9-818
2

DESCRIPTION AND TABULATED DATA

construction. The windshield is of one-piece pattern and has shatterproof glass. With the windshield, the canvas top provides complete enclosure, or it may be arranged as a roof only, or for other partial protection. For ventilation, the windshield is hinged at the top, and cab doors have hooks for holding them partly open. Top and windshield fold down for open-seat arrangement. Seat cushions are of the spring type, and the driver's seat is adjustable. A round rear-vision mirror is mounted at the base of the windshield post on each side of the cab (figs. 17, 18, 19, 20 and 21).

d. Body. The body is of wood reinforced with steel, and is secured by clamps and at the front spring-tension fastenings to the chassis frame on continuous, wooden sills. It has a tail gate, tarpaulin bows

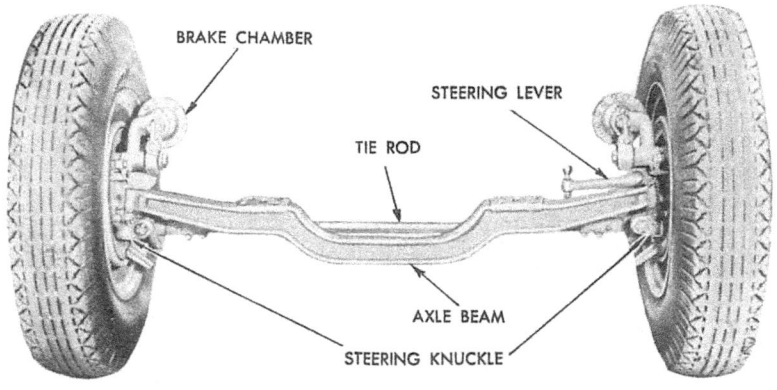

RA PD 310505

Figure 9—Front Axle

and tarpaulin and troop seats. Tail gate and tarpaulin provide full enclosure. Tie fastenings secure the tarpaulin. There are splash aprons ahead of and behind the rear wheels (figs. 1 and 2).

e. Identification.

(1) CHASSIS SERIAL NUMBER (figs. 22 and 23). The chassis serial number, which incorporates model designation (NR), is stamped on the left side frame member just ahead of the steering gear arm. Model and chassis serial number are also shown on the chassis name plate, located on the right-hand seat.

(2) ENGINE SERIAL NUMBER (fig. 24). The engine serial number, which incorporates engine model designation (ED), is stamped on the right side of the timing gear housing at the front of the engine.

15

10-TON 6 x 4 TRUCK (MACK MODEL NR)

(3) PUBLICATION DATA PLATE (fig. 25). The publication plate bears the number of Technical Manuals and Parts List which apply to the vehicle. This plate is attached to the door of the compartment which is built into the instrument panel.

(4) IDENTIFICATION PLATE. There is an identification plate on the door of the instrument panel compartment.

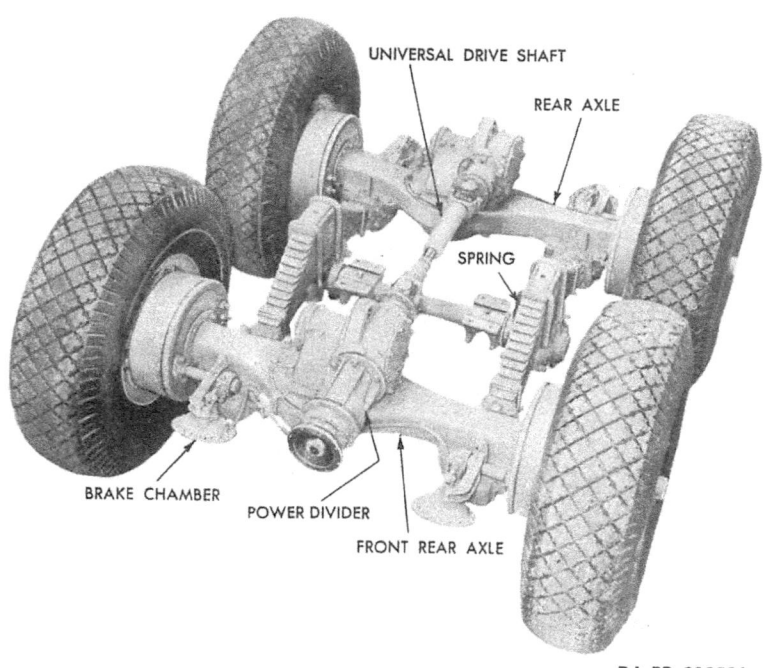

RA PD 310506

Figure 10—Rear Axles Assembly

f. **Mobility.** This cargo truck is essentially a road vehicle, but operation need not be confined to first-class roads, as its four-wheel rear drive gives it tractive ability to travel well where there is mud, sand or snow, provided conditions are not extreme. No attempt should be made to operate them cross-country, or where conditions are so severe that use of a front-drive or a track-type vehicle is indicated.

DESCRIPTION AND TABULATED DATA

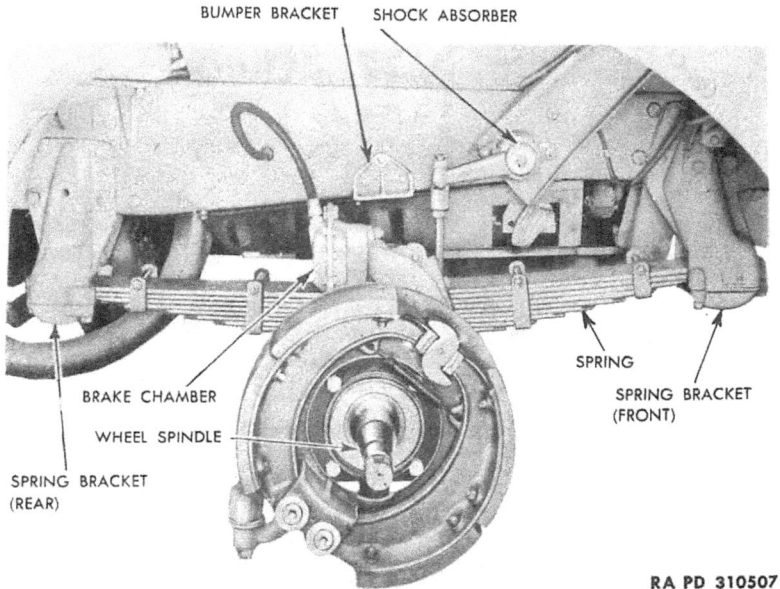

Figure 11—Front Spring

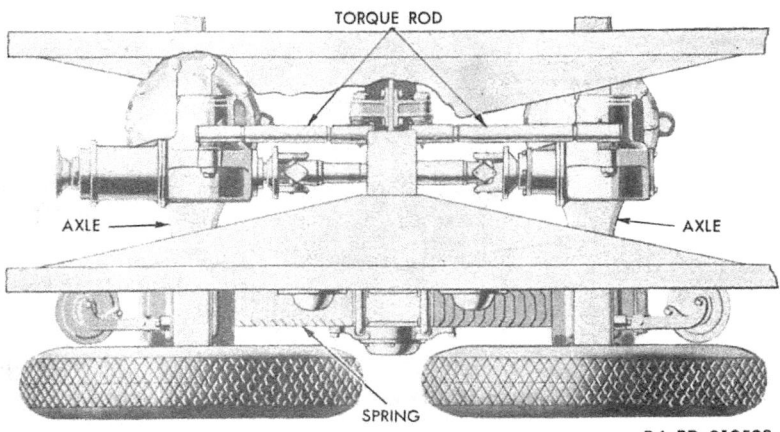

Figure 12—Top View of Rear Suspension

10-TON 6 x 4 TRUCK (MACK MODEL NR)

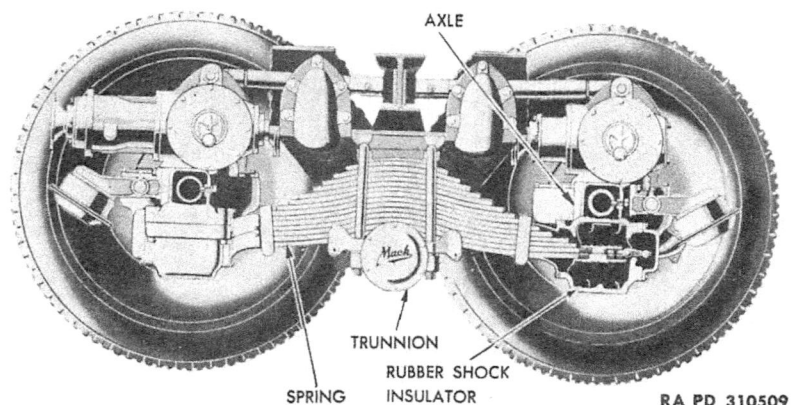

Figure 13—Left Side View of Rear Suspension

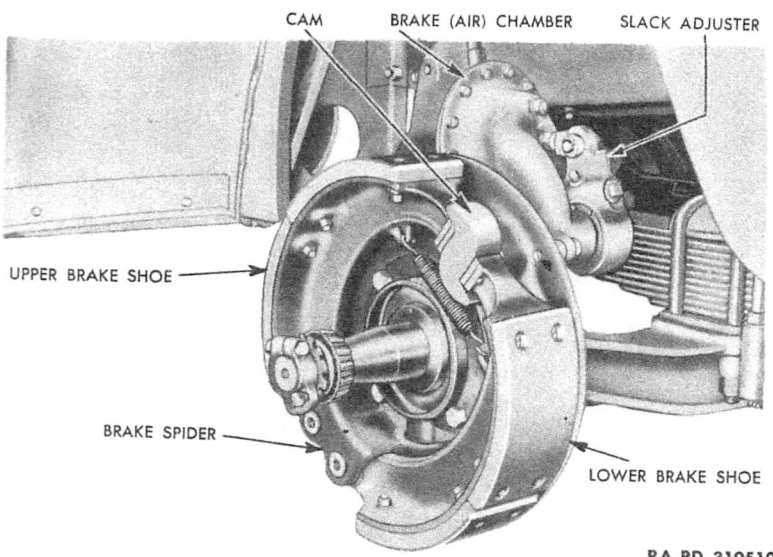

Figure 14—Front Wheel Brake

TM 9-818
2

DESCRIPTION AND TABULATED DATA

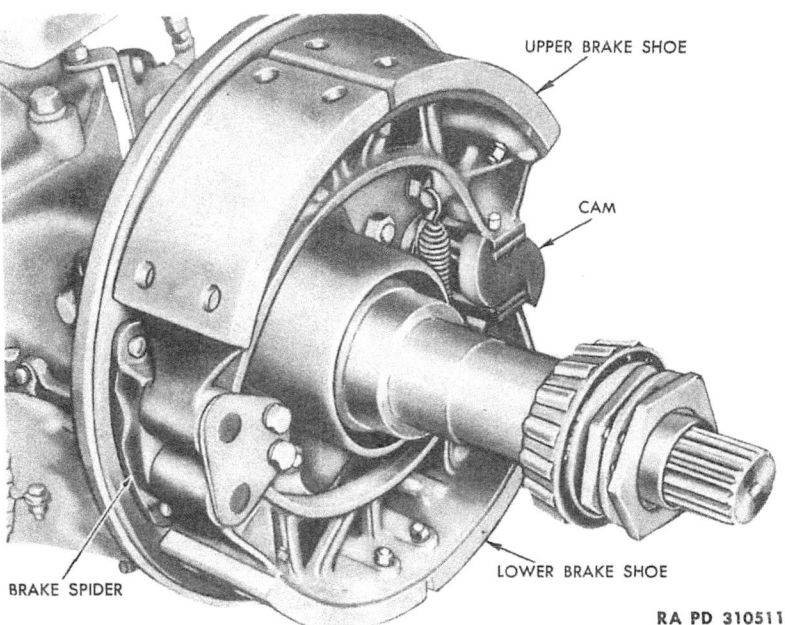

Figure 15 — Rear Wheel Brake

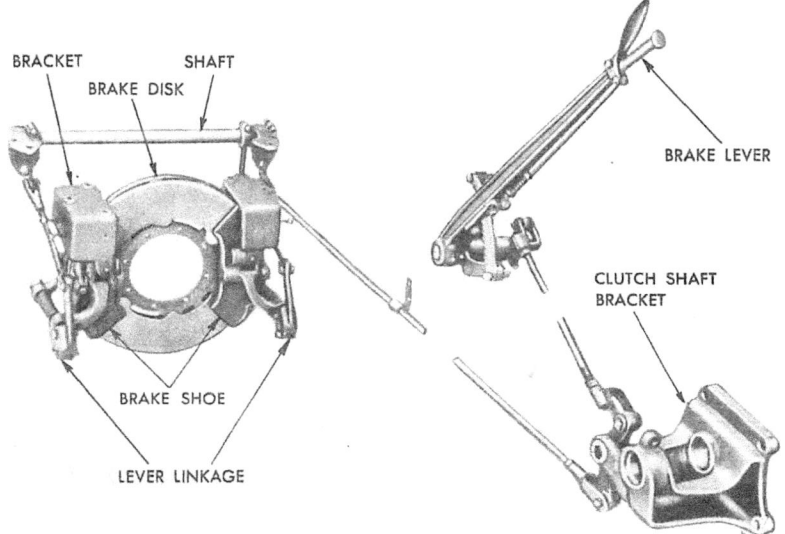

Figure 16 — Parking Brake Parts

19

TM 9-818
2
10-TON 6 x 4 TRUCK (MACK MODEL NR)

RA PD 310513
Figure 17—Cab, Enclosed—Left Rear View

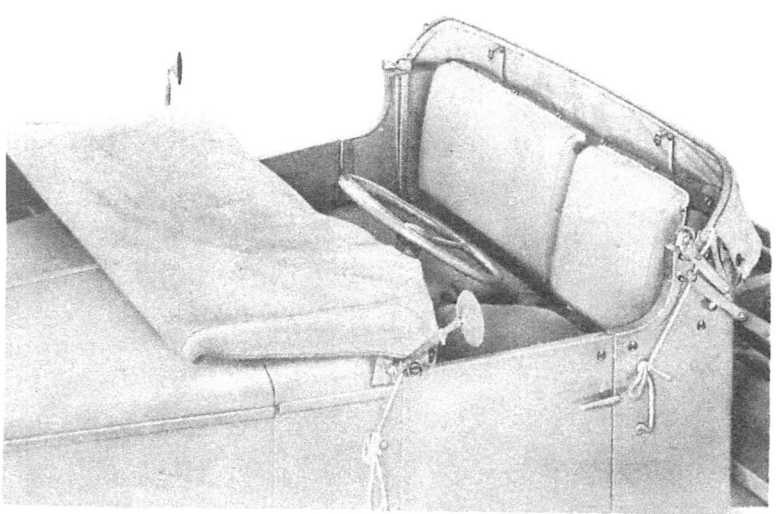

RA PD 310514
Figure 18—Cab with Top and Windshield Folded Down

DESCRIPTION AND TABULATED DATA

Figure 19 — Cab with Top Up, and Windshield Open

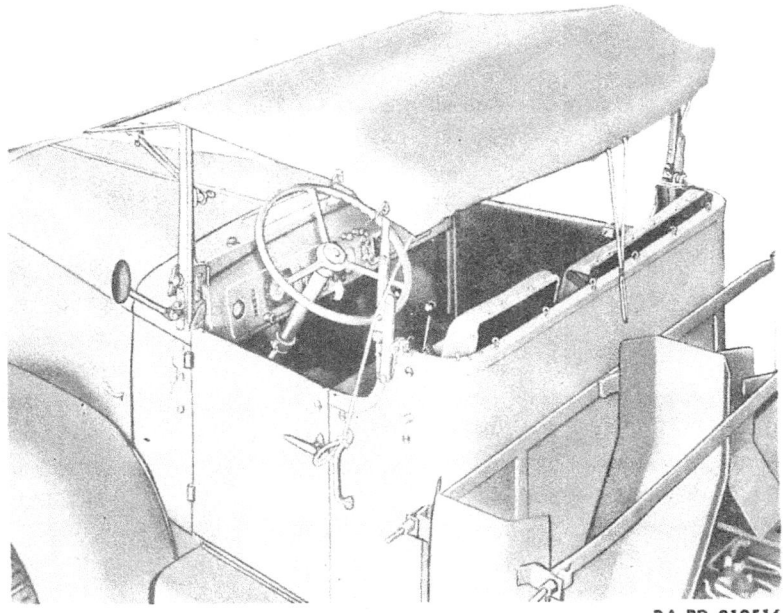

Figure 20 — Cab with Rear Curtain Rolled Up, and Windshield Open

TM 9-818
3

10-TON 6 x 4 TRUCK (MACK MODEL NR)

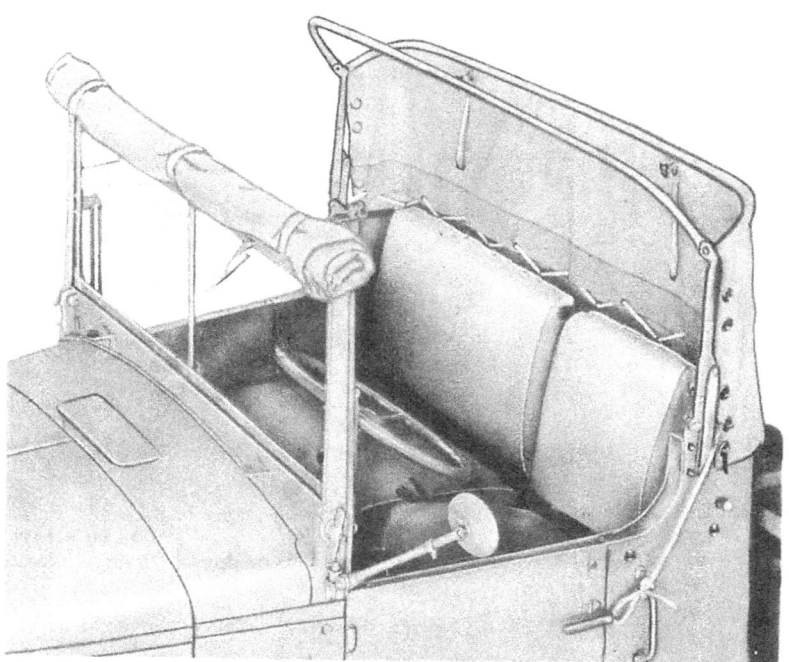

Figure 21—Cab with Bows for Top Raised

RA PD 310517

3. DATA.

a. Vehicle Specifications.

Wheelbase—Front axle to front rear axle .... 14 ft 5 in.
            Between rear axles .... 4 ft 7 in.
Length, over-all .... 26 ft 10⅛ in.
Width, over-all .... 8 ft 7 in.
Height, over-all .... 10 ft 2¾ in.
Tread (center to center) Front .... 79½ in.
                            Rear .... 75¼ in.
Weight of vehicle—(empty) .... 20,750 lb
Weight of vehicle—(loaded) .... 42,750 lb
Ground pressure:
    Gross { Front .... 67.2 psi
             { Rear .... 86.3 psi
    Net   { Front .... 96.15 psi
             { Rear .... 123.13 psi

TM 9-818
3

DESCRIPTION AND TABULATED DATA

Figure 22—Chassis Name Plate

Figure 23—Chassis Serial Number Location

Ground contact:
Gross { Front ....74.3 si
       { Rear ....95.0 si
Net   { Front ....52.0 si
      { Rear ....66.5 si
Height—under side of body at trunnion center line (empty) 56$\frac{13}{16}$ in.
Ground clearance—rear axle at center ....13 in.

TM 9-818
3

10-TON 6 x 4 TRUCK (MACK MODEL NR)

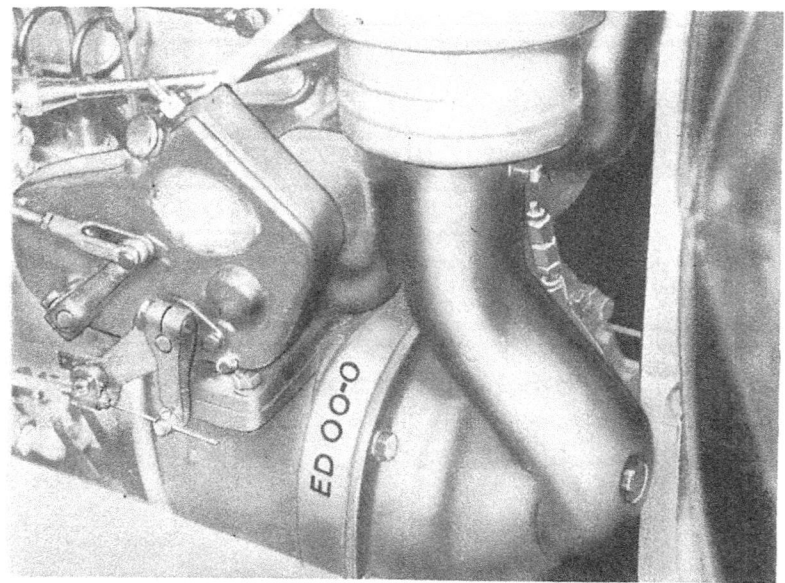

RA PD 310520

Figure 24—Engine Serial Number Location

Pintle height—to center line .... 38¾ in.
Kind and grade of fuel .... U.S.A. Spec. No. 2-102B

  b. **Performance.**

	Gear	Fast Ratios	Slow Ratios
Vehicle speeds in miles per hour....	1st	4.37	3.17
	2nd	7.70	5.58
	3rd	15.32	11.10
	4th	29.35	21.30
	5th	37.70	27.25
	Reverse	4.32	3.13

Approach angle .... 36°
Departure angle .... 38°
Minimum turning radius (at tread center):
  Right .... 39 ft 6 in.
  Left .... 40 ft 6 in.
Towing facilities:
  Front .... Tow hooks
  Rear .... Tow hooks and pintle hook

TM 9-818
3

DESCRIPTION AND TABULATED DATA

Maximum allowable engine (governed) speed .... 2,000 rpm
Maximum allowable speed .... 37.7 mph

   c. **Capacities.**

Transmission capacity .... 18 qt
Rear axle capacity (each) .... 4 qt
Fuel capacity (two tanks) (each) .... 75 gal
Cooling system capacity .... 15 gal
Crankcase capacity .... 15 qt

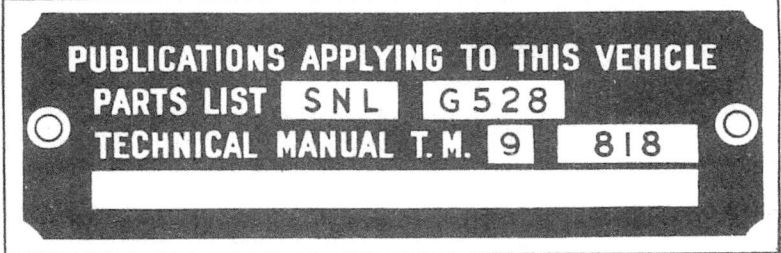

Figure 25—Publication Plate

   d. **Wheels and Tires.**

Wheel size:
  Front .... 24 in.
  Rear .... 20 in.
Tire size:
  Front .... 11.00 x 24
  Rear .... 14.00 x 20
Air pressure:
  Front .... 70 lb
  Rear .... 90 lb

TM 9-818
4

10-TON 6 x 4 TRUCK (MACK MODEL NR)

Section III

DRIVING CONTROLS AND OPERATION

	Paragraph
Controls	4
Operation of the vehicle	5
Towing the vehicle	6

4. CONTROLS.

a. General. Figures 26 and 27 illustrate all the instrument controls necessary for the operation of the vehicle.

b. Driving Controls. The controls by which the truck is driven are as follows:

(1) STARTER SWITCH AND HAND THROTTLE CONTROL (fig. 27). Both the starter switch and the hand throttle control are on the instrument panel, and are specifically covered under their subject names (steps c (1) and (3) below).

(2) STEERING (fig. 26). Steering control of the front wheels, through the steering gear and its linkage, is by the steering wheel at the top of the inclined steering column.

(3) CLUTCH PEDAL (fig. 26). This is the pedal located to the left of the steering column, operable by the driver's left foot. Pressure on the pedal disengages the clutch for stopping the truck, and for allowing the shifting of gears. Release of the pedal engages the clutch which is the drive between the engine and transmission.

(4) PARKING BRAKE LEVER (fig. 26). This lever is at the right of the driver's seat. It is used to apply the disk brake on the propeller shaft when the truck is stopped and parked. It must always be in full-release position (forward) when truck is traveling.

(5) TRANSMISSION SHIFTER LEVER (fig. 26). This lever to the right of the driver is the means for selectively engaging any of the five speeds forward and one reverse provided in the main portion of the transmission. It is vertical when in neutral. A separate lever (step (6) below) controls the auxiliary gears through which two groups of speeds are obtained, a total of ten forward speeds and two reverse. Diagram showing shifting positions of this and the auxiliary lever is attached to the instrument panel (fig. 27). Diagram plate is shown in figure 28.

(6) AUXILIARY TRANSMISSION SHIFTER LEVER (fig. 26). The lever just to the right of the main shift lever is for shifting the auxiliary gears. This is a two-position lever for selection of the fast group of speeds, of which the fifth is overdrive—or the slower group, of which the fifth is direct. Diagram showing shifting positions of this and the main shift lever is attached to the instrument panel (fig. 27). Diagram plate is shown in figure 28.

TM 9-818
4

DRIVING CONTROLS AND OPERATION

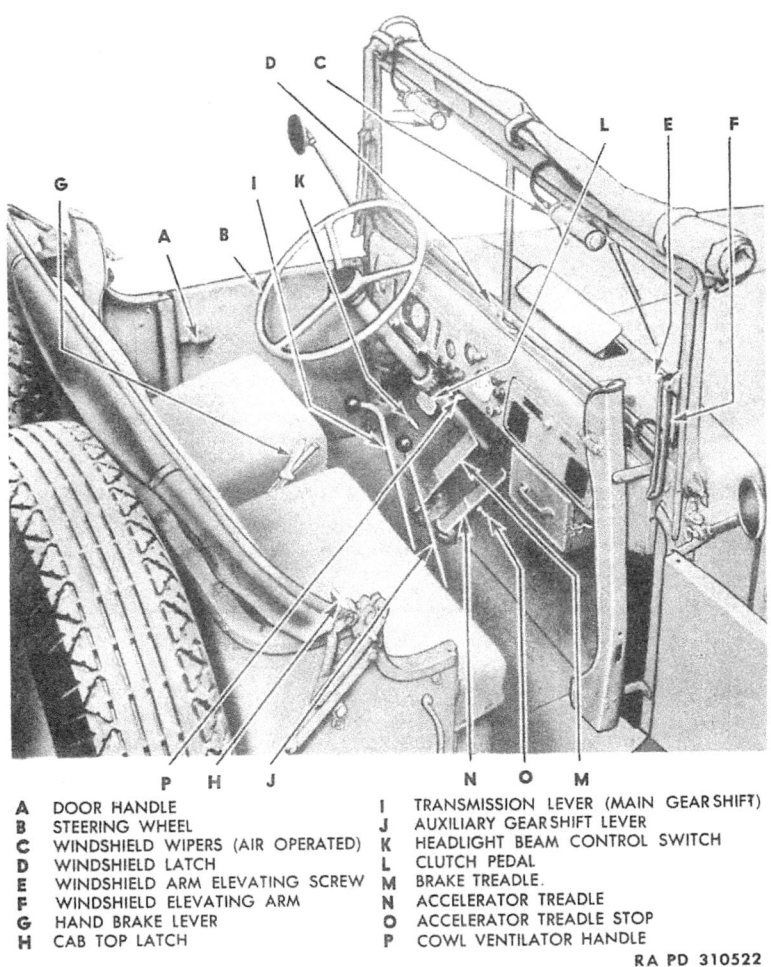

A	DOOR HANDLE	I	TRANSMISSION LEVER (MAIN GEAR SHIFT)
B	STEERING WHEEL	J	AUXILIARY GEARSHIFT LEVER
C	WINDSHIELD WIPERS (AIR OPERATED)	K	HEADLIGHT BEAM CONTROL SWITCH
D	WINDSHIELD LATCH	L	CLUTCH PEDAL
E	WINDSHIELD ARM ELEVATING SCREW	M	BRAKE TREADLE.
F	WINDSHIELD ELEVATING ARM	N	ACCELERATOR TREADLE
G	HAND BRAKE LEVER	O	ACCELERATOR TREADLE STOP
H	CAB TOP LATCH	P	COWL VENTILATOR HANDLE

RA PD 310522

Figure 26—Controls in Cab

(7) ACCELERATOR PEDAL (fig. 26). The accelerator pedal is located at the right of the steering post, beside and to the right of the brake pedal. It is operated by the driver's right foot, and controls the speed and power of the engine.

(8) ACCELERATOR PEDAL STOP (fig. 26). Located beneath the accelerator pedal, this stop limits the downward travel of the pedal to prevent damage to the accelerator linkage.

(9) BRAKE PEDAL (fig. 26). Control of the service brakes is by a pedal which is located at the right of the steering post, between the

TM 9-818
4

10-TON 6 x 4 TRUCK (MACK MODEL NR)

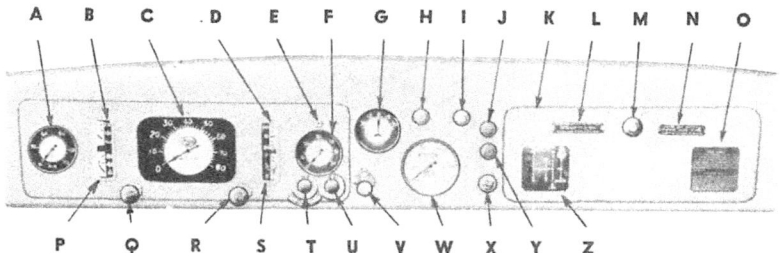

- A AIR PRESSURE GAGE
- B WATER TEMPERATURE GAGE
- C SPEEDOMETER
- D MAIN AMMETER
- E HEADLIGHT BEAM INDICATOR
- F FUEL OIL PRESSURE GAGE
- G AUXILIARY AMMETER
- H WINDSHIELD WIPER VALVE (L.H.)
- I WINDSHIELD WIPER VALVE (R.H.)
- J INSTRUMENT PANEL LIGHT SWITCH
- K COMPARTMENT DOOR
- L PUBLICATION DATA PLATE
- M DOOR LATCH BUTTON
- N IDENTIFICATION PLATE
- O DIESEL INSTRUCTION PLATE
- P OIL PRESSURE GAGE
- Q THROTTLE CONTROL
- R ENGINE STOP CONTROL
- S FUEL LEVEL GAGE
- T STARTING HEATERS SWITCH
- U STARTER SWITCH
- V BLACKOUT LIGHTING SWITCH
- W TACHOMETER
- X ENGINE EMERGENCY STOP CONTROL
- Y FUEL GAGE SWITCH BUTTON
- Z GEAR SHIFT INSTRUCTION PLATE

RA PD 332258

Figure 27 — Instrument Panel

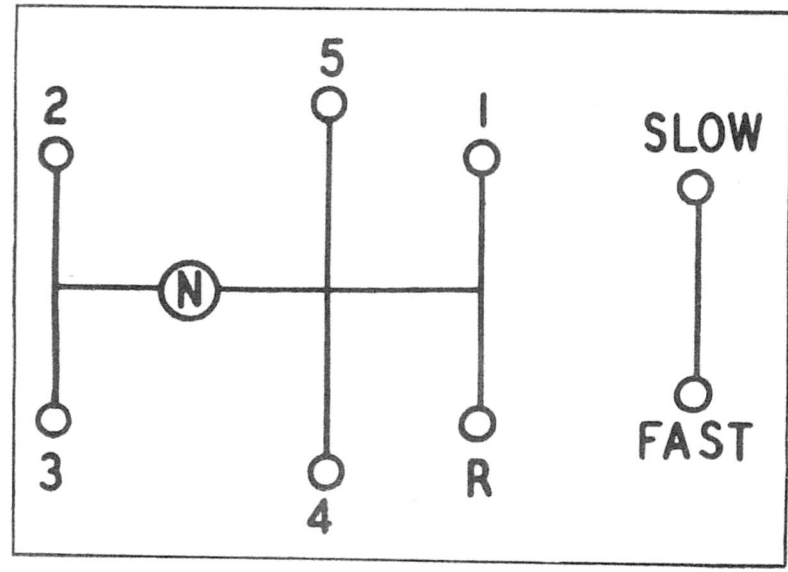

RA PD 310524

Figure 28 — Gearshift Diagram

28

DRIVING CONTROLS AND OPERATION

post and the accelerator pedal, and is operated by the driver's right foot. It is hinged directly on the top of the brake application (air) valve, and the brakes are applied by depressing the pedal.

(10) HEADLIGHT BEAM CONTROL (fig. 26). The switch which controls the headlight beam is to the left of the clutch pedal, and is operated by the driver's left foot. Depressing and then releasing this foot button changes the light beam from high to low, and vice versa. When the red light shows in the beam indicator on the instrument panel, headlight beam is high.

(11) SEAT ADJUSTMENT LEVER (fig. 26). Adjustments of the driver's seat are made by pulling upward, and holding the lever at the front of the seat frame. The seat can then be moved forward or backward.

(12) WINDSHIELD ADJUSTMENT (fig. 26). The elevating arms are the supports for adjustable opening of the windshield from the bottom. When the windshield is in the desired open position it must be secured by tightening the clamping thumb screws. In order to avoid breakage of the glass from twisting, both arms must slide freely for simultaneous operation. In order to open the windshield from the bottom, the thumb screws must be loosened. When the windshield latch is raised, the windshield can be opened. When windshield is entirely closed it should always be secured by this latch.

(13) WINDSHIELD WIPERS (fig. 26). There are two inside-mounted wiper units at the top of the windshield. These are operated by air, and are independently controlled by two valves on the instrument panel. Control by these valves includes speed regulation.

(14) COWL VENTILATOR HANDLE (fig. 26). This handle, under the instrument panel, moved forward or backward, respectively opens and closes the ventilator at the top and center of the cowl.

(15) DOOR HANDLE (fig. 26). Moving this handle downward unlatches the door.

(16) CAB TOP LATCH (fig. 26). This latch, when swung outward, provides a positive lock for keeping the cab top in the up position. In order to lower the top, the latch must first be swung inward to allow the cab top bows to pass.

(17) ENGINE STOP CONTROL AND ENGINE EMERGENCY STOP CONTROL (fig. 27). Both the regular stop control and the emergency stop control for the engine are on the instrument panel, and are specifically covered under their subject names (steps c (10) and (11) below).

c. **Instrument Panel—Controls, Switches, Gages and Instruments.**

(1) STARTER SWITCH (fig. 27). This is a push-button type of switch near the middle of the panel.

(2) STARTING HEATER SWITCH (fig. 27). This is a push-button type switch located just to the left of the starter button. Pressing this button operates six heater grids at the branches of the intake manifold.

TM 9-818
10-TON 6 x 4 TRUCK (MACK MODEL NR)

In cold weather, this button should be held in for thirty seconds, immediately before pressing starter switch button.

(3) HAND THROTTLE CONTROL (fig. 27). This control is a pull-knob located at the left. It may be used when starting, or in other cases when a setting for greater than idling speed is wanted. When pushed in, the engine has idling speed. This is its normal position when truck is traveling, and speed and power of engine is controlled by the accelerator pedal.

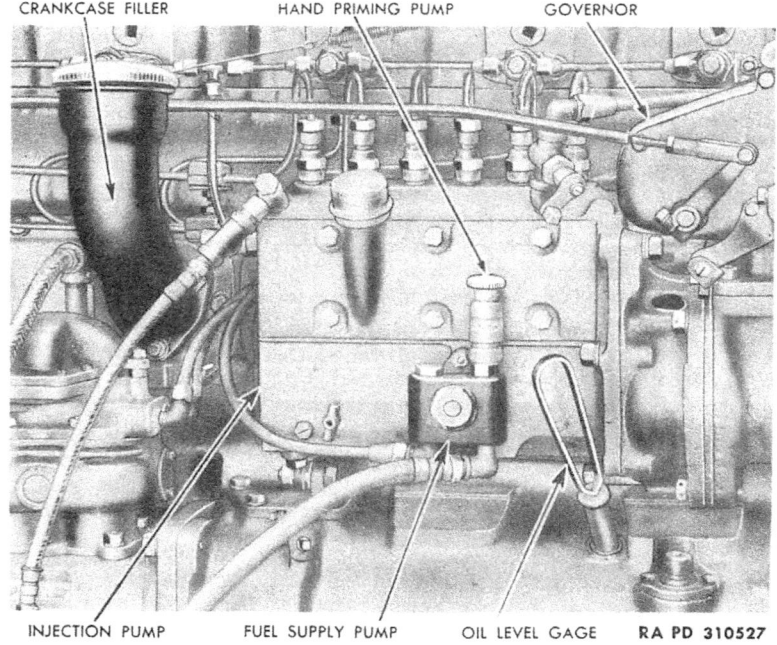

Figure 29—Injection Pump

(4) FUEL LEVEL GAGE (fig. 27). At the left center is the fuel level gage, an electrically operated unit which registers the level of fuel in each of the two tanks.

(5) FUEL GAGE SWITCH (fig. 27). This pull-button, which is at right center on the panel, has two gage-control positions—first, to give the reading for the left-hand tank, and second, to give the reading for the right-hand tank.

(6) FUEL PRESSURE GAGE (fig. 27). This gage is slightly to the left of the panel center, and indicates the pressure in the fuel line between the fuel supply pump and the injection pump. It should show between 3 and 18 pounds during operation.

TM 9-818
4

DRIVING CONTROLS AND OPERATION

(7) OIL PRESSURE GAGE (fig. 27). The pressure gage is near the left end of the panel, the lower unit of a two-unit gage of which the upper is the water temperature gage. The oil pressure gage indicates the pressure of the engine lubricating oil. It should show 45 to 60 pounds at 2100 revolutions per minute maximum speed, and 10 to 20 pounds at 500 revolutions per minute idling speed. This gage shows pressure only; it does not indicate the amount of oil in the lower crankcase.

(8) AIR PRESSURE GAGE (fig. 27). The air pressure gage, which is the last to the left on the panel, indicates the air pressure in the reservoirs and pipe lines. This gage should show a pressure of between 85 and 105 pounds (the "cut-in" and "cut-out" pressures, respectively) when the engine has run a short time. Applying the brakes will cause a slight drop in pressure, but with the air compressor and air system functioning properly pressure will be restored. Avoid "fanning brakes," that is, making numerous brake applications within a short time, as the pressure may drop to a point where brake applications will not be powerful enough for safety. The low-pressure indicator buzzer sounds as an alarm when the pressure drops to 60 pounds.

(9) WATER TEMPERATURE GAGE (fig. 27). This gage, which indicates the temperature of the water in the cooling system, is near the left end of the panel, the upper unit of a two-unit gage, of which the lower is the pressure gage for the lubricating oil. The normal operating temperature range is 165°F to 185°F. The temperature should rise to within this range when the engine has run a short time, and it should remain within this range during operation.

(10) ENGINE STOP CONTROL (fig. 27). This is a pull-knob control by which the engine is stopped. It is to the left center of the panel. Pulling out this knob stops the engine. The knob should be pushed in immediately after the engine has stopped, so that it will not prevent starting.

(11) ENGINE EMERGENCY STOP CONTROL (fig. 27). This is a pull-knob control which operates a valve in the fuel line. It is at the right center of the panel. Pulling out this knob stops the engine by shutting off the fuel at the feed line. It must be pushed in before engine is to be started.

(12) TACHOMETER (fig. 27). The tachometer indicates the engine speed in revolutions per minute (rpm) and also the number of hours of engine operation. The engine must not be run at over 2100 revolutions per minute (the nominal governed speed is 2000 rpm). The recording of hours of operation of the engine is for guidance of service and maintenance.

(13) GEARSHIFT DIAGRAM PLATE (figs. 27 and 28). This plate, attached to the door of the glove and map compartment, diagrammatically gives the shifting positions for the main and auxiliary shift levers.

(14) SPEEDOMETER (fig. 27). The speedometer indicates the speed in miles per hour (mph) at which the vehicle is traveling. Reference

TM 9-818
10-TON 6 x 4 TRUCK (MACK MODEL NR)

to it should be a driving habit to eliminate guessing at the road speeds.

(15) MAIN AMMETER (fig. 27). This is one of two ammeters, and is the upper unit of a two-unit component to the left center of the panel, the lower unit being the fuel gage. The main ammeter indicates in amperes the current received by the two sets of batteries, in parallel, when engine is running. When the engine is stopped, this ammeter will show the amount of current being drawn from the batteries by any lights which are turned on.

(16) AUXILIARY AMMETER (fig. 27). The auxiliary ammeter, located near the center of the panel, is provided to show that the second pair of batteries, which are connected in parallel with the first, is receiving a charge. Since this set of batteries receives its charge through two sets of contacts in the series-parallel starting switch and the circuits (through two fuses), this ammeter will im-

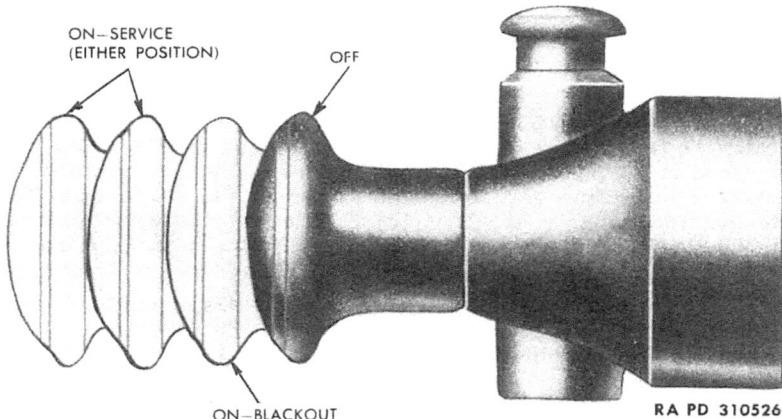

Figure 30—Light Switch

mediately disclose, through absence of charge, if any of these contacts are not properly closed.

(17) WINDSHIELD WIPER CONTROL VALVES (fig. 27). There are two knob-type valves at the middle of the panel for separate control of the two windshield wipers. Their control includes speed regulation. The left valve controls the left wiper.

(18) BLACKOUT LIGHTING SWITCH (figs. 27 and 30). The switch which controls the headlight, stop and taillights is a blackout type. It is of the pull type, and is just to the left of the middle of the panel. It has one "OFF" and three "ON" positions.

(a) *Off Position.* With the knob pushed all the way in, all lights are off and stop light will not operate.

(b) *Blackout (First "ON") Position.* Pulling the knob to the first "ON" position lights the blackout parking and blackout taillights,

DRIVING CONTROLS AND OPERATION

and energizes the circuits to the stop light operating switch and to the trailer taillight contact in the trailer receptacle. When the stop light operating switch closes, the blackout stop lights go on, and the stop light contact in the trailer receptacle is energized.

(c) Service (Second "ON") Position. The latch or locking button on the switch body must be pressed down before switch can be pulled to the second "ON" position. When the switch is in this position it lights the blackout headlight, the instrument panel lights, and blackout taillights. It also energizes the wire to the stop light operating switch (thus making stop light operable) as well as the contacts for the stop and taillights in the trailer connection receptacle.

(d) Third "ON" Position. As used on this truck, the switch produces the same results in the third "ON" position as it does in the second "ON" position.

(19) PANEL LIGHT SWITCH (fig. 27). A switch for separate control of the instrument panel lights makes it possible to turn on these lights when the main light switch is in the first "ON" position; in this position the main switch does not light unless the instrument panel light switch is "ON."

d. Headlight Beam Indicator (fig. 27). To the left of the panel center, just above the fuel oil pressure gage, is the headlight beam indicator. It shows a red light when the beam is high. No light shows when the headlight is turned off, or when the beam is low.

e. Instrument Panel—Miscellaneous (fig. 27).

(1) GLOVE AND MAP COMPARTMENT. For carrying small articles there is a compartment with a flush door at the right end of the instrument panel.

(2) COMPARTMENT LATCH BUTTON. The compartment door, which swings downward, is unlatched by pressing the button near its top edge at the center.

(3) DIESEL INSTRUCTION PLATE. This plate, attached to the compartment door at the left, gives recommendations relative to fuel and lubricating oils, also oil filters and air cleaners.

(4) GEARSHIFT DIAGRAM PLATE (figs. 27 and 28). This plate which diagrammatically shows the shifting positions of both the main and auxiliary shift levers is attached to the compartment door. It is mentioned in conjunction with transmission shifter lever (par. 4 b (5)), and auxiliary transmission shift lever (par. 4 b (6)).

f. Hand Priming Pump (fig. 29). This is a plunger type pump which is located on the fuel injection pump and is accessible when the right side of the hood is opened. While this hand pump is not regularly used when starting the engine it is necessary occasionally for expelling air from fuel lines or priming them. To use this pump the plunger must be released for action by turning the knurled knob to the left. After using this pump the knurled knob must be screwed down (to the right) to reasonable tightness.

10-TON 6 x 4 TRUCK (MACK MODEL NR)

5. OPERATION OF THE VEHICLE.

a. Starting the Engine.

(1) BEFORE-OPERATION SERVICE. Before the engine is started the prestarting inspection outlined in paragraph 14 must be performed.

(2) BEFORE STARTING THE ENGINE.

(a) Try the hand brake lever, and if it is not pulled up well, apply the brake firmly.

(b) Push the clutch pedal down all the way. This disengages the clutch. Keep this pedal depressed until the engine is started.

(c) Try the main transmission shifter lever, and if it is not at neutral, move it to the neutral position.

(3) ENGINE STARTING PROCEDURE—MILD WEATHER.

(a) Be sure the "stop" control button and the emergency stop control button are both in running position—all the way in.

(b) Press down the accelerator pedal—all the way.

(c) Press the starter switch button.

(d) Release the starter switch button as soon as the engine starts.

(e) Begin to release the accelerator pedal as soon as the engine starts, keeping the engine running at a fast idling speed for several minutes, but do not race the engine to "warm up."

(f) Allow the engine to become warm before putting the truck under way. CAUTION: *No attempt should be made to start this Diesel engine with a hand crank. A cranking speed favorable to starting cannot be attained.* If the starter system fails, the truck may be towed to crank the engine. A hand crank is furnished, but it is for turning over the engine when making adjustments, etc.

(4) STARTING THE ENGINE IN COLD WEATHER.

(a) This engine is equipped with manifold heaters (electrical, grid type) and in cold weather should be preheated by holding in the heater switch button for thirty seconds before operating the cranking motor. If the engine does not start in fifteen or twenty seconds of cranking, by the cranking motor, release the starter switch and wait two minutes (for benefit of the batteries and cranking motor) before repeating the preheating and cranking.

(b) Starting will be facilitated by use of winter engine oil of low cold test (par. 21).

(c) During severe (low-temperature) weather, it may be necessary in an emergency to perform with caution the following:

1. Remove the plug from the central branch of the intake manifold of the engine, and direct the flame of an ordinary blowtorch into the manifold opening, at the same time having an assistant open the throttle all the way and press the starter switch button. Withdraw the torch as quickly as the engine fires evenly, and install plug.

2. In extreme cases it may be necessary to drain the cooling system and fill it with hot water.

TM 9-818
5

DRIVING CONTROLS AND OPERATION

3. A further aid is to drain the engine oil, warm it, and pour it back into the crankcase.

4. When conditions indicate that there must be no risk of delay in starting, the engine should be started and run for several minutes at intervals of an hour or so, in order to keep the engine heat at or near running temperatures.

5. Do not overtax the cranking motor and batteries by cranking longer than thirty seconds without interruption; and wait about two minutes before making another attempt to start the engine. This permits the cranking motor to cool, and the batteries to "pick-up."

(d) Do Not Race the Engine.

1. Never race the engine during warm-up periods, nor while tuning engine—running without load.

2. Never operate the engine beyond the governed speed (2000 rpm).

3. When driving the truck, select the correct gear to hold the vehicle back when descending grades, using brakes when necessary, as it is possible for the vehicle to drive the engine beyond its governed speed even though the engine is equipped with a governor.

(e) Checking Instruments. The dash instruments should be observed when starting and occasionally throughout travel.

1. *Oil pressure gage.* This gage should register pressure of from 10 to 20 at idling speed; 45 to 60 at maximum speed. Engine should be stopped if no pressure is indicated.

2. *Ammeters.* The ammeters should normally show charge. If battery is fully charged, ammeter will show little or perhaps no charge at any engine speed. If no charge is shown, report battery for hydrometer test.

3. *Water temperature gage.* This gage should show between 165°F and 185°F. If engine overheats, stop and investigate. Do not pour cold water into an overheated engine; after cooling, add water slowly with engine idling, so that the water pump will prevent the cold water from entering the cylinder head directly. A frozen cooling system must be allowed to thaw out slowly in a warm place if damage is to be averted. (If the vehicle freezes en route and cannot be taken into a building, cover the radiator and hood with a blanket, and let the engine run slowly until the blanket steams. Shut off the engine, wait a few minutes and repeat the operation, three or four times if necessary, until the heat gradually thaws it out.) Do not use boiling water to thaw a freeze-up.

4. *Fuel pressure gage.* This gage should not register in the red section at the beginning or the end of the scale. If indication is in the 0-3 section, the element of No. 2 filter requires renewal; if in the 18-60 section the trouble is No. 3 filter.

5. *Air pressure gage.* The pressure gage for the air brakes should register at least 60 pounds and preferably 85 pounds. Should the gage indicate a rapid drop in pressure, the vehicle should be stopped

10-TON 6 x 4 TRUCK (MACK MODEL NR)

immediately and the trouble corrected before proceeding further, as without adequate air pressure the braking action would be unsafe.

6. *Tachometer.* The tachometer should be watched habitually so that proper engine speeds are maintained during all load and road conditions. The engine (working) speed range is between a minimum of 1000 revolutions per minute and the governed speed of 2000. Proper selection of the transmission gears will keep the engine in this "power range" and achieve proper performance. The tachometer should also be watched when descending grades, and the proper gears and the brakes should be used to prevent overspeeding the engine.

b. **Driving the Vehicle.**

(1) STARTING THE VEHICLE.

(a) Release the hand brake lever. (If necessary, hold the vehicle with the service brakes—using brake pedal.)

(b) Depress clutch pedal.

(c) Keeping clutch pedal down, move main gearshift lever to the position used for starting. On hard ground that is level or nearly level, and with vehicle loaded, the second speed of the slower group of ratios should be used. Do not use higher speeds for starting except on down grades where vehicle will roll forward freely when hand brake is released. Starting in too high a gear requires excessive slipping of the clutch, a practice that is extremely harmful to the clutch disk. The lowest gear of the slower group of ratios is for starting under severe conditions where slow speed with plenty of power is the requirement. Figure 28 shows diagrammatically the shift positions for both the main and auxiliary gear levers.

(d) While speeding up the engine slightly, begin to release the clutch pedal and engage the clutch gradually but progressively, increasing the fuel feed, as the clutch takes hold, so as to neither stall the engine nor jerk the vehicle.

(e) With the clutch fully engaged, increase the vehicle speed until it approaches the speed shown for the gear engaged (and at engine governed speed) in the following tabulation.

ROAD SPEEDS
(for engine-governed speed of 2000 rpm)

SLOW RATIOS		FAST RATIOS	
Gear	Miles per Hour	Gear	Miles per Hour
1st	3.17	1st	4.37
2nd	5.58	2nd	7.70
3rd	11.10	3rd	15.32
4th	21.30	4th	29.35
5th	27.25	5th	37.70
Reverse	3.14	Reverse	4.32

(f) Again depress clutch pedal, and simultaneously decrease fuel feed (by releasing accelerator pedal) and with clutch pedal held

TM 9-818

DRIVING CONTROLS AND OPERATION

down, move gearshift lever to the position for the next faster gear, following this immediately with reengagement of the clutch (step b (2) above), and increased fuel feed.

(g) Repeat shifting procedure as given in preceding step (f) until advance has been made to the highest gear, or the highest suitable for the grade.

(2) ENGAGING THE CLUTCH. When shifting gears, the engine speed should be synchronized with the vehicle speed before engaging the clutch. When engaging the clutch, release the pedal slowly at first, synchronizing the movement with the depressing of the accelerator pedal, but toward the end of the engagement action allow clutch to take hold quickly; do not restrain pedal unnecessarily long when the clutch approaches full engagement. Never allow the weight of the foot to rest on the clutch pedal when the clutch is not being disengaged, as this will result in decreasing the life of the release or throw-out bearing, and a "heavy" foot may even cause clutch slippage, which shortens the life of the clutch disk. *Avoid the habit of "riding" the clutch.*

(3) SHIFTING THE TRANSMISSION.

(a) The clutch must always be disengaged when changing gears. In shifting, begin with the starting ratio, then shift to the next faster one, and so on to the highest (fifth) unless the grade or the road conditions require staying in a lower gear (fig. 28).

(b) Vehicle must always be brought to a full stop before shifting from forward to reverse, or from reverse to forward.

(c) "Double-clutch" when shifting "down," that is, from any gear to a slower one. "Double-clutching" is a procedure for making such shifts without clashing gears and it comprises the following:

1. Disengage the clutch, move the main shift lever to its neutral position, and reengage the clutch.

2. Accelerate the engine to approximately the speed of the vehicle for the lower gear to be engaged, disengage the clutch, move the lever to the selected lower-gear position, and engage the clutch.

(d) Learn to shift expertly, and never hesitate to shift whenever necessary to prevent engine from "lugging." By keeping the engine speed up, grades will be negotiated quicker, the trip made in less time, economy will be improved, and the engine subjected to less strain and wear.

(4) SHIFTING AUXILIARY TRANSMISSION. The lever to the right of the main shift lever is for shifting the range gear from "fast" to "slow," and vice versa, that is, to make the entire set of main transmission ratios either "fast" or "slow."

(a) This lever and the main lever may be shifted in any sequence, except that the combined ratios should be progressively higher as the vehicle is accelerated. Should load and grade, or road conditions, require the use of slow-low to start, both levers should be shifted to the low positions before starting. The two levers give ten forward speeds (and two reverse), and where certain difficulty in grades

10-TON 6 x 4 TRUCK (MACK MODEL NR)

are negotiated on a certain route, experience will determine an advantageous combination for each grade. It is important to use the half-steps between the main speeds, provided by the range gear.

(b) When Shifting Either Main Speeds or Range Gear to:

1. Higher ratio: Accelerate the vehicle to goverened speed, and time shifting the shift to avoid clashing.

2. Lower ratio: "Double-clutch" as explained in step b (3) (c) above.

(c) When Shifting Both Main Speeds and Range Gear:

1. From "Fast" to "Slow" range and to next higher main speed: Shift the range gear first, "double-clutch" (step b (3) (c) above), and shift into "slow" range.

2. Engine speed. Allow the engine to reach governed speed before making each shift to a higher ratio, and make each shift to a lower ratio before the engine speed drops to lugging conditions. By keeping the engine speed up, when engine is under load, proper combustion is assured, and performance and economy will be at their maximum.

3. Gear combinations selection. Proper combination of the two shifts gives ratios between the main speeds which will be favorable for pulling certain grades. It is important to understand that if a certain grade cannot quite be pulled in, say, third gear with the "fast" range, it may not be necessary to drop down to second gear, but rather only to shift the range gear to "slow" position, which gives a ratio between the two main speed reductions. Every effort should be made to take advantage of the greater flexibility offered by the dual-range (Duplex) transmission.

(5) PROCEEDING IN HIGH GEAR. Operate the vehicle so as to hold the speed as even as possible, increasing the speed on the open highway and decreasing the speed in congested areas. Accomplish this by gradual power application. "Fanning" the accelerator constantly and needlessly increasing and decreasing the speed, will cause shock and excessive wear, especially to the propelling units, from the engine straight through to the driving wheels. Do not approach intersections and signals at high speeds and then apply the brakes harshly, as to do so is wasteful of fuel and causes rapid wear of the brakes and driving mechanism. Quick stops must be reserved for emergencies, for to constantly make such stops may cause failure of braking and driving parts. On slippery pavement it is absolutely necessary from a safety standpoint to increase and decrease the power slowly to prevent spinning of wheels. If skidding is evident, the power should be decreased, but the clutch must not be disengaged or skidding will be aggravated. If the vehicle skids, immediately turn the front wheels in the direction of the skid. This will minimize the tendency to skid, and will assist in regaining movement in the original direction. Slow, intermittent brake applications will minimize skidding.

(6) BRAKING.

(a) All brake applications should be made early enough whenever possible, so that moderate braking will achieve the desired stop.

DRIVING CONTROLS AND OPERATION

(b) On slippery pavements, apply and release brakes alternately instead of holding them on continuously. When descending a long grade, select the proper gear to hold back the vehicle, and supplement with intermittent light brake applications, if necessary.

(c) With the air brakes on this truck, the pressure is applied in proportion to the depressing of the brake pedal. If the pedal is partly released the braking pressure is decreased in proportion. Never completely depress the pedal unless an emergency stop must be made. The pedal should be depressed fairly well at first and then the application should be diminished as the speed is reduced so as to accomplish a smooth stop. Do not "fan" the pedal unnecessarily as this gives poor brake performance, wastes air and drops the air pressure.

(d) The engine should be utilized to assist in retarding the vehicle when possible, when braking action is of reasonable duration, by leaving the clutch engaged until the vehicle speed has been well reduced.

(e) CAUTION: *The parking brake is intended for parking. It must not be used as a service brake; the service (air) brakes have ample braking power.* Apply the parking brake after vehicle has been brought to a stop. Avoid harsh application with vehicle moving except as required in an emergency.

(7) CAUTION AGAINST COASTING. Never coast with the clutch disengaged or with the transmission in neutral, as this is extremely unsafe. Should reengagement of the transmission gears and clutch be attempted while coasting downgrade at high speeds, the engine, propeller shafts, axles and other units will be damaged, and control of the vehicle lost.

c. **Stopping the Vehicle.**

(1) Release accelerator pedal and apply service brakes by depressing the brake pedal. To avoid skidding on slippery surfaces, and to utilize the compression of the engine to save brakes, do not disengage the clutch (do not depress the clutch pedal) until the vehicle has slowed down to near idle speed for the engine, or just before the engine would stall; after which continue the brake application (step b (5) above) until the vehicle stops.

(2) Move the main gearshift lever to its neutral position, and allow the clutch to reengage by releasing the clutch pedal.

(3) Apply the parking brake firmly.

d. **Stopping the Engine.**

(1) With accelerator pedal completely released (and with hand throttle control pushed all the way in) pull out engine stop control.

(2) ENGINE EMERGENCY STOP CONTROL. Should the engine continue to run after the regular stop control has been pulled out to the limit of its travel, pull out the emergency stop control. It is a rare occurrence when this control is necessary. It will only be necessary

10-TON 6 x 4 TRUCK (MACK MODEL NR)

when some disorder prevents the regular control from accomplishing full shut-off in the injection pump. The emergency stop control closes a valve in the feed line outside of the pump.

6. TOWING THE VEHICLE.

a. Towing to Start Vehicle.

(1) Place a loop from one end of the tow cable on a tow hook on the front of the towed vehicle, and with a similar attachment at the rear of the towing unit, carefully take up the slack of the tow connection. Hold the clutch disengaged, and place the towed vehicle in third or fourth gear. Tow the disabled vehicle until a reasonable speed is attained, then engage the clutch gradually and smoothly.

(2) Once the engine has been started, do not stop it if the trouble is with the starting system, until the destination has been reached or repairs are possible.

(3) CAUTION. Pushing with another truck should not be attempted because of the possibility of damage to the vehicles. When towing, avoid sudden, forceful tightening of the towing connection; keep the connection taut and pull smoothly, as otherwise it will be subjected to severe shock strains.

b. Towing Disabled Vehicle.

(1) When a vehicle has become disabled but can still be moved by towing, the most satisfactory immediate method of moving it is to place the loop on each end of the tow cable over a tow hook on each side of the vehicle at the front and place the cable loop thus formed in the tow hook, or preferably the pintle hook of the towing vehicle, and, should the emergency require it, tow the disabled vehicle away.

(2) In case of disablement with less urgency of removal, remove the vehicle with a wrecking truck, or the method outlined above may be utilized, with the exception that an improvised spreader bar is used to lessen or eliminate lateral strain between the tow hooks when the cable is used. Any material of sufficient strength may be used, if its length be equal or nearly equal to the distance between the tow hooks, approximately 37 inches, so that the pull exerted will be directly in line with the side members.

Section IV
OPERATION UNDER UNUSUAL CONDITIONS

	Paragraph
Operation in high temperature	7
Operation in low temperature	8
Operation in sandy or desert terrain	9
Operation in mud	10
Operation in snow and ice	11
Operation in flood	12

7. OPERATION IN HIGH TEMPERATURE.

a. Fan belt should be inspected at frequent intervals and adjusted if necessary. Water pump must be kept in good operating condition (par. 83 a).

b. Keep the brush guard free from foreign matter which may obstruct air circulation through radiator.

c. Watch engine oil temperature gage. If engine overheats, stop the engine and investigate.

d. Use correct grade of lubricant (par. 19).

e. Check the tire pressures frequently, and where necessary reduce them to the recommended pressures (par. 147 b). At the end of runs after tires have cooled and accordingly have insufficient pressure, check them again and inflate to the recommended pressures.

f. Be guided by the fuel specifications for various temperatures as noted in paragraph 63 b.

8. OPERATION IN LOW TEMPERATURE.

a. Fuel.

(1) Use Diesel fuel oil procured under U. S. Army Specification 2-102 (grade X), latest revision, at temperatures below 0°F.

(2) Due to condensation of moisture in the air, water will accumulate in tanks, drums and containers. At low temperatures, this water will form ice crystals that will clog fuel lines and stop fuel flow to the fuel pump, unless the following precautions are taken:

(a) Strain the fuel through filter paper or any other type of strainer that will prevent the passage of water.

(b) Keep tank full, if possible. The more fuel there is in a tank, the smaller will be the volume of air from which moisture can be condensed.

(c) Add 1 quart of grade 3 denatured alcohol to the fuel tank at start of winter season, and 1 pint per month thereafter. This will reduce the hazard of ice formation in the fuel.

(d) Be sure all containers are thoroughly clean and free from rust before storing fuel in them.

(e) If possible, after filling or moving a container, allow the fuel to settle 24 hours before filling the vehicle tank from it.

10-TON 6 x 4 TRUCK (MACK MODEL NR)

(f) Keep all closures of containers tight to prevent snow, ice, dirt, and other foreign matter from entering.

(g) Wipe all snow or ice from dispensing equipment, and from around fuel tank fill cap before removing cap to refuel vehicle.

b. Antifreeze.

(1) Protect the cooling system with antifreeze compound for operation below $+32°F$.

(2) Before adding antifreeze compound, be sure the cooling system is clean, and completely free from rust.

(3) Inspect all hose and replace if deteriorated. Inspect hose clamps, plugs, and petcocks, and tighten if necessary. Make sure radiator does not leak before adding antifreeze compound, and that no exhaust gas or air leaks into the cooling system.

(4) After the cooling system is clean and tight, fill with water to about one-third capacity. Then add antifreeze compound, using the proportion of antifreeze compound to the cooling system capacity indicated below.

ANTIFREEZE TABLE

Lowest expected Temperature	Pints, compound antifreeze per gallon of cooling system capacity
$+10°F$	2
$0°F$	$2\frac{1}{2}$
$-10°F$	3
$-20°F$	$3\frac{1}{2}$
$-30°F$	4
$-40°F$	$4\frac{1}{2}$
$-50°F$	5

(5) After adding antifreeze compound, fill with water to slightly below the filler neck; then start and warm the engine to normal operating temperature.

(6) The engine should then be stopped and the solution checked with a hydrometer, adding antifreeze compound if required.

(7) Inspect the coolant weekly for strength and color. Rusty solution must be drained, the cooling system thoroughly cleaned, and new solution of the required strength added. CAUTION: *Use an accurate hydrometer. To test a hydrometer use one part antifreeze compound to two parts water. This solution should produce a hydrometer reading of $0°F$.*

c. Lubrication.

(1) Drain the lubricant from the crankcase while the engine is warm.

(2) Fill the crankcase to the "FULL" mark on the oil gage with engine oil prescribed for use between $+32°F$ and $0°F$ below $0°F$, dilute with grade X Diesel fuel oil in the proportion of $1\frac{1}{2}$ quarts of grade X Diesel fuel oil to 5 quarts of engine oil.

OPERATION UNDER UNUSUAL CONDITIONS

(3) Check oil level frequently. Use the grade of engine oil prescribed for use between +32°F and 0°F, diluted as prescribed in step (2) above, to maintain oil level to "FULL" mark on the gage during operation.

(4) Universal gear lubricant SAE 80, is suitable for use at temperatures as low as −20°F.

(5) If consistent temperature below 0°F is anticipated, drain the gear cases while warm and refill with universal gear lubricant, grade 75, which is suitable for operation at all temperatures below 0°F. If grade 75 universal gear lubricant is not available, use universal gear lubricant, SAE 80, diluted with the fuel used in the engine in the proportion of one part fuel to six parts universal gear lubricant. Make-up oil must be diluted in the same proportion before it is added to gear case.

d. **Starting Instructions.**

(1) This engine is equipped with manifold heaters, and should be preheated by pressing heater switch for 30 seconds before the engine is cranked. If the engine does not start in 15 or 20 seconds of cranking by the cranking motor release cranking motor switch for 2 minutes before repeating the preheating cranking.

(2) During severe weather the following may be performed in an emergency to facilitate quick starting: Remove plug from the central branch of the inlet manifold and direct the flame of a blowtorch into the manifold opening, at the same time having an assistant open the throttle completely and close the cranking motor switch; withdraw the torch as soon as engine fires evenly and install plug.

(3) In extreme cases it may be necessary to drain the cooling system and fill it with hot water, and/or drain the engine oil, warm it and pour it back into the crankcase.

(4) Do not overtax cranking motor and batteries by cranking longer than 30 seconds without interruption.

(5) When conditions indicate that there must be no risk of delay in starting, the engine should be started and run for several minutes at intervals of an hour or so in order to keep the engine heat at or near running temperatures. CAUTION: *Do not race the engine after starting to increase heat rapidly, as the oil will not flow freely enough to lubricate parts, resulting in unnecessary wear and perhaps more immediate damage.*

e. **Initial Movement of Vehicle.**

(1) After engine has been warmed up, engage clutch and maintain engine speed at fast idle for 5 minutes or until gears can be engaged. Put transmission in low (first) gear, and drive vehicle for 100 yards, being careful not to stall engine. This will heat gear lubricants to the point where normal operation can be expected.

(2) Brake bands, particularly on new vehicles, have a tendency to bind when they are very cold. Always have a blowtorch handy to warm up these parts if they bind when an attempt is made to move the vehicle. Parking the vehicle with the brake released will eliminate

10-TON 6 x 4 TRUCK (MACK MODEL NR)

most of the binding. Precaution must be taken, under these circumstances, to block the wheels or otherwise prevent movement of the vehicle.

f. Effects of Low Temperatures on Metals. Inspect the vehicle frequently. Shock resistance of metals, or resistance against breaking, is greatly reduced at extremely low temperatures. Operation of vehicles on hard, frozen ground causes strain and jolting which will result in screws breaking or nuts jarring loose.

9. OPERATION IN SANDY OR DESERT TERRAIN.

a. Reference. Refer to Training Circular No. 2 (6 January 1943) entitled Desert Operation of Motor Vehicles, as well as FM 31-25.

b. Tires. For better traction in sand, reduce tire pressures somewhat to gain more tire surface on which to travel, but do not soften tires to an extent that they develop excessive heat; and for the same reason travel at slower speeds when tire pressures are reduced. As soon as the sandy ground has been traversed, do not fail to inflate to proper pressures.

c. Air Cleaners. Be sure that engine air cleaners, breathers and breather caps are kept clean at all times. Daily inspection and cleaning is recommended, with even more frequent attention when air is especially sand- or dust-laden.

d. Oil Filter. Examine engine oil every day. Wipe engine dip stick on white or light-colored clean rag. Oil will be absorbed and dirt particles will be visible if present. If the oil seems gritty, or shows dirt particles on the rag, report the condition immediately, as it is an indication that oil filter is not functioning properly. The filter should have attention, and fresh oil should be supplied.

e. General Hints on Desert Operation.

(1) Select and use a gear that will keep the vehicle steadily under way with sufficient power to keep it moving, but avoid the risk of stalling the engine.

(2) When the vehicles are in column, it is better to follow in the tracks of the leading vehicle. Careful reconnaissance will determine the best route for traversing sandy terrain.

(3) The instruction for operation in extreme heat (par. 7) will apply in cases where this condition prevails.

10. OPERATION IN MUD.

a. Do not spin the wheels in an effort to get out of muddy places. Pull out slowly in low gear. If this does not succeed, obtain the help of another vehicle, using a tow connection. Do not "rock out." This puts a severe strain on the vehicle, and even if it is successful the wheels will have dug in so that following vehicles will have greater difficulty.

b. Reinforce a soft surface by placing brushwood, timbers, stones and other material under the wheels. Keep the vehicle in motion and

OPERATION UNDER UNUSUAL CONDITIONS

avoid stopping, starting, or shifting gears, as these are all factors contributing to loss of traction.

11. OPERATION IN SNOW AND ICE.

a. General Operation and Safety Hints.

(1) When halting on wet, slushy, snowy or icy surface, place weeds, brush, boards, or other material on the ground in proper position, and drive the truck upon them to prevent its freezing while parked.

(2) Use customary precautions when driving on ice so that skidding does not occur. Descend grades and hills slowly, use care on turns, and if necessary, partially deflate tires when travel is especially slippery, to give more traction surface. As soon as conditions are better, inflate the tires to proper pressures.

(3) When stuck in snow, try to "rock out" by alternate forward and rearward motion without spinning the wheels, keeping front wheels straight. If rocking fails, use the pioneer tool equipment to dig out the truck.

12. OPERATION IN FLOOD.

a. When a vehicle has been submerged by flood or other cause, it is necessary that complete overhaul of all units be performed, therefore notify higher authority. CAUTION: *No starting of the engine or other running should be permitted until all units have been completely dried, cleaned and overhauled.*

TM 9-818
13

10-TON 6 x 4 TRUCK (MACK MODEL NR)

Section V
INSPECTION AND PREVENTIVE MAINTENANCE SERVICE

	Paragraph
Purpose	13
Before-operation service	14
During-operation service	15
At-halt service	16
After-operation service and weekly service	17

13. PURPOSE.

a. To insure mechanical efficiency it is necessary that the vehicle be systematically inspected at intervals each day it is operated, also weekly, so that defects may be discovered and corrected before they result in serious damage or failure. Certain scheduled maintenance services will be performed at these designated intervals. The services set forth in this section are those performed by driver before operation, during operation, at halt, after operation and weekly.

b. Driver preventive maintenance services are listed on the back of "Driver's Trip Ticket and Preventive Maintenance Service Record," W.D. Form No. 48, to cover vehicles of all types and models. Items peculiar to specific vehicles, but which are not listed on W.D. Form No. 48 are covered in manual procedures under the items with which they are related. Certain items listed on the form that do not pertain to the vehicle involved are eliminated from the procedures as written into the manual. Every organization must thoroughly school each driver in performing the maintenance procedures set forth in manuals whether or not they are listed specifically on W.D. Form No. 48.

c. The items listed on W.D. Form No. 48 that apply to this vehicle are expanded in this manual to provide specific procedures for accomplishment of the inspections and services. These services are arranged to facilitate inspection, and to conserve the time of the driver, and are not necessarily in the same numerical order as shown on W.D. Form No. 48. The item numbers, however, are identical with those shown on that form.

d. The general inspection of each item applies also to any supporting member or connection, and generally includes a check to see whether the item is in good condition, correctly assembled, secure, or excessively worn.

(1) The inspection for "good condition" is usually an external visual inspection to determine whether the unit is damaged beyond safe or serviceable limits. The term "good condition" is explained further by the following: not bent or twisted, not dented or collapsed, not torn or cut.

(2) The inspection of a unit to see that it is "correctly assembled" is usually an external visual inspection to see whether it is in its normal assembled position in the vehicle.

TM 9-818
13-14

INSPECTION AND PREVENTIVE MAINTENANCE SERVICE

(3) The inspection of a unit to determine if it is "secure" is usually an external visual examination, a hand-feel, or a pry-bar check for looseness. Such an inspection should include any brackets, lock washers, lock nuts, locking wires, or cotter pins used in assembly.

(4) "Excessively worn" will be understood to mean worn close to, or beyond, serviceable limits, and likely to result in a failure if not replaced before the next scheduled inspection.

e. Any defects or unsatisfactory operating characteristics beyond the scope of first echelon to correct, must be reported at the earliest opportunity to the designated individual in authority.

14. BEFORE-OPERATION SERVICE.

a. This inspection schedule is designed primarily as a check to see that the vehicle has not been tampered with or sabotaged since the After-operation Service was performed. Various combat conditions may have rendered the vehicle unsafe for operation, and it is the duty of the driver to determine whether or not the vehicle is in condition to carry out any mission to which it is assigned. This operation will not be entirely omitted, even in extreme tactical situations.

b. Procedures. Before-operation Service consists of inspecting items listed below according to the procedure described, and correcting or reporting any deficiencies. Upon completion of the service, results should be reported promptly to the designated individual in authority.

(1) ITEM 1, TAMPERING AND DAMAGE. Look around the vehicle and examine wheels, tires, bogie assembly, and engine for evidence of tampering or damage since parking vehicle. Inspect load for shifting and make sure that load fastenings and tarpaulins are fastened securely.

(2) ITEM 2, FIRE EXTINGUISHER. Inspect portable fire extinguisher to see that it is in place and securely mounted. Shake to determine contents.

(3) ITEM 3, FUEL, OIL, AND WATER. Inspect fuel tank, lines, radiator, hose connections and oil pan for evidence of leaks or tampering. Read fuel level on instrument panel gage, engine oil level on bayonet gage, fuel injection pump level at pump, and water level in radiator; add if necessary.

(4) ITEM 4, ACCESSORIES AND DRIVES. Inspect generator and fan belts for condition and alinement. Belts must show one-half inch deflection under hand pressure on the long side of the drive.

(5) ITEM 5, AIR BRAKE TANKS. Examine air reservoir tanks, lines, and connections for evidence of tampering or damage. Drain condensation from the air tanks.

(6) ITEM 6, LEAKS: GENERAL. Look on ground under vehicle, around hubs, axles, transmission, and radiator for evidence of leaks. Trace all leaks to their source and correct or report them.

(7) ITEM 7, ENGINE WARM-UP. Push in emergency stop control, push starting button, and start engine. Note action of starting mecha-

10-TON 6 x 4 TRUCK (MACK MODEL NR)

nism, particularly whether cranking motor has adequate cranking speed and engages and disengages properly without unusual noise. Run the engine at 800 to 1000 revolutions per minute until temperature reaches at least 165°F before moving the vehicle.

(8) ITEM 9, INSTRUMENTS.

(a) Oil Pressure. Observe oil pressure gage. Normal pressure with engine idling is 10 to 20 pounds. Normal pressure at 1,800 to 2,100 revolutions per minute is 45 to 60 pounds. Stop engine immediately when oil pressure does not register.

(b) Ammeter. Ammeter should show high (+) charging rate for first few minutes after starting, until generator has restored to battery current used in starting engine. High charging rate for an extended period may indicate a dangerously low battery, or a faulty regulator.

(c) Auxiliary Ammeter. Ammeter should show high (+) charging rate for first few minutes after starting, until generator has restored to battery current used in starting engine. High charging rate for an extended period may indicate a dangerously low battery, or a faulty regulator.

(d) Engine Temperature Gage. Engine temperature should rise slowly during warm-up period until 165°F is reached. Stop engine if temperature exceeds 212°F.

(e) Tachometer. Tachometer should indicate engine revolutions per minute without needle fluctuation or grind, and record the accumulated engine hours.

(f) Fuel Pressure Gage. The pressure indicator should read between 3 and 18 pounds at all times while engine is running. If above or below this reading, stop engine and investigate.

(g) Air Pressure Gage. During warm-up period gage should show a steady build-up until 80 to 105 pounds pressure is reached. If pressure exceeds 105 pounds, stop engine and investigate. Pressure should not drop excessively when brakes are applied.

(h) Fuel Gage. Fuel gage should register approximate amount of fuel in tank.

(9) ITEM 10, HORN AND WINDSHIELD WIPERS. Sound horn for proper tone, if tactical situation permits. Examine windshield wipers for proper operation and good blades.

(10) ITEM 11, GLASS AND REAR-VIEW MIRRORS. Inspect all glass for cracks or breaks, clean windshield, and adjust rear-view mirrors.

(11) ITEM 12, LAMPS. With all light switches in "ON" position, including stop light, inspect all lights to see that they are clean, securely mounted, operating properly, and that they go out when switches are turned "OFF."

(12) ITEM 13, WHEEL AND FLANGE NUTS. Inspect all wheel and flange nuts to see that they are present and secure.

(13) ITEM 14, TIRES. Inspect for flat or low-pressure, tires. Front tires should have 70 pounds pressure; rear tires, 90 pounds pressure. Examine tires, wheels, rims and valve caps for good condition. After operations, note any evidence of damage since previous service.

TM 9-818
14-15

INSPECTION AND PREVENTIVE MAINTENANCE SERVICE

(14) ITEM 15, SPRINGS AND SUSPENSIONS. Inspect springs and bogie suspensions for evidence of tampering or damage since parking the vehicle.

(15) ITEM 16, STEERING LINKAGE. Examine housing for leaks or loose mounting bolts, and all linkage for looseness or damaged condition.

(16) ITEM 17, FENDERS AND BUMPERS. Make sure that fenders and bumpers are securely mounted and have not been damaged since parking the vehicle.

(17) ITEM 18, TOWING CONNECTIONS. Towing and pintle hooks must be in serviceable condition, latch must operate freely and pintle pin must be attached to chain.

(18) ITEM 19, BODY, LOAD AND TARPAULINS. Inspect body for good condition, secure mounting, for any evidence of tampering or shifting of the load, and make sure that tarpaulin is in good condition and securely fastened.

(19) ITEM 20, DECONTAMINATOR. Make sure decontaminator is in place and securely mounted. Shake to determine contents.

(20) ITEM 21, TOOLS AND EQUIPMENT. Inspect all tools and equipment for condition, proper stowage, and serviceability, using the vehicle stowage list.

(21) ITEM 22, ENGINE OPERATION. Engine should idle smoothly. After normal operating temperature has been reached, accelerate engine momentarily and note any misfiring, unusual noise, or excessive smoking.

(22) ITEM 23, DRIVER'S PERMIT AND STANDARD ACCIDENT REPORT FORM No. 26. Driver must have his operator's permit on his person. Check to see that Form No. 26, Form No. 478, vehicle manuals, and Lubrication Guides are present in vehicle, legible, and properly stowed.

(23) ITEM 25, DURING-OPERATION CHECK. Immediately after putting the vehicle in motion, start the During-operation Service, in the nature of a road test.

15. DURING-OPERATION SERVICE.

a. While vehicle is in motion, listen for any sounds such as rattles, knocks, squeals, or hums that may indicate trouble. Look for indications of trouble in cooling system, and for smoke from any part of the vehicle. Be on the alert to detect any odor of overheated components or units such as generator, brakes or clutch, fuel vapor from a leak in fuel system, exhaust gas, or other signs of trouble. Any time the brakes are used, gears shifted, or vehicle turned, consider this a test, and notice any unsatisfactory or unusual performance. Watch the instruments constantly. Notice promptly any unusual instrument indication that may signify possible trouble in system to which the instrument applies.

10-TON 6 x 4 TRUCK (MACK MODEL NR)

b. **Procedures.** During-operation Service consists of observing items listed below according to the procedures following each item, and investigating any indications of serious trouble. Notice minor deficiencies to be corrected or reported at earliest opportunity, usually at next scheduled halt.

(1) ITEM 27, FOOT AND HAND BRAKES. With vehicle in motion apply foot brake; observe for smooth and effective braking, and see that air pressure does not drop excessively. With vehicle on a grade, apply the hand brake. Brake should hold vehicle stationary with one-third ratchet travel in reserve.

(2) ITEM 28, CLUTCH. While shifting gears note any chatter, squealing or slipping of the clutch; free travel of pedal should be 2 inches.

(3) ITEM 29, TRANSMISSION. Transmission gears should shift smoothly, operate without unusual noise, and not slip out of mesh. Stop vehicle in case of unusual noise.

(4) ITEM 30, ENGINE AND CONTROLS. Driver should always be on the alert for any deficiencies in engine operation such as lack of power on acceleration, excessive smoking, misfiring, or overheating. Note any binding or unusual operation of the engine control linkage.

(5) ITEM 32, INSTRUMENTS. Observe reading of all instruments frequently during operation for proper functioning.

(a) Oil Pressure. Oil pressure should be 10 to 20 pounds at idle speed, and 45 to 60 pounds at maximum speed.

(b) Ammeter. A high (+) charging rate for first few minutes after starting is normal; a high charging rate for an extended period may indicate a dangerously low battery, or a faulty regulator.

(c) Auxiliary Ammeter. A high (+) charging rate for first few minutes after starting is normal; a high charging rate for an extended period may indicate a dangerously low battery, or a faulty regulator.

(d) Engine Temperature Gage. Engine temperature should not exceed 212°F nor fall below 165°F.

(e) Tachometer. Tachometer should indicate engine revolutions per minute without needle fluctuation or grind, and record accumulated engine hours.

(f) Fuel Pressure Gage. The pressure indicator should read between 3 and 18 pounds at all times while engine is running.

(g) Air Pressure Gage. Air pressure should be between 80 and 105 pounds at all times during operation. Pressure should not drop excessively when brakes are applied; low pressure indicator should function at 60 pounds pressure.

(h) Speedometer. Speedometer should register speed of vehicle without fluctuation or noise, and odometer should record accumulated mileage.

(i) Fuel Gage. Fuel gage should register approximate amount of fuel in tank.

TM 9-818
15-16

INSPECTION AND PREVENTIVE MAINTENANCE SERVICE

(6) ITEM 33. STEERING GEAR. Note any binding of steering gear; observe any pulling to one side, wandering, or shimmy of vehicle.

(7) ITEM 34. RUNNING GEAR. Listen for any unusual noise in the wheels, axles, spring, and bogie that might indicate loose or damaged units.

(8) ITEM 35. BODY. Note any unusual noise that would indicate loose body mountings or attachments.

16. AT-HALT SERVICE.

a. At-halt Service may be regarded as minimum maintenance procedures and should be performed under all tactical conditions, even though more extensive maintenance services must be slighted or omitted altogether.

b. **Procedures.** At-halt Service consists of investigating any deficiencies noted during operation, inspecting items listed below according to the procedures following the items, and correcting any deficiencies found. Deficiencies not corrected should be reported promptly to the designated individual in authority.

(1) ITEM 38. FUEL, OIL, AND WATER. Determine quantity of fuel, oil, and water. Make sure supply is adequate for mission.

(2) ITEM 39. TEMPERATURES (HUBS, BRAKE DRUMS, TRANSMISSION AND AXLES). Hand-feel wheel hubs and brake drums for abnormal temperatures. If any wheel hub is too hot to touch, bearings may be inadequately lubricated, damaged, or improperly adjusted. Abnormal temperature at brake drums may be caused by dragging or improperly adjusted brake. Examine transmission and rear axle for overheating or excessive oil leaks.

(3) ITEM 40. AXLE AND TRANSFER VENTS. Inspect vents to see that they are present, secure and not clogged. Remove and clean if necessary.

(4) ITEM 41. PROPELLER SHAFTS. Inspect shafts and universal joints for looseness, damage, or excessive lubricant leaks.

(5) ITEM 42. SPRINGS AND BOGIES. Examine springs and bogies for loose, broken, or shifted leaves, and loose mounting bolts.

(6) ITEM 43. STEERING LINKAGE. Inspect Pitman arm and linkage for looseness or damage. Inspect for grease leaks at steering housing.

(7) ITEM 44. WHEEL AND FLANGE NUTS. All wheel and flange nuts must be present and secure.

(8) ITEM 45. TIRES. Check tire pressure. (Front tires should have 70 pounds pressure; rear tires, 90 pounds pressure.) Look for missing valve caps and examine tires for cuts or bruises.

(9) ITEM 46. LEAKS. Look on ground under vehicle and around radiator, axles and transmission for evidence of fuel oil or water leaks.

(10) ITEM 47. ACCESSORIES AND BELTS. Examine accessories for loose mountings and frayed or broken belts. Belts must show $\frac{1}{2}$-inch deflection under hand pressure, on the long side of the drive.

TM 9-818
16-17

10-TON 6 x 4 TRUCK (MACK MODEL NR)

(11) ITEM 48, AIR CLEANER. Inspect air cleaner for good condition and secure mounting. When operating under extreme dust or sand conditions, make sure filter element is not clogged; remove and clean if necessary.

(12) ITEM 49, FENDERS AND BUMPERS. Inspect fenders and bumpers for loose mountings or damaged condition.

(13) ITEM 50, TOWING CONNECTIONS. Towing and pintle hooks must be in serviceable condition; pintle latch must operate freely and lock pin must be attached to chain.

(14) ITEM 51, BODY, LOAD AND TARPAULINS. Inspect load for shifting and see that all load and tarpaulin fastenings and hold-downs are in good condition and securely mounted.

(15) ITEM 52, APPEARANCE AND GLASS. Clean windshield rearview mirror and all light lenses. Inspect for good condition and secure mounting.

17. AFTER-OPERATION SERVICE AND WEEKLY SERVICE.

a. After-operation Service is particularly important, because at this time the driver inspects his vehicle to detect any deficiencies that may have developed and corrects those he is permitted to handle. He should report promptly, to the designated individual in authority, the results of his inspection. If this schedule is performed thoroughly, the vehicle should be ready to roll again on a moment's notice. The Before-operation Service, with a few exceptions, is then necessary only to ascertain whether the vehicle is in the same condition in which it was left upon completion of the After-operation Service. The After-operation Service should never be entirely omitted, even in extreme tactical situations, but may be reduced to the bare fundamental services outlined for the At-halt Service if necessary.

b. **Procedures.** When performing the After-operation Service, the driver must remember and consider any irregularities noticed during the day in the Before-operation, During-operation, and At-halt Services. The After-operation Service consists of inspecting and servicing the following items. Those items of the After-operation Service which are marked by an asterisk (*) require additional weekly services, the procedures for which are indicated in step *(b)* of each applicable item.

(1) ITEM 54, FUEL, OIL, AND WATER. Fill the fuel tank; check the crankcase oil level on the bayonet gage, the fuel injection pump level at pump, and water level in radiator; add supplies as necessary to bring to proper levels.

(2) ITEM 55, ENGINE OPERATION. Observe engine for smooth idle, without stalling. Accelerate and decelerate engine and listen for any unusual noise or excessive smoking of the engine. Investigate any deficiencies observed during operation.

(3) ITEM 56, INSTRUMENTS. Before stopping engine, inspect instruments for secure mounting, good condition, and proper readings.

TM 9-818
17

INSPECTION AND PREVENTIVE MAINTENANCE SERVICE

(a) *Oil Pressure Gage*. Normal oil pressure is 10 to 20 pounds at engine idle, and 45 to 60 pounds at 1,800 to 2,100 revolutions per minute.

(b) *Ammeter*. Normal ammeter reading is zero or slight (+) charge at idle speed, with light switches "OFF."

(c) *Auxiliary Ammeter*. Normal ammeter reading is zero or slight (+) charge at idle speed, with light switches "OFF."

(d) *Engine Temperature Gage*. Engine temperature should not exceed 212°F, nor fall below 165°F.

(e) *Fuel Pressure Gage*. The fuel pressure indicator should register between 3 and 18 pounds at all times while engine is running.

(f) *Air Pressure Gage*. Air pressure should be between 80 and 105 pounds; pressure must not drop excessively when brakes are applied.

(g) *Fuel Gage*. Fuel gage should register approximate amount of fuel in tank; recheck after filling tank.

(h) *Tachometer*. Tachometer should indicate engine revolutions per minute without fluctuation of needle or noise, and should record accumulated engine hours.

(4) ITEM 57, HORN AND WINDSHIELD WIPERS. Sound horn for proper tone if tactical situation permits. Examine windshield wipers for proper operation and good blades.

(5) ITEM 58, GLASS AND REAR-VIEW MIRRORS. Inspect all glass for cracks or breaks. Clean the windshield and adjust rear-view mirror.

(6) ITEM 59, LAMPS AND REFLECTORS. With all light switches in the "ON" position, including stop light, inspect all lights to see that they are in good condition, securely mounted, clean, burning, and that they go out when switches are turned "OFF."

(7) ITEM 60, FIRE EXTINGUISHERS. Inspect portable fire extinguisher to see that it is in place and securely mounted. Shake to determine contents.

(8) ITEM 61, DECONTAMINATOR. Make sure decontaminator is in place and securely mounted. Shake to determine contents. Decontaminator must be recharged every 90 days, as contents deteriorate.

(9) ITEM 62*, BATTERY.

(a) Battery connections and mountings should be kept clean and tight. Electrolyte level should be three-eighths inch above top of plates. Add fresh, clean water if necessary.

(b) *Weekly*. Examine battery for cracks and leaks, tighten all terminals and mountings, clean corroded terminals, and apply coating of grease.

(10) ITEM 63, ACCESSORIES AND BELTS.

(a) Inspect generator fan and belts to see that they are securely mounted and properly alined. Belts must show ½-inch deflection, under hand pressure on the long side of the drive.

10-TON 6 x 4 TRUCK (MACK MODEL NR)

(b) *Weekly*. Examine fuel pump, generator, regulator, starter and fan for loose mounting bolts. Inspect belts for good condition and proper adjustment.

(11) ITEM 64*, ELECTRICAL WIRING.

(a) With engine stopped and all switches in "OFF" position, ammeter should read zero. Examine loom and wiring for damaged or frayed condition, and make certain that all electrical connections are secure.

(b) *Weekly*. Wipe oil and dirt from wiring and inspect for chafed, cracked, or deteriorated condition. Make sure all connections are secure.

(12) ITEM 65*, AIR CLEANERS AND BREATHER CAPS.

(a) When vehicle has been operated under extreme dust or sand conditions make sure filter element is not clogged; remove and clean if necessary.

(b) *Weekly*. Remove and thoroughly clean elements and sumps. Refill sumps to proper level with engine oil. Reassemble and tighten all mountings securely.

(13) ITEM 66*, FUEL FILTERS.

(a) Remove top cap on primary filter and turn the scraper one full turn.

(b) *Weekly*. Remove bottom plug on primary and each secondary filter, and drain off approximately one-quarter pint.

(14) ITEM 67, ENGINE CONTROLS. Examine the engine and accessory controls for loose, worn, or binding linkage.

(15) ITEM 68*, TIRES.

(a) Check tire pressure (front, 70 lbs; rear, 90 lbs). Replace missing valve caps, examine tires for cuts, bruises, or fractures, and remove any foreign objects such as stones, nails, or glass from the tread.

(b) *Weekly*. Match tires according to tread and over-all circumference.

(16) ITEM 69*, SPRINGS AND SUSPENSIONS.

(a) Inspect springs for sag, broken or shifted leaves, loose or missing U-bolts, eyebolts and rebound clips.

(b) *Weekly*. Inspect torque rods for alinement, bushings for deterioration, trunnion cap for grease leaks, and spring housing cap for looseness.

(17) ITEM 70, STEERING LINKAGE. Examine Pitman arm, drag link, and tie rod for loose connections or damaged condition. Steering stop screws must be in position and lock nut must be tight.

(18) ITEM 71, PROPELLER SHAFTS. Examine propeller shafts for alinement, and universal joints for loose mountings or excessive grease leaks.

TM 9-818
17

INSPECTION AND PREVENTIVE MAINTENANCE SERVICE

(19) Item 72*, Axle and Transfer Vents.

(a) Wipe off excessive dirt or grease from vents and examine for clogged condition. Remove and clean if necessary.

(b) *Weekly.* Clean vents thoroughly.

(20) Item 73, Leaks and (General). Start engine and inspect all fuel, oil, and water connections for evidence of leaks. Look on ground under vehicle and around axles and transmission for grease leaks.

(21) Item 74, Gear Oil Levels. Inspect transmission and driving axles for proper oil level. Oil must not be more than one-half inch below filler hole when cold, and not above bottom edge of filler hole when hot.

(22) Item 75*, Air Brakes. Inspect all tank and air line connections for good condition, secure mounting and leaks. Open tank pet cocks and drain off condensation.

(23) Item 76, Fenders and Bumpers. Inspect fenders and bumpers for good condition and secure mounting. Tighten all loose mounting bolts. Report to designated authority if in damaged condition.

(24) Item 77*, Towing Connections.

(a) Towing and pintle hooks must be in serviceable condition, and lock pin must be attached to chain.

(b) *Weekly.* Lubricate and free-up any binding condition of pintle hook or latch.

(25) Item 78, Body, Load and Tarpaulins. Inspect body for loose mountings, load for shifting and tarpaulin for tears, loose or missing hold-down ropes and fastenings.

(26) Item 82*, Tighten (Wheel Rim and Axle Drive Flange).

(a) Inspect to see that all wheel rim and axle drive flange bolts and nuts are present and tightened securely.

(b) *Weekly.* Inspect all units for missing or loose mounting bolts and nuts; replace or tighten as necessary.

(27) Item 83*, Lubricate as Needed.

(a) Oil and lubricate all parts as required when performing After-operation Service. For specific intervals and lubricants to be used, refer to Lubrication Guide (par. 19).

(b) *Weekly.* Lubricate as inspections reveal the need of lubrication at this time.

(28) Item 84*, Clean Engine and Vehicle.

(a) Clean all dirt from inside of cab and body. Remove excessive dirt and grease from the engine and exterior of vehicle. Make sure identification markings are visible.

(b) *Weekly.* Wash vehicle if possible.

(29) Item 85*, Tools and Equipment.

(a) Inspect all tools and equipment for condition, proper stowage and serviceability, using the vehicle stowage list.

(b) *Weekly.* Clean, put in good condition, and stow all tools and equipment. Replace or report missing tools or equipment.

TM 9-818
18-19

10-TON 6 x 4 TRUCK (MACK MODEL NR)

Section VI

LUBRICATION

	Paragraph
Introduction	18
Lubrication Guide	19

18. INTRODUCTION.

a. Lubrication is an essential part of preventive maintenance, determining to a great extent the serviceability of parts and assemblies.

19. LUBRICATION GUIDE.

a. **General.** Lubrication instructions for this materiel are consolidated in a Lubrication Guide (fig. 31). These specify the points to be lubricated, the periods of lubrication, and the lubricant to be used. In addition to the items on the Guide, other small moving parts, such as hinges and latches, must be lubricated at frequent intervals.

b. **Supplies.** In the field it may not be possible to supply a complete assortment of lubricants called for by the Lubrication Guide to meet the recommendations. It will be necessary to make the best use of those available, subject to inspection by the officer concerned, in consultation with responsible ordnance personnel.

c. **Lubrication Notes.** The following notes apply to the Lubrication Guide (fig. 31). All note references in the Guide itself are to the steps below having the corresponding numbers.

(1) FITTINGS. Clean before applying lubricant. Lubricate until new lubricant is forced from the bearing, unless otherwise specified. CAUTION: *Lubricate chassis points after washing truck.*

(2) INTERVALS. Intervals indicated are for normal service. For extreme conditions of speed, heat, water, sand, mud, snow, rough roads, dust, etc., reduce interval by $\frac{1}{3}$ or $\frac{1}{2}$, or more if conditions warrant.

(3) AIR CLEANERS. Daily, check level and refill oil reservoir of engine air cleaner to bead level with used crankcase oil or OIL, engine, crankcase grade. Every 2,000 miles, daily under extreme dust conditions, remove and wash all parts. Every 250 miles or more often in dusty operation, drain crankcase breather and compressor air cleaner, and refill to bead level with used crankcase oil or OIL, engine, crankcase grade. Every 1,000 miles, remove breather and air cleaner, and wash all parts. Every 6,000 miles, remove and wash brake governor strainer located in small housing at base of governor. Proper maintenance of air cleaners is essential to prolonged engine life.

(4) CRANKCASE. Drain only when engine is hot. Refill to "FULL" mark on gage. Run engine a few minutes and recheck oil level. CAUTION: *Be sure pressure gage indicates oil is circulating.*

56

LUBRICATION

(5) OIL FILTER. Every 1,000 miles, drain oil filter housing to remove accumulated sediment. Every 6,000 miles or more often if filter becomes clogged, remove filter element, clean case and renew element. After renewing element, refill crankcase to "FULL" mark on gage. Run engine a few minutes, recheck level and add oil to "FULL" mark. Before reinstalling oil filter cover, make sure cover gasket is not damaged.

(6) GEAR CASES. Weekly, check level with truck on level ground and, if necessary, add lubricant to within $\frac{1}{2}$ inch of plug level when cold, or to plug level when hot. Drain, flush and refill at intervals indicated on Guide. When draining, drain immediately after operation. Keep housing vents clean. When filling rear axle differentials, first fill the power divider housing and bevel gear housings to level of filler plug. Allow a few minutes for excess lubricant to drain into differential main case, then fill main case to proper level. A drain plug is provided for the power divider housing and bevel gear housings. These should be removed when draining the differentials. To flush, fill cases to about one-half capacity with OIL, engine, SAE 10. Operate mechanism within cases slowly for several minutes and redrain. Replace drain plugs and refill cases to correct level with lubricant specified on Guide.

(7) FUEL FILTERS. Daily, remove either upper or lower cap of No. 1 filter. Insert suitable tool in the hole in the shaft and turn shaft clockwise two revolutions. Make sure that the packing gland at ends of shaft and cap gasket prevent fuel leakage. Every 6,000 miles, remove drain plug to drain accumulated water and sediment. CAUTION: *Do not run engine with drain plug removed.* Keep vent holes in fuel tank filler cap open. The element of No. 2 filter should be replaced when during operation the fuel pressure drops to 0-3 pounds, as indicated by the fuel pressure gage. The element of No. 3 filter should be replaced when fuel pressure increases to 18-60 pounds Install new cover gaskets and tighten retaining nuts uniformly to prevent leakage. Every 6,000 miles, remove drain plugs to drain accumulated water and sediment. When air enters the fuel system as may be caused by replacement of the element in No. 2 or 3 fuel filter, running out of fuel, or the loosening of a line between the transfer pump and the injection pump, the fuel system must be primed. To prime the fuel system, unscrew the plunger knob of the priming pump located on the fuel injection pump. Operate the priming pump until the fuel pressure gage indicates 15 pounds, then continue pumping a few strokes to insure expulsion of all air. Screw priming pump plunger in place. In instances where priming as described does not prove effective, refer to the technical manual for additional information on bleeding and priming the fuel system.

(8) WHEEL BEARINGS (FRONT AND REAR). To clean and pack wheel bearings properly, they must be removed from the hub. Follow the procedure below:

TM 9-818
19

10-TON 6 x 4 TRUCK (MACK MODEL NR)

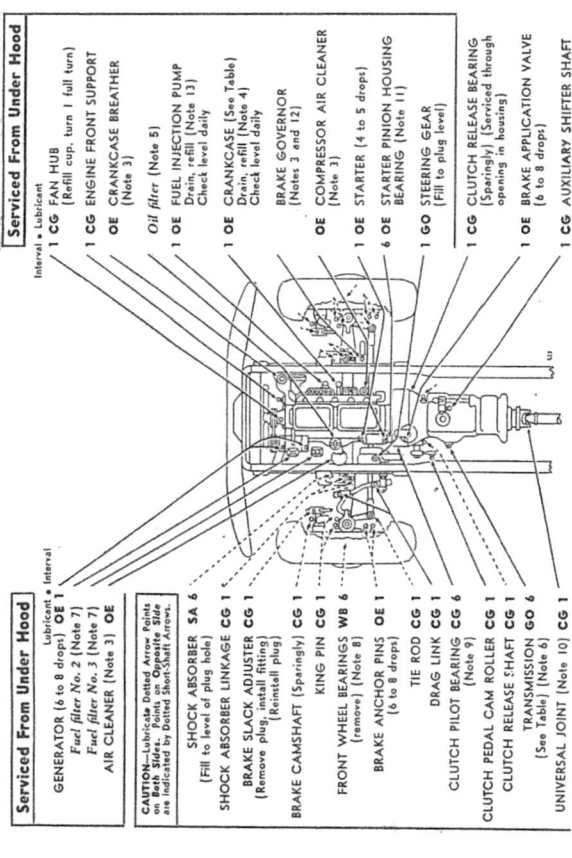

TM 9-818
19

LUBRICATION

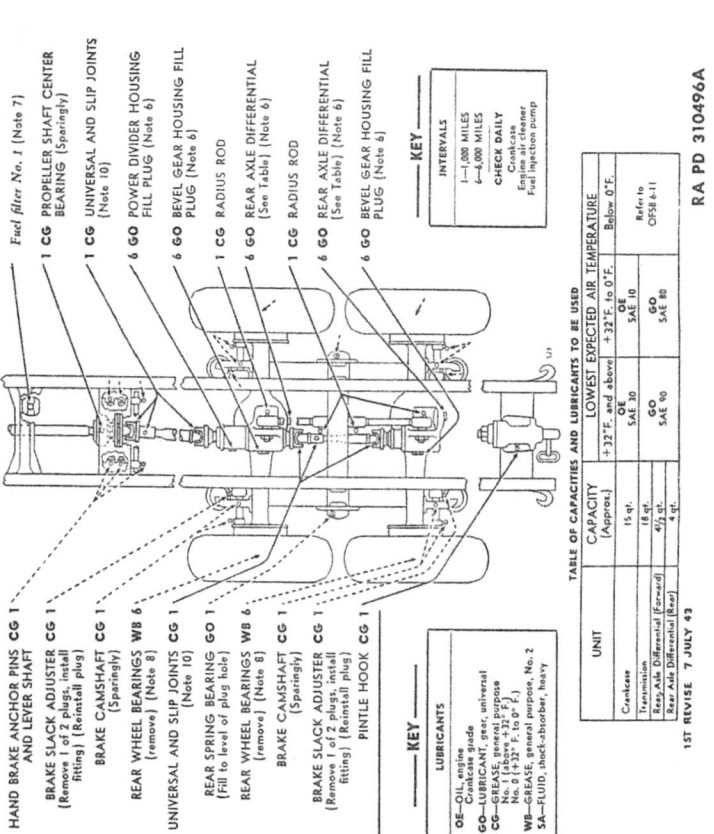

Figure 31—Lubrication Guide

TM 9-818
19

10-TON 6 x 4 TRUCK (MACK MODEL NR)

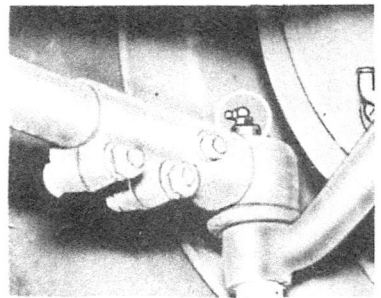

TIE ROD END — CG
Two fittings (one each end). Pressure gun. Apply lubricant until new grease begins to come out of joint.

DRAG LINK — CG
Two fittings (one each end). Pressure gun. Apply lubricant until new grease begins to come out of connection.

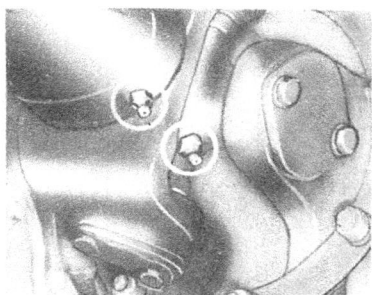

UNIVERSAL JOINT, NEEDLE BEARINGS CG
Five fittings (one each in five joints in propeller shaft drive line). Pressure gun. Apply lubricant until it begins to show at relief valve at center of cross. (See Note 10).

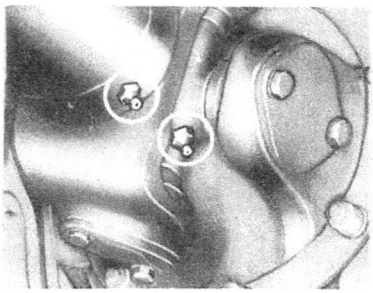

UNIVERSAL SLIP JOINT — CG
Two fittings (propeller shaft — one at disk brake and one to rear of front rear axle). Pressure gun. Apply lubricant until it begins to show at relief opening. (See Note 10).

PROPELLER SHAFT CENTER BEARING — CG
One fitting. Pressure gun. Add lubricant until pressure builds up.

LEVER SHAFT (FOR DISK BRAKE) — CG
Two fittings (one each end). Pressure gun. Apply lubricant until new grease begins to show between bearing brackets and levers.

RA PD 310528

Figure 32 — Localized Lubrication Views

TM 9-818
19

LUBRICATION

RADIUS ROD, AXLE END — CG
Two fittings (one at each rear axle). Presssure gun. Apply lubricant until new grease begins to come out of connection.

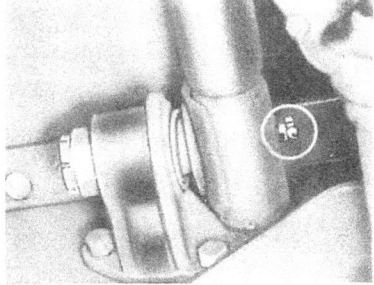

RADIUS ROD, CROSS-MEMBER END — CG
Two fittings (one for each rod). Pressure gun. Apply lubricant until new grease begins to come out of connection.

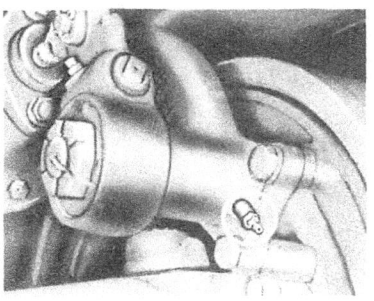

SLACK ADJUSTER — CG
Six plugs (one at each wheel — fitting temporarily installed). Pressure gun. Remove plug and install fitting, add lubricant until pressure builds up, remove fitting and reinstall plug.

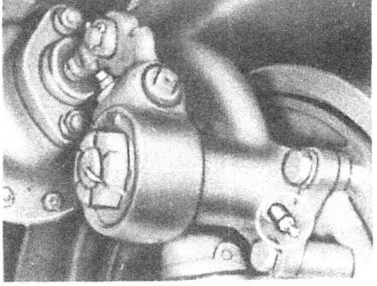

BRAKE CAMSHAFT, FRONT — CG
Two fittings (one at each front wheel). Hand grease gun. Apply lubricant until pressure increase is felt. Do not over-lubricate.

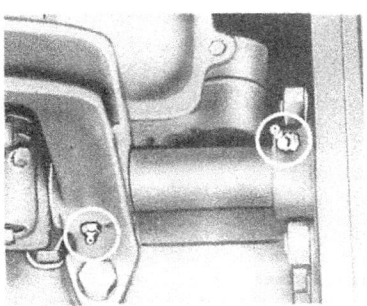

BRAKE CAMSHAFT, REAR — CG
Eight fittings (two at each rear wheel). Hand grease gun. Apply lubricant until pressure increase is felt. Do not over-lubricate.

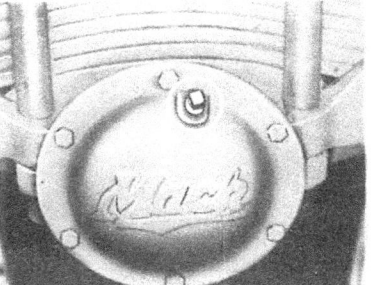

REAR SPRING BEARING — GO
Two filler plugs (one for right and one for left spring). Pump gun. Remove plug and add lubricant to level of filling hole. Reinstall plug.

RA PD 310529

Figure 33—Localized Lubrication Views

TM 9-818
19

10-TON 6 x 4 TRUCK (MACK MODEL NR)

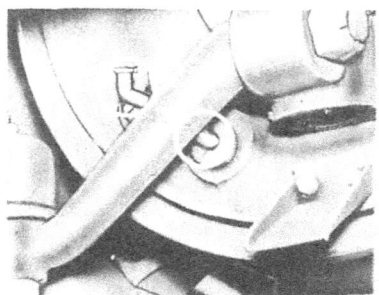

BRAKE ANCHOR PINS, FRONT SERVICE BRAKES — OE
Four oil cups (two at each front wheel). Oil can. Place 6 to 8 drops of oil in each oil cup.

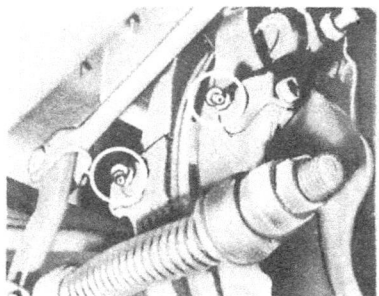

BRAKE ANCHOR PINS, HAND (DISK) BRAKE — CG
Four fittings (two on each side). Pressure gun. Apply lubricant until new grease begins to show between shoes and brackets.

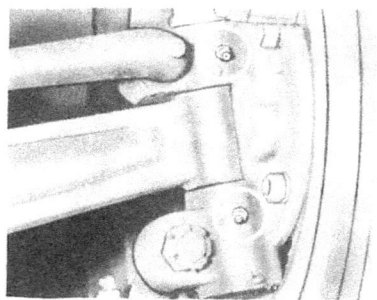

KING PIN BEARINGS — CG
Four fittings (two at each front wheel). Pressure gun. Apply lubricant until it begins to come out between axle end and knuckle.

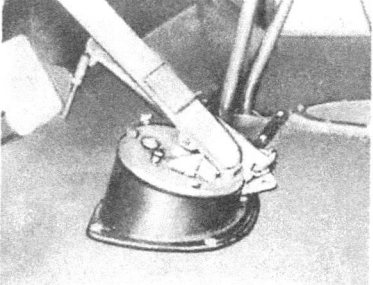

BRAKE APPLICATION VALVE — OE (SAE 10)
One place. Oil can. Apply 6 to 8 drops of oil.

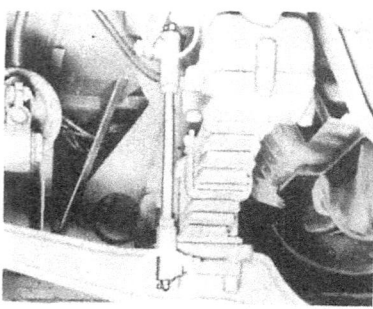

SHOCK ABSORBER LINK — CG
Four fittings (one at each end of link for left shock absorber and one at each end of link on right side). Pressure gun. Apply lubricant until new grease begins to come out of connections.

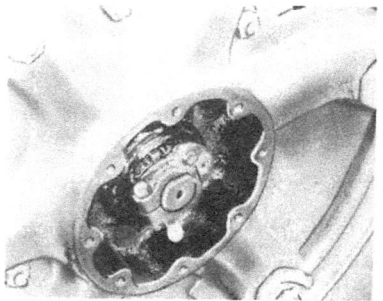

WHEEL BEARING, FRONT — WB
Two places (hub of each front wheel). Hand pack. Remove wheel. Thoroughly clean bearings and cavity. If solvent and air are used do not spin bearings with air blast. Inspect bearings and make necessary renewals. Pack bearings and pack cavity ½ full; do not overfill; Mount wheel and adjust bearings. (See Note B).

RA PD 310530

Figure 34—Localized Lubrication Views

TM 9-818
19

LUBRICATION

WHEEL BEARINGS, REAR — WB
Four places (hub of each rear wheel). Hand Pack. Remove Wheel. Thoroughly clean bearings and cavity. If solvent and air are used do not spin bearings with air blast. Inspect bearings and make necessary renewals. Pack bearings and pack cavity ½ full; do not overfill. Mount wheel and adjust bearings. (See Note 8).

STEERING GEAR HOUSING — GO
One filler plug. Pump gun. Remove plug and add lubricant to level of filling hole. Reinstall plug.

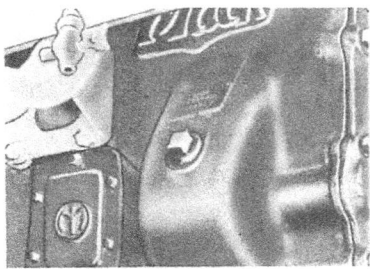

TRANSMISSION — GO
One filler plug. Pump gun. Remove plug. Check oil level; if low, add oil to within ½ in. of plug level when cold; to plug level when hot. Reinstall plug. Clean vent. Drain, flush and refill at intervals indicated on guide. (See Table) (See Note 6).

POWER DIVIDER HOUSING — GO
One filler plug (front rear axle). Pump gun. Remove plug and check oil level; if low, add oil to within ½ in. of plug level when cold, or to plug level when hot. Reinstall plug. Clean vent. Drain, flush and refill at intervals indicated on guide. (See Table) (See Note 6).

BEVEL GEAR HOUSING — GO
Two filler plugs (one on each rear axle). Pump gun. Remove plug and check oil level; if low, add oil to within ½ in. of plug level when cold, or to plug level when hot. Reinstall plug. Clean vent. Drain, flush and refill at intervals indicated on guide. (See Table) (Note 6).

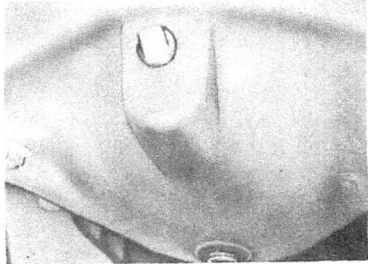

DIFFERENTIAL HOUSING — GO
Two filler plugs (one on each rear axle). Pump gun. Remove plug and check oil level; if low, add oil to within ½ in. of plug level when cold, or to plug level when hot. Reinstall plug. Clean vent. Drain, flush and refill at intervals indicated on guide. When draining also remove drain plugs from POWER DIVIDER HOUSING and BEVEL GEAR HOUSING. Refill after filling POWER DIVIDER HOUSING and BEVEL GEAR HOUSING. (See table) (See Note 6).

RA PD 310531

Figure 35 — Localized Lubrication Views

TM 9-818
19

10-TON 6 x 4 TRUCK (MACK MODEL NR)

GENERATOR — OE
Two oil cups (one at each end of generator). Oil can. Place 6 to 8 drops of oil in each oil cup.

STARTING MOTOR — OE
Three oil cups (one at each end of starting motor and one on inside, at Bendix drive shaft outer bearing). Oil can. Place 4 to 5 drops of oil in each outside oil cup (every 1,000 miles). Remove starting motor to oil inner cup, 6 to 8 drops of oil (every 6,000 miles).

CRANKCASE — OE
One place. Fill from container. Check level daily; if low, add oil to high level mark on oil level gage. Drain and refill at intervals indicated on guide. Drain when hot waiting long enough for thorough drainage. (See Note 4).

ENGINE AIR CLEANER — OE
One place. Fill from Container. Disassemble, clean and refill at intervals indicated on guide. (See Note 3).

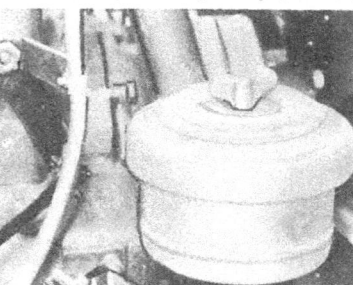

CRANKCASE BREATHER — OE
One place. Fill from container. Remove oil and renew to level of bead. Drain and refill at intervals indicated on guide. When it becomes necessary to wash the element use thin oil, not gasoline. (See Note 3).

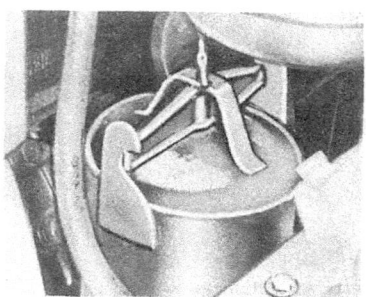

OIL FILTER — OE
One place. Fill from container. Remove drain plug from bottom of filter and draw off quart of oil while engine is running. Remove cover and filter element. Install new element and reinstall drain plug. Add oil to bring level to FULL mark on oil level gage of engine, recheck level and add more oil if necessary. (See Note 5).

RA PD 310532

Figure 36 — Localized Lubrication Views

TM 9-818
19

LUBRICATION

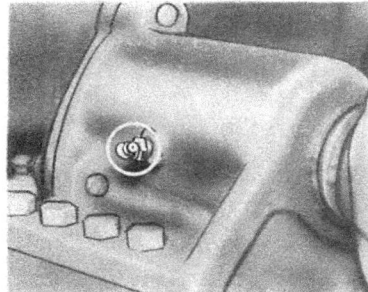

PINTLE HOOK — CG
One fitting. Pressure gun. Apply lubricant until housing is about ½ full.

FAN HUB — CG
One fitting. Screw-cap grease cup. Keep cup filled and screw down cap one turn.

ENGINE FRONT SUPPORT — CG
One fitting. Pressure gun. Apply lubricant until pressure is felt, or new grease begins to come out at side of trunnion.

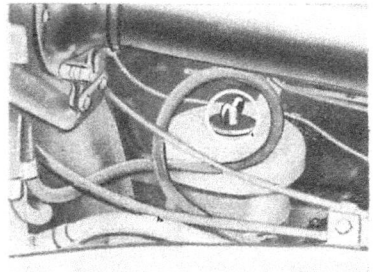

COMPRESSOR AIR CLEANER — OE
One place. Fill from container. Remove oil and refill at intervals indicated on guide. When it becomes necessary to wash the element use thin oill, not gasoline. (See Note 3).

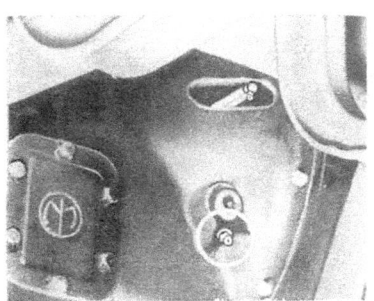

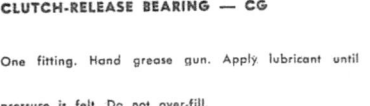

CLUTCH-RELEASE BEARING — CG

One fitting. Hand grease gun. Apply lubricant until pressure is felt. Do not over-fill.

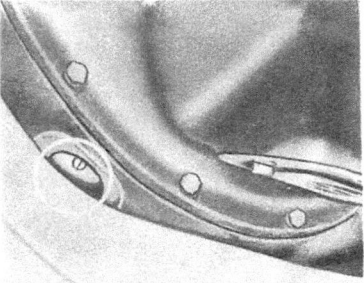

CLUTCH PILOT BEARING—CG
One plug (fitting temporarily installed). Hand grease gun. Remove cover plate from flywheel housing. Crank engine until plug in flywheel is accessible. Install fitting. Appy lubricant until pressure is felt. Do not over-fill. Remove fitting and re-install plug and cover plate. (See Note 9).

RA PD 310533

Figure 37 — Localized Lubrication Views

65

TM 9-818
19

10-TON 6 x 4 TRUCK (MACK MODEL NR)

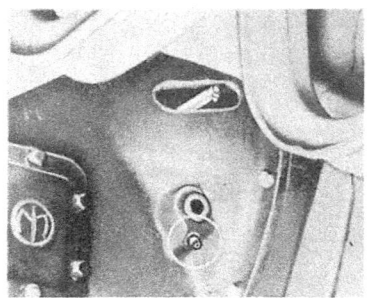

CLUTCH-RELEASE SHAFT, R.H. — CG
One fitting. Pressure gun. Apply lubricant until new grease begins to come out of bearing, or until pressure builds up.

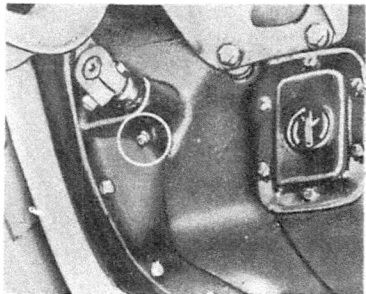

CLUTCH-RELEASE SHAFT, L.H. — CG
One fitting. Pressure gun. Apply lubricant until new grease begins to come out of bearing, or until pressure builds up.

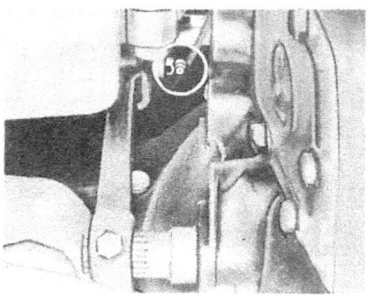

CLUTCH PEDAL CAM ROLLER — CG
One fitting. Pressure gun. Apply lubricant until new grease comes out between roller and yoke.

FUEL INJECTION PUMP — OE
One place. Fill from container. Check oil level in pump sump daily. If level has risen, drain excess at drain cock. Change oil when changing engine crankcase oil. (See Note 13).

No. 1 FUEL FILTER   SHAFT CAP

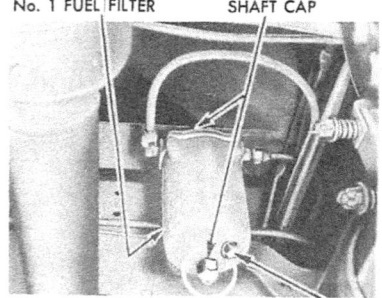

NO. 1 FUEL FILTER   DRAIN PLUG
One place (filter on chassis frame). Remove either upper or lower cap, insert tool through end of shaft and turn shaft clockwise two revolutions, daily. Remove drain plug to drain water and sediment, every 6,000 miles. (See Note 7).

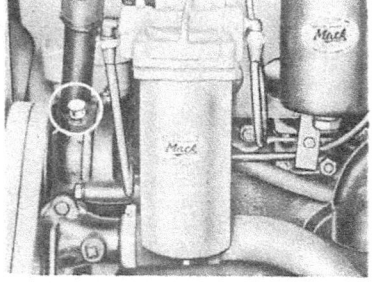

NO. 2 FUEL FILTER
One place (filter on engine). Renew element when fuel pressure drops to 0-3 lb. Remove drain plug to drain water and sediment, every 6,000 miles. (See Note 7).

RA PD 310534

Figure 38—Localized Lubrication Views

TM 9-818
19

LUBRICATION

(a) Remove bearings from hub and wash in SOLVENT, dry-cleaning, until all the old lubricant is removed from both inside and outside of cage.

(b) Lay bearings aside to dry and wash inside of hub and spindle with SOLVENT, dry-cleaning.

(c) When bearings are thoroughly dry, pack races with GREASE, general purpose, No. 2 and reassemble in hub. To satisfactorily pack a bearing it is necessary to knead lubricant by hand into the space between cage and inner race. Coat inside of hub and spindle with a thin coat of grease to prevent rusting. Do not fill hub. The lubricant packed in bearing races is sufficient to provide lubrication until the next service period. An excess may result in leakage of lubricant into the brake drum.

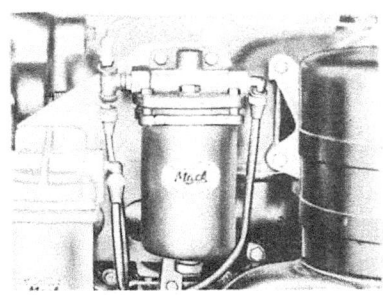

NO. 3 FUEL FILTER
One place (filter on engine). Renew element when fuel pressure increases to 18-60 lb. Remove drain plug to drain water and sediment, every 6,000 miles. (See Note 7).

MISCELLANEOUS LUBRICATION — OE
Places as noted. Oil can. Apply several drops of oil to all clevises, compressor rocker arm shaft and pintle hook latch. Oil all door hinges, hood hinges, hood locks and door catches. Oil every 1,000 miles. (See Note 14).

PRE-LUBRICATED PARTS — CG
Places at noted. Hand Coat. Clutch pedal: when disassembled from bracket for other reasons, coat pin and bushing lightly with lubricant before reassembling rear brake anchor pins; when rear brake shoes are removed for relining or other reasons coat anchor pins lightly with lubricant before reassembling. (See Note 16).

RA PD 310535

Figure 39—Localized Lubrication Views

(d) Replace wheel and adjust bearings according to instructions in technical manual.

(9) CLUTCH PILOT BEARING. Remove small plate on left side of flywheel housing. Crank engine until plug in flywheel is visible. Remove plug, install fitting and lubricate sparingly—do not overfill. Reinstall plug and plate.

(10) UNIVERSAL AND SLIP JOINTS. Apply GREASE, general purpose, seasonal grade, to universal joint until it overflows at relief valve, and to slip joint until lubricant is forced from vent at universal joint end of spline.

(11) STARTER PINION HOUSING BEARING. Every 6,000 miles, remove, clean and lubricate Bendix drive with 6 to 8 drops of OIL, lubricating, preservative, special. Lubricate Bendix drive shaft outer bearing through oiler with 6 to 8 drops of OIL, engine, crankcase grade.

10-TON 6 x 4 TRUCK (MACK MODEL NR)

(12) BRAKE GOVERNOR. Every 6,000 miles, remove governor housing cover and lubricate upper valve stem with OIL, lubricating, preservative, special.

(13) FUEL INJECTION PUMP. If oil in housing tends to rise, open drain cock to allow excess to drain. Keep filler cap element clean.

(14) OILCAN POINTS. Every 1,000 miles, lubricate throttle control linkage, hinges, clutch pedal shaft, latches, clevises, pintle hook latch and pin, compressor rocker arm shaft and hand brake control linkage with OIL, engine, crankcase grade.

(15) POINTS REQUIRING NO LUBRICATION SERVICE. These are the engine governor, water pump, compressor crankcase and springs.

(16) POINTS TO BE SERVICED AND/OR LUBRICATED BY ORDNANCE MAINTENANCE PERSONNEL. These are the rear wheel brake anchor pins, engine breather, flywheel clutch, .tachometer and speedometer cables.

TM 9-818
20

Section VII

TOOLS AND EQUIPMENT STOWAGE ON THE VEHICLE

	Paragraph
Tools	20
Equipment	21
Spare parts	22

20. TOOLS.

a. The following tools are carried on each truck for making minor adjustments, and for performing those operations which are inherent to driving operation, such as mounting a spare tire or giving immediate attention to a wheel bearing or a loose fastening. The tools are stowed under the right-hand seat in the cab except the oiler, which is carried in the bracket on the dash sheet, fire extinguisher which is inside of the cowl on the left-hand side, and the starting crank which is back of the driver's seat.

Tool	Federal Stock No.	Number Carried
Bag, tool	41-B-15	1
Chisel, mach's., hand, cold, alloy-S., ⅝-in. (on Models NR8 and NR9)	41-C-1116	1
Hammer, mach's., ball peen, 1 lb	41-H-523	1
Handle, wrench and tool, rim	....	1
Pliers, comb., slip-jt., 6 in.	41-P-1650	1
Punch, drive pin, solid, alloy-S., $\frac{5}{32}$-in. point, 5 in. long (on Models NR8 and NR9)	41-P-1650	1
Screwdriver, comm., hv.-duty, integral handle, 6 in.	41-S-1076	1
Screwdriver, comm., normal duty, 3-in. (on Models NR8 and NR9)	41-S-1101	1
Screwdriver, cross recess hd., No. 1, 3-in. blade	41-S-1636	1
Screwdriver, cross recessed hd., No. 2, 4-in. blade	41-S-1638	1
Screwdriver, cross recess-hd., No. 3, 6-in. blade (on Models NR8 and NR9)	41-S-1640	1
Wrench, adj., single-end, 12-in. ($1\frac{5}{16}$-in. cap)	41-W-488	1
Wrench, engr's., dble-hd. alloy-S., ⅜ and $\frac{7}{16}$	41-W-991	1
Wrench, engr's., dble-hd. alloy-S., ½ and $\frac{19}{32}$	41-W-1003	1
Wrench, engr's., dble-hd. alloy-S., $\frac{9}{16}$ and $\frac{11}{16}$	41-W-1005-5	1

TM 9-818
20-21

10-TON 6 x 4 TRUCK (MACK MODEL NR)

Tool	Federal Stock No.	Number Carried
Wrench, engr's., dble-hd. alloy-S., $\frac{5}{8}$ and $\frac{25}{32}$	41-W-1008-10	1
Wrench, engr's., dble-hd. alloy-S., $\frac{3}{4}$ and $\frac{7}{8}$	41-W-1012-5	1
Wrench, open-end, $\frac{15}{16}$ and $1\frac{1}{16}$	41-W-1021-10	1
Wrench, rear wheel bearing nut ($4\frac{7}{32}$-in. hex.)	....	1
Wrench, wheel rim nut	....	1
Wrench, screw., adj., auto., 15-in.	41-W-450	1
Wrench, wheel stud nut, dble-end, $14\frac{7}{8}$ in. long, $1\frac{3}{32}$ and $\frac{29}{32}$-in. hex.	41-W-450	1

21. EQUIPMENT.

Equipment	Federal Stock No.	Number Carried
Oiler	13-O-1530	1
Holder, oiler	....	1
Chuck, air	....	1
Jack, hydr., 12-ton, w/handle	41-J-73-5	1
Gun, lubr. pres., push type hydr.	....	1
Nozzle, 16-oz	41-G-1344-40	1
Gage, tire pressure, general service type	8-G-615	1
Extinguisher, fire, 1 qt	58-E-202	1
Crank, starting	....	1
Book, Standard Nomenclature List No. G-528	....	1
Book, Technical Manual, Truck, 10-ton 6 x 4 (Mack)	....	1
Cable, towing, S., hemp-center, $\frac{5}{8}$ in. diam. x 15 ft	....	1
Guide, Lubrication, War Dept. No. 523	....	1
Gun, lubr., pres., push-type, hydr. nozzle, 9-oz (on Models NR8 and NR9)	41-G-1344-40	1
Hose, tire-inflation, single-foot chuck, 22 ft long	....	1
Plate, jack mounting, 14 x 14 x $\frac{1}{2}$ in.	....	1

TM 9-818
22

TOOLS AND EQUIPMENT STOWAGE ON THE VEHICLE

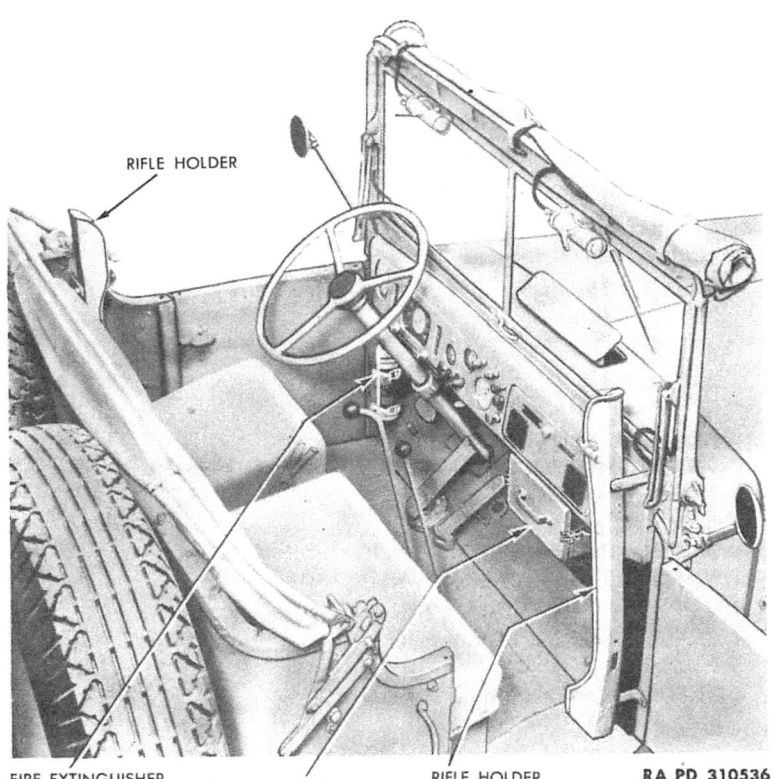

Figure 40—Inside of Cab

22. SPARE PARTS.

Item	Federal Stock No.	Number Carried
Bulb, 3-cp No. 67	. . . .	. .
Unit, sealed, blackout, service stop, left upper	17-L-5230	2
Unit, sealed, blackout, stop, right upper	8-L-419-10	1
Unit, sealed, blackout, tail, left and right lower	8-L-421-10	1
Cap, valve, tire, screwdriver type, olive drab	8-L-417	1
Core, valve, inside, tire (box of 5 cores)	8-C-650	2

71

TM 9-818

22

10-TON 6 x 4 TRUCK (MACK MODEL NR)

Item	Federal Stock No.	Number Carried
Tape, friction (8-oz roll) ¾ in. wide	22-W-650	1
Spark plug, with gasket	17-T-805	1
Fan belt, dual, Dayton No. 839	None	..
Fuse, 10-amp.	33-B-137	2
Shear pin, winch	....	..
Cotter pin (package assortment)	42-P-5347	1
Including:		
20—1/16 x 1½ in.		
25—3/32 x 2 in.		
20—1/8 x 2 in.		
5—5/32 x 2½ in.		
5—3/16 x 2½ in.		
METAL SPARE PARTS KIT		
CONTAINER (Ord-095334Y)	8-C-5498-50	1
Attaching parts:		
4 BBKX2C nut, screw, hex., S., No. 10-24NC-1		
4 BCNX2AP screw, mach., rd-hd., S., No. 10-24NC-2 x 1¾		
4 BECX1E washer, lock, reg., S., No. 10.		
4 BEBX1E washer, plain, S., SAE std., No. 10.		
Wire, iron, annealed, 22-gage (¼-lb spool).	22-W-650	1

TM 9-818
23

PART TWO—VEHICLE MAINTENANCE INSTRUCTIONS

Section VIII

MODIFICATION RECORDS

Paragraph

FSMWO and major unit assembly replacement record....... 23

23. FSMWO AND MAJOR UNIT ASSEMBLY REPLACEMENT RECORD.

　　a. **Description.** Every vehicle is supplied with a copy of AGO Form No. 478 which provides a means of keeping a record of each FSMWO completed or major unit assembly replaced. This form includes spaces for the vehicle name and U. S. A. Registration No., instructions for use, and information pertinent to the work accomplished. It is very important that the form be used as directed and that it remain with the vehicle until the vehicle is removed from service.

　　b. **Instructions for Use.** Personnel performing modifications or major unit assembly replacements must record clearly on the form a description of the work completed and must initial the form in the columns provided. When each modification is completed, record the date, hours and/or mileage, and FSMWO number. When major unit assemblies, such as engines, transmissions, transfer cases, are replaced, record the date, hours and/or mileage and nomenclature of the unit assembly. Minor repairs and minor parts and accessory replacements need not be recorded.

　　c. **Early Modifications.** Upon receipt by a third or fourth echelon repair facility of a vehicle for modification or repair, maintenance personnel will record the FSMWO numbers of modifications applied prior to the date of AGO Form No. 478.

TM 9-818
24-26

10-TON 6 x 4 TRUCK (MACK MODEL NR)

Section IX

NEW VEHICLE RUN-IN TEST

	Paragraph
Purpose	24
Correction of deficiencies	25
Run-in test procedures	26

24. PURPOSE.

a. When a new or reconditioned vehicle is first received at the using organization, it is necessary for second echelon personnel to determine whether or not the vehicle will operate satisfactorily when placed in service. For this purpose, inspect all accessories, subassemblies, assemblies, tools, and equipment to see that they are in place and correctly adjusted. In addition, they will perform a run-in test of at least 50 miles as directed in AR 850-15, paragraph 25, table III, according to procedures in paragraph 26 below.

25. CORRECTION OF DEFICIENCIES.

a. Deficiencies disclosed during the course of the run-in test will be treated as follows:

(1) Correct any deficiencies within the scope of the maintenance echelon of the using organization before the vehicle is placed in service.

(2) Refer deficiencies beyond the scope of the maintenance echelon of the using organization to a higher echelon for correction.

(3) Bring deficiencies of serious nature to the attention of the supplying organization.

26. RUN-IN TEST PROCEDURES.

a. **Preliminary Service.**

(1) FIRE EXTINGUISHER. See that portable extinguisher is present and in good condition. Test it momentarily for proper operation, and mount it securely.

(2) FUEL, OIL, AND WATER. Fill fuel tanks. Check engine crankcase, fuel injection pump oil, and coolant supply. Add oil or coolant as necessary to bring to correct levels. Allow room for expansion in fuel tanks and radiator. During freezing weather, test value of antifreeze, and add as necessary to protect cooling system against freezing. CAUTION: *If there is a tag attached to filler cap concerning engine oil in crankcase, follow instructions on tag before driving the vehicle.*

(3) FUEL FILTERS. Inspect all fuel filters for leaks, damage, and secure mountings and connections. Remove filter drain plugs (side plug on No. 1 filter) and drain sediment bowls. NOTE: *Turn No. 1 filter shaft clockwise two complete revolutions before draining.* If an

NEW VEHICLE RUN-IN TEST

appreciable amount of dirt or water is evident in No. 2 filter, inspect elements for excessive dirt. Change if necessary according to instructions in Lubrication Guide, paragraph 19.

(4) BATTERIES. Make hydrometer and voltage test of all four batteries; if necessary bring electrolyte up to full level with distilled or clean water.

(5) AIR BRAKE TANKS. Drain water from all air brake reservoir tanks and close drain cocks.

(6) AIR CLEANERS AND BREATHER CAPS. Examine carburetor and air compressor air cleaners and crankcase breather and outlet tube to see if they are in good condition and secure. Remove elements, and wash thoroughly in dry-cleaning solvent. Fill reservoirs to bead level with fresh oil, and reinstall securely. Be sure all gaskets are in good condition, and that ducts and air horn connections are tight.

(7) ACCESSORIES AND BELTS. See that accessories such as carburetor, generator, regulator, cranking motor, water pump, fan, oil filter, and air compressor, are securely mounted. Make sure that fan and generator drive belts are properly adjusted to have $\frac{1}{2}$-inch finger-pressure deflection.

(8) ELECTRICAL WIRING. Examine all accessible wiring and conduits to see if they are in good condition, securely connected, and properly supported.

(9) TIRES. See that all tires including spares are properly inflated (70 lb front, 90 lb rear), cool; that stems are in correct position, all valve caps present and fingertight. Remove objects lodged in treads or carcass and between duals, and inspect for damage. See that spares are secure in carriers.

(10) WHEEL AND FLANGE NUTS. See that all wheel mounting lug nuts and axle flange nuts are present and secure.

(11) FENDERS AND BUMPERS. Examine front fenders and bumper, and rear splash guards, brush guards and radiator grille, for looseness and damage.

(12) TOWING CONNECTIONS. Examine all towing shackles and pintle hook for looseness and damage, and see that pintle latch operates properly and locks securely.

(13) BODY AND TARPAULIN. See that all cab and body mountings are secure. Inspect attachments, hardware, glass, seats, doors, tail gate, chain and latches, to see if they are in good condition and secure. Examine cab top, top frame, and cargo body bows for looseness and damage. See that cab soft top, and cargo body tarpaulins and curtains are properly installed, securely fastened, and not damaged.

(14) LUBRICATE. Perform a complete lubrication of the vehicle, covering all intervals, according to the instructions on the Lubrication Guide, paragraph 19, except gear cases, wheel bearings, and other units covered in preceding procedures. Check all gear case oil levels, and add as necessary to bring to correct level. Change only if

TM 9-818
10-TON 6 x 4 TRUCK (MACK MODEL NR)

condition of oil indicates the necessity, or if gear oil is not of proper grade for existing atmospheric temperature. NOTE: *Perform items (15) to (18) during lubrication.*

(15) SPRINGS AND SUSPENSIONS. Inspect front shock absorbers, front and rear springs, rear bogie suspension spring seats, and torque rods to see if all are in good condition, correctly assembled, and secure, and not leaking excessively.

(16) STEERING LINKAGE. See that all steering arms, rods, and connections are in good condition and secure; and that gear case is securely mounted and not leaking excessively.

(17) PROPELLER SHAFTS, CENTER BEARING, VENTS. Inspect all shafts, universal joints, and center bearing (pillow block) to see if they are in good condition, correctly assembled, alined, secure, and not leaking excessively at seals or vents. Be sure vent passages are not clogged.

(18) AXLE AND TRANSFER VENTS. See that axle housing and transfer case vents are present, in good condition, and not clogged.

(19) ENGINE WARM-UP. During cold weather, when starting engine. test manifold preheating unit for proper operation. Start engine, and note if cranking motor action is satisfactory, and engine has any tendency toward hard starting. Set hand throttle to run engine at fast idle during warm-up period.

(20) INSTRUMENTS.

(a) Oil Gage. Immediately after engine starts, observe if oil pressure is satisfactory. (Normal pressures, hot, are 10 to 20 lb at idle speed, 45 to 60 lb at 1800-2000 rpm). Stop engine if oil pressure is not indicated in 30 seconds.

(b) Ammeters. Main ammeter should show slight positive (+) charge. High charge may be indicated until generator restores to batteries the current used in starting. Auxiliary ammeter also should show positive (+) charge indicating that all contacts through series-parallel starting switch and fuses are properly closed, and that second pair of batteries are receiving a charge.

(c) Temperature Gage. Engine temperature gage should rise gradually during warm-up, to normal operating range (165°F to 185°F).

(d) Tachometer. Tachometer should indicate the engine revolutions per minute, and also register the total hours of engine operation.

(e) Fuel Pressure Gage. Fuel pressure of 3 to 18 pounds should be indicated continuously when engine is running.

(f) Fuel Gage and Selector Switch. With selector switch in relative position, fuel gage should register "FULL" if tanks have been filled.

(g) Air Pressure Gage. During warm-up, air pressure should build up to 105 pounds. Governor should cut off air from compressor at this pressure, and should again cut in if pressure in lines and

NEW VEHICLE RUN-IN TEST

reservoirs is reduced to 85 pounds. Warning indicator (buzzer), should sound at pressures below 60 pounds.

(21) ENGINE CONTROLS. Observe if engine responds properly to controls and if controls operate without excessive looseness or binding.

(22) HORN AND WINDSHIELD WIPERS. See that these items are in good condition and secure. If tactical situation permits, test horn for proper operation and tone. See if wiper arms will operate through their full range and that blade contacts glass evenly and firmly.

(23) GLASS AND REAR-VIEW MIRRORS. Clean all body glass and mirrors, and inspect for looseness and damage. Adjust rear-view mirrors for correct vision.

(24) LAMPS (LIGHTS). Clean lenses and inspect all units for looseness and damage. If tactical situation permits, open and close all light switches to see if lamps respond properly.

(25) LEAKS, GENERAL. Look under vehicle, and within engine compartment, for indications of fuel, oil, and coolant leaks. Trace any leaks found to source, and correct or report them to designated authority.

(26) TOOLS AND EQUIPMENT. Check tools and On Vehicle Stowage Lists, paragraphs 20, 21, and 22, to be sure all items are present, and see that they are serviceable and properly mounted or stowed.

b. Run-in Test. Perform the following procedures (1) to (11) inclusive during the road test of the vehicle. On vehicles which have been driven 50 miles or more in the course of delivery from the supplying to the using organization, reduce the length of the road test to the least mileage necessary to make observations listed below. CAUTION: *Continuous operation of the vehicle at speeds approaching the maximums indicated on the caution plate should be avoided during the test.*

(1) DASH INSTRUMENTS AND GAGES. Do not move vehicle until engine temperature reaches 135°F. Maximum safe operating temperature is 200°F. Observe readings of ammeter, oil, temperature, and fuel gages to be sure they are indicating the proper function of the units to which they apply. Also see that speedometer registers the vehicle speed, and that odometer registers accumulating mileage.

(2) BRAKES (FOOT AND HAND). Test service air brakes to see if they stop vehicle effectively without side pull, chatter, or squealing. Be sure application valve closes when treadle pressure is released, and that there is not an excessive drop in air pressure when brakes are applied. Parking brake should hold vehicle on a reasonable incline, with $\frac{1}{3}$ ratchet travel in reserve and should lock securely in applied position. CAUTION: *Avoid long application of brakes until shoes have become seated.* Do not apply full treadle pressure except for an emergency stop.

(3) CLUTCH. Observe if clutch operates smoothly without grab, chatter or squeal on engagement, or slippage (under load) when fully engaged. See that pedal has 1-inch free travel before meeting

10-TON 6 x 4 TRUCK (MACK MODEL NR)

resistance. CAUTION: *Do not ride clutch pedal at any time, and do not engage and disengage new clutch severely or unnecessarily until driven and driving disks have become properly worn in.*

(4) TRANSMISSION. Gearshift mechanism should operate easily and smoothly, and gears should operate without undue noise, and not slip out of mesh.

(5) STEERING. Observe steering action for binding or looseness, and note any excessive pull to one side, wander, shimmy, or wheel tramp. See that column bracket and wheel are secure.

(6) ENGINE. Be on the alert for any abnormal engine operating characteristics or unusual noise, such as lack of pulling power or acceleration; backfiring, misfiring, stalling, overheating, or excessive exhaust smoke. Observe if engine responds properly to all controls.

(7) UNUSUAL NOISE. Be on the alert throughout road test for any unusual noise from body and attachments, running gear, suspensions, or wheels, that might indicate looseness, damage, wear, inadequate lubrication, or underinflated tires.

(8) HALT VEHICLE AT 10-MILE INTERVALS FOR SERVICES (9) TO (11) BELOW.

(9) AIR BRAKE SYSTEM LEAKS. With air pressure at governed maximum (105 lb) and brakes applied, stop engine. There should not be a noticeable drop in pressure in 1 minute.

(10) TEMPERATURES. Cautiously hand-feel each brake drum and wheel hub for abnormal temperatures. Examine transmission and rear axle housings for indications of overheating, or excessive lubricant leaks at seals, gaskets, or vents.

(11) LEAKS. With engine running, and fuel, engine oil, and cooling systems under pressure, look within engine compartment and under vehicle for indications of leaks.

c. Upon completion of run-in test, correct or report any deficiencies noted. Report general condition of vehicle to designated individual in authority.

TM 9-818
27

Section X

ORGANIZATION PREVENTIVE MAINTENANCE SERVICE

	Paragraph
Second echelon preventive maintenance service	27

27. SECOND ECHELON PREVENTIVE MAINTENANCE SERVICE.

a. Regular scheduled maintenance inspections and services are a preventive maintenance function of the using arm, and are the responsibility of commanders of operating organizations.

(1) FREQUENCY. The frequencies of the preventive maintenance services outlined herein are considered a minimum requirement for normal operation of vehicles. Under unusual operating conditions such as extreme temperatures, and dusty or sandy terrain, it may be necessary to perform certain maintenance services more frequently.

(2) FIRST ECHELON PARTICIPATION. The drivers should accompany their vehicles, and should assist the mechanics while periodic second echelon preventive maintenance services are performed. Ordinarily the driver should present the vehicle for a scheduled preventive maintenance service in a reasonably clean condition: that is, it should be dry and not caked with mud or grease to such an extent that inspection and servicing will be seriously hampered. However, the vehicle should not be washed or wiped *thoroughly* clean, since certain types of defects, such as cracks, leaks, and loose or shifted parts or assemblies, are more evident if the surfaces are slightly soiled or dusty.

(3) If instructions other than those contained in the general procedures ((4) below), or the specific procedures ((5) below), are required for the correct performance of a preventive maintenance service, or for correction of a deficiency, other sections of the vehicle operator's manual pertaining to the item involved, or a designated individual in authority, should be consulted.

(4) GENERAL PROCEDURES. These general procedures are basic instructions which are to be followed when performing the services on the items listed in the specific procedures. NOTE: *The second echelon personnel must be thoroughly trained in these procedures, so that they will apply them automatically.*

(a) When new or overhauled subassemblies are installed to correct deficiencies, care should be taken to see that they are clean, correctly installed, properly lubricated, and adjusted.

(b) When installing new lubricant retainer seals, a coating of the lubricant should be wiped over the sealing surface of the lip of the seal. When the new seal is a leather seal, it should be soaked in SAE 10 engine oil (warm if practicable), for at least 30 minutes. The leather lip should then be worked carefully by hand before installing the seal. The lip must not be scratched or marred.

(c) The general inspection of each item applies also to any supporting member or connection, and usually includes a check to see whether the item is in good condition, correctly assembled, secure,

10-TON 6 x 4 TRUCK (MACK MODEL NR)

or excessively worn. The mechanics must be thoroughly trained in the following explanations of these terms.

1. The inspection for "good condition" is usually an external visual inspection to determine whether the unit is damaged beyond safe or serviceable limits. The term "good condition" is explained further by the following: not bent or twisted, not chafed or burned, not broken or cracked, not bare or frayed, not dented or collapsed, not torn or cut.

2. The inspection of a unit to see that it is "correctly assembled" is usually an external visual inspection to see whether it is in its normal assembled position in the vehicle.

3. The inspection of a unit to determine if it is "secure" is usually an external visual examination, a hand-feel, or a pry-bar check for looseness. Such an inspection should include any brackets, lock washers, lock nuts, locking wires, or cotter pins used in assembly.

4. "Excessively worn" will be understood to mean worn close to, or beyond, serviceable limits, and likely to result in a failure if not replaced before the next scheduled inspection.

(d) Special services. These are indicated by repeating the item numbers in the columns which show the interval at which the services are to be performed, and show that the parts or assemblies are to receive certain mandatory services. For example, an item number in one or both columns opposite a procedure, means that the actual tightening of the object must be performed. The special services include:

1. Adjust. Make all necessary adjustments in accordance with the pertinent section of the vehicle operator's manual, special bulletins, or other current directives.

2. Clean. Clean units of the vehicle with dry-cleaning solvent to remove excess lubricant, dirt, and other foreign material. After the parts are cleaned, rinse them in clean fluid, and dry them thoroughly. Take care to keep the parts clean until reassembled, and be certain to keep cleaning fluid away from rubber or other material which it will damage. Clean the protective grease coating from new parts, since this material is not a good lubricant.

3. Special lubrication. This applies both to lubrication operations that do not appear on the vehicle Lubrication Guide, and to items that do appear on such charts, but which should be performed in connection with the maintenance operations if parts have to be disassembled for inspection or service.

4. Serve. This usually consists of performing special operations, such as replenishing battery water, draining and refilling units with oil, and changing the oil filter cartridge.

5. Tighten. All tightening operations should be performed with sufficient wrench-torque (force on the wrench handle) to tighten the unit according to good mechanical practice. Use torque-indicating wrench where specified. Do not overtighten, as this may strip threads or cause distortion. Tightening will always be understood to include the correct installation of lock washers, lock nuts, and cotter pins provided to secure the tightening.

TM 9-818
ORGANIZATION PREVENTIVE MAINTENANCE SERVICE

(e) When conditions make it difficult to perform the complete preventive maintenance procedures at one time, they can sometimes be handled in sections, but all operations should be completed within the week if possible. All available time at halts and in bivouac areas must be utilized if necessary to assure that maintenance operations are completed. When limited by the tactical situation, items with special services in the columns, should be given first consideration.

(f) The numbers of the preventive maintenance procedures that follow are identical with those outlined on W.D. A.G.O. Form No. 461, which is the Preventive Maintenance Service Work Sheet for Wheeled and Half-track Vehicles. Certain items on the work sheet that do not apply to this vehicle are not included in the procedures in this manual. In general, the numerical sequence of items on the work sheet is followed in the manual procedures, but in some instances there is deviation for conservation of the mechanic's time and effort.

(5) SPECIFIC PROCEDURES. The procedures for performing each item in the 1,000-mile (monthly) and 6,000-mile (6-month) maintenance procedures are described in the following chart. Each page of the chart has two columns at its left edge corresponding to the 6,000 mile and the 1,000 mile maintenance respectively. Very often it will be found that a particular procedure does not apply to both scheduled maintenances. In order to determine which procedure to follow, look down the column corresponding to the maintenance due and, wherever an item number appears, perform the operations indicated opposite the number.

MAINTENANCE		ROAD TEST
6000 Mile	1000 Mile	
		NOTE: When the tactical situation does not permit a full road test, perform those items which require little or no movement of the vehicle. When a road test is possible, it should be for preferably 3 miles, and not over 5 miles.
1	1	Before-operation Service. When vehicle is to be tested, perform Before-operation Service (par. 14) as a check to determine if the vehicle is in satisfactory condition to make the road test.
2	2	Air Pressure. Run engine at fast idle speed and observe whether the air pressure builds up at a steady rate to 105 pounds, and if the governor cuts off compressor action at this pressure. Bleed the air pressure down by application of the brake, and observe if governor cuts in the compressor at 85 pounds. Again bleed down the air pressure and observe if the low-pressure buzzer signals at 60 pounds.
3	3	Dash Instruments and Gages. Observe reading of all instruments frequently during operation for proper functioning.

10-TON 6 x 4 TRUCK (MACK MODEL NR)

MAINTENANCE	
6000 Mile	1000 Mile
4	4
5	5
6	6

OIL PRESSURE. Normal oil pressure is 10 to 20 pounds with engine idling, and maximum pressure is 45 to 60 pounds with engine operating at 2100 revolutions per minute.

AMMETER. A high charging rate for first few minutes after starting engine is normal; a high charging rate for an extended period with all light and accessory switches "OFF," may indicate a dangerously low battery, or a faulty regulator.

AUXILIARY AMMETER. A high charging rate for first few minutes after starting engine is normal; a high charging rate for an extended period with all light and accessory switches "OFF," may indicate a dangerously low battery, or a faulty regulator.

ENGINE TEMPERATURE. Engine temperature should not exceed 212°F, nor fall below 165°F.

TACHOMETER. Tachometer should indicate engine revolutions per minute without needle fluctuation or noise, and record accumulated engine hours.

FUEL PRESSURE GAGE. The fuel pressure indicator should read between 3 and 18 pounds at all times while engine is running.

AIR PRESSURE GAGE. Air pressure should be between 80 and 105 pounds; pressure should not drop excessively when brakes are applied. Low-pressure indicator buzzer should signal at 60 pounds.

SPEEDOMETER. Speedometer should register speed without needle fluctuation or noise, and odometer should record accumulated mileage.

FUEL GAGE. Fuel gage should register approximate amount of fuel in tank; recheck gage after filling tank.

Horns, Mirrors, and Windshield Wipers. Sound horn for proper tone, tactical situation permitting. Observe windshield wiper for proper operation, and adjust rear-view mirrors.

Brakes. Hand brake must lock securely in applied position and hold vehicle on grade with one-third reserve ratchet travel. Apply foot brake during operation to test for smooth and effective braking.

Clutch. While shifting gears note any chatter, squealing, or grabbing of the clutch. Observe clutch action for

TM 9-818
27

ORGANIZATION PREVENTIVE MAINTENANCE SERVICE

MAINTENANCE		
6000 Mile	1000 Mile	
		slipping while under load; free pedal travel should be two inches.
7	7	**Transmission.** Transmission gears should shift smoothly, without unusual noise, and not slip out of mesh. Stop vehicle in case of unusual noise and investigate.
8	8	**Steering.** Note any binding of steering gears, also note any pulling to one side, wandering, or shimmy while driving the vehicle. When the vehicle is stopped, inspect the steering wheel and column for good condition and secure mounting.
9	9	**Engine.** Engine should run smoothly at idle speed of 500 revolutions per minute. Gradually increase engine speed to maximum governed speed of 2,100 revolutions per minute. Observe any misfiring, detonation, unusual noise, or excessive smoke. Test vehicle for power and acceleration under load, and maximum governed speed.
10	10	**Unusual Noises.** Be on the alert constantly for any unusual noise or vibrations that would indicate worn or defective units or lack of lubrication.
11	11	**Air Brake System Leaks.** With air pressure at 105 pounds, apply the foot brake and stop the engine; there should be no noticeable drop of pressure in one minute. If pressure drops, test system for leaks with soap suds.
13	13	**Temperatures (Hubs, Brake Drums, Transmission and Axles).** Hand-feel these assemblies for abnormal temperatures that would indicate lack of lubrication, improper adjustment or defective units.
14	14	**Leaks.** Inspect engine and look on ground under vehicle and around radiator, axles, and transmission for evidence of fuel, oil, or water leaks.
16	16	**Gear Oil Level.** Inspect transmission and driving axles for proper oil level. Oil must not be more than one-half inch below filler hole when cold, nor above bottom edge of filler hole when hot.
		RAISE VEHICLE—BLOCK SAFELY
17	17	**Unusual Noises (Engine, Belts, Accessories, Transmission, Propeller Shafts and Joints, and Axles).** Run vehicle in gear slightly above idle speed, and listen for any unusual or excessive vibrations that would indicate worn, damaged or underlubricated units.
		ENGINE AND ACCESSORIES
18	18	**Cylinder Head and Gasket.** Examine cylinder head for cracks, oil, water, or compression leaks around studs

83

10-TON 6 x 4 TRUCK (MACK MODEL NR)

MAINTENANCE	
6000 Mile	1000 Mile
	19
19	
22	22
22	
23	23
24	24
25	25

or gaskets. Cylinder heads must not be tightened unless there is definite evidence of looseness or leaks. If necessary to tighten cylinder head or replace gasket, studs must be tightened in sequence to 130 foot-pounds, using a torque-indicating wrench.

Valve Mechanism. If valve tappets are noisy or cause engine to run unevenly, adjust as outlined in paragraph 49. Exhaust valve clearance is 0.015 to 0.017 inch; intake valve clearance is 0.010 to 0.012 inch, when hot.

SERVE. Warm up engine to 165°F. Remove rocker arm cover, inspect rocker arms and shafts for excessive wear, and valve springs for breakage. Inspect all rocker arms and push rods to determine proper flow of oil through all moving parts. Adjust valves: intake valve clearances, to 0.010-0.012 inch, hot; exhaust valve clearances, to 0.015-0.017 inch. Replace cover, using new gasket.

Battery. Inspect battery case for cracks and leaks. Clean top of battery. Inspect cables, terminals, bolts, posts, straps and hold-downs for good condition. Test specific gravity and voltage, and record on W.D. A.G.O. Form No. 461. Specific gravity readings below 1.225 indicate battery should be recharged or replaced. Electrolyte level should be three-eighths inch above plates. Replenish by adding distilled or clean, fresh water.

SERVE. Perform high-rate discharge test according to instructions for "condition" test which accompany test instrument and record voltage on W.D. A.G.O. Form No. 461. Cell variation should not be more than 30 percent. NOTE: *Specific gravity must be above 1.225 to make this test.*

CLEAN. Clean entire battery and carrier; repaint carrier if corroded. Clean battery cable terminals, terminal bolts and nuts, and battery posts; grease lightly and inspect bolts for serviceability. Tighten terminals and hold-downs carefully to avoid damage to battery.

Crankcase. With engine running at idle speed inspect oil pan, valve covers, timing gear cover and clutch housing for oil leaks. Stop engine and check to see whether oil is at proper level. If engine oil change is due, remove crankcase drain plug and drain oil. Refill to proper level with specified oil.

Oil Filters and Lines. Drain oil filter, and thoroughly clean housing with dry-cleaning solvent. Replace filter element and gaskets. Tighten all mountings and connections securely.

Radiator (Core, Shell, Mountings, Hose, Cap and Gasket, Antifreeze) Record. Inspect radiator core for

ORGANIZATION PREVENTIVE MAINTENANCE SERVICE

MAINTENANCE	
6000 Mile	1000 Mile
26	26
27	27
27	
28	28
28	
29	29
30	30
33	33
34	34

leaks, loose mounting bolts or obstructions. Inspect the shell for loose mountings or damaged condition. Examine radiator hose and cap for evidence of leaks. Test the antifreeze for temperatures anticipated, and record on back of W.D. A.G.O. Form No. 461. Clean dirt and insects from exterior of core, and tighten all mounting bolts and connections securely.

Water Pump, Fan and Shroud. Examine water pump, gasket, and connections for leaks and loose mounting bolts. Inspect fan and shroud for alinement and loose attaching bolts. Tighten bolts and replace gaskets as necessary.

Generator, Cranking Motor. Inspect generator and cranking motor for loose mounting bolts or loose wiring connections.

Clean and Tighten. Remove commutator cover band and inspect brush holders to see that brushes are free and not excessively worn, and that brush leads are not damaged; clean out commutator end of generator and cranking motor with compressed air, and tighten all mounting bolts and electrical connections securely.

Air Compressor. Inspect compressor for loose mountings, also for air, water, or oil leaks.

ADJUST. Check the clearance of the unloader valve and the discharge valve. Unloader must have clearance of 0.010-0.015 inch; discharge valve must have clearance of 0.042-0.075 inch. Adjust valves if not within these limits.

TEST GOVERNOR OPERATION. Air compressor governor should cut in at 85 pounds, and cut out at 105 pounds. Adjust and lubricate as necessary.

Drive Belts and Pulleys. Inspect fan, generator, and drive pulleys for cracks or fractures; check belts for adjustment and wear. Adjust belts to show ½-inch deflection, under hand pressure on the long side of the drive.

Tachometer Drive and Adapter. Inspect shaft housing for good condition, and make sure all clamps and connections are securely tightened.

Manifolds and Heat Controls. Inspect intake and exhaust manifolds for cracks, loose mounting bolts, or gasket leaks. Make sure all heat control electrodes are in place and tightened securely.

Air Cleaners. Remove air cleaners and disassemble. Inspect all gaskets and seals for evidence of leaks. Clean

TM 9-818
27
10-TON 6 x 4 TRUCK (MACK MODEL NR)

MAINTENANCE	
6000 Mile	1000 Mile
35	35
37	37
37	
39	39
40	40
42	42
43	43
45	45
45	
46	46
46	

element and oil reservoir thoroughly with dry-cleaning solvent. Refill reservoir with engine oil to "FULL" mark on body and reassemble, using new gaskets and seals if necessary.

Breather Caps. Remove the breather and clean the element and sump thoroughly with dry-cleaning solvent; refill sump to proper level with engine oil, and reassemble.

Fuel Filters, Screen, and Lines. Remove top cap on primary filter, and turn the scraper one full turn. Remove bottom plug on primary and on each secondary filter, and drain off approximately one-quarter pint.

SERVE. When fuel pressure is below three pounds, remove the filter elements from the secondary filters (Nos. 1 and 2) and replace them. When fuel pressure is above 18 pounds, remove the element from primary filter (No. 3) and replace it.

Cranking Motor. Operate cranking motor to make sure it engages freely and produces adequate cranking speed without unusual noise or grind. Tighten wires and mountings as necessary.

Leaks. Inspect all units serviced for fuel, oil or water leaks, while engine is running. Trace any leaks to their source and correct the cause.

Engine Idle. Adjust the engine idle speed to 500 revolutions per minute. Engine must run smoothly without stalling.

Regulator Unit. Inspect regulator mounting and electrical connections to see that they are tight. Regulator must be properly grounded.

TEST. With regulator at normal operating temperature, make voltage and amperage test, using low-voltage tester, to determine the proper functioning of the regulator as outlined in test set instructions.

Diesel Fuel Injector Pump. Examine the pump for loose mounting bolts and lines, and gaskets for evidence of leaks. Open pet cock and check oil level; add oil if needed.

DRAIN. Drain oil from pump, and refill to proper level with engine oil. Check injector pump timing markings, and adjust if necessary.

Diesel Fuel Nozzles and Lines. Inspect injector fuel lines and connections for leaks. Make sure the injectors are securely mounted.

INSPECT AND TIGHTEN. Remove injectors and inspect tips for evidence of burning. Test each injector for

86

ORGANIZATION PREVENTIVE MAINTENANCE SERVICE

MAINTENANCE	
6000 Mile	1000 Mile

proper spray. Install injectors, and tighten all connections and mounting bolts securely.

CHASSIS, BODY AND ATTACHMENTS.

47 | 47 — **Tires and Rims (Including Spare).** Examine tires for cuts, bruises, breaks, blisters, and irregular tread wear. Remove imbedded glass, nails, or stones. Replace leaky valve cores or missing caps. Inspect rims for damage and tighten wheel-lug nuts. Inflate tires to 70 pounds front, 90 pounds rear pressure. Match tires according to over-all circumference and type of tread.

48 — **Rear Brakes.** Remove the wheels and drums, inspect and service. The several wheel bearing and brake items up to and including 52 are group services in which there will be overlap. Perform these services in the best order for economy of mechanic's time and for orderly assembly.

48 — CLEAN (DRUMS AND SUPPORTS). Remove dirt and grease, and inspect brake drums to see if they are in good condition, securely mounted, and are not excessively worn or scored.

49 — **Rear Brake Shoes.** Examine linings through inspection holes to see whether they are worn to the rivet heads. If brake shoes are not worn excessively, adjust at slack adjuster. Brake chamber push-rod travel should be $\frac{3}{4}$ to $\frac{7}{8}$ inch.

49 — SERVE. Thoroughly clean brake shoes, links, guides, and anchors; inspect for worn linings, rusted or binding linkage, and worn or broken retraction springs. If linings are excessively worn, they should be replaced.

50 | 50 — **Torque Rods.** Examine torque rods for good condition, proper alinement, and secure mountings. Inspect bushings for deterioration.

50 — SERVE. Tighten all nuts securely.

51 | 51 — **Rear Spring Seats and Bearings.** Inspect spring seat for loose U-bolts or leaking grease seals. Remove filler plugs from rear spring-trunnion bearing caps and fill to level with specified lubricant.

52 — **Rear Wheels.** Revolve the wheel and observe for run-out. Listen for any indications of damaged wheel bearings. Force wheel in and out to test for end play. Tighten and adjust if necessary.

52 — SERVE. Remove wheels, bearings and seals. Clean all parts thoroughly in dry-cleaning solvent. Inspect bear-

10-TON 6 x 4 TRUCK (MACK MODEL NR)

MAINTENANCE	
6000 Mile	1000 Mile
53	
53	
	54
54	
55	55
56	56
57	57
58	58
58	
60	60

ings and races for cracks, pits, or fractures. Pack bearings with specified lubricant and reinstall, using new grease seals.

Front Brakes. Remove the wheels and drums, inspect and service. The several wheel bearing and brake items up to and including 60 are group services in which there will be overlap. Perform these services in the best order for economy of mechanic's time and for orderly assembly.

CLEAN (DRUMS AND SUPPORTS). Remove dirt and grease and inspect to see that they are in good condition, securely mounted, and not excessively worn or scored.

Front Brake Shoes. Examine linings through inspection holes to see whether they are worn to the rivet heads. If brake shoes are not worn excessively, adjust at slack adjuster. Brake chamber push-rod travel should be $\frac{5}{8}$ to $\frac{3}{4}$ inch.

SERVE. Thoroughly clean brake shoes, links, guides, and anchors; inspect for worn linings, rusted or binding linkage, and worn or broken retraction spring. If linings are excessively worn, they should be replaced.

Steering Knuckles. Inspect to see whether steering knuckles are in good condition and properly secured by the king pins; and determine whether the king pins and their bushings are excessively worn.

Front Springs. Inspect springs for sag, broken or shifted leaves, loose or missing U-bolts, eyebolts and rebound clips. Tighten all mounting and assembly bolts and nuts.

Steering. Examine Pitman arm, drag link, and tie rod for loose mounting, worn, or damaged condition. Inspect steering gear housing for oil level, loose mountings, or leaks. Lubricate and tighten mountings and connections. CAUTION: *Loosen the steering column bracket when tightening the steering case mounting nuts so as not to distort the column.*

Front Shock Absorbers and Links. Examine shock absorbers for wear, loose mountings, or fluid leaks.

SERVE. Remove filler plug and fill reservoir with shock absorber fluid; disconnect link, and test shock absorber action; shock absorbers should have resistance in both directions.

Front Wheels. Revolve the wheel and observe for runout. Listen for any indications of damaged wheel bear-

TM 9-818
27

ORGANIZATION PREVENTIVE MAINTENANCE SERVICE

MAINTENANCE		
6000 Mile	1000 Mile	
		ings. Force wheel in and out to test bearing adjustment. Tighten and adjust if necessary.
60		SERVE. Remove wheels, bearings, and seals; clean all parts thoroughly in dry-cleaning solvent. Inspect bearings and races for cracks, pits, or fractures. Pack bearings with specified lubricant and reinstall, using new grease seals.
61	61	**Front Axle.** Examine front axle for twists and bends, and determine whether it is in proper alinement.
63	63	**Engine Mountings.** Inspect engine mountings, ground straps, and side pans to see if they are in good condition and securely mounted. Tighten all loose mountings and connections.
64	64	**Hand Brake.** Examine ratchet, pawl, and linkage for excessive wear or loose mountings. Inspect linings for wear or oil-soaked condition, and examine disk for scores.
		ADJUST. Hand brake should be adjusted so that there is one-third reserve ratchet travel when brake is fully applied.
64		REPLACEMENT. If brake linings are excessively worn or oil-soaked, they should be replaced.
65	65	**Clutch Pedal.** Test clutch pedal for free travel. Examine linkage for wear or missing cotter pins, and pedal-return spring for broken or stretched spring coils. Adjust free pedal travel to 2 inches (par. 60). Tighten linkage and replace cotter pins, if necessary.
69	69	**Air Brake Application Valve.** Examine air brake to see that the application valve operates freely and is securely mounted, and that the valve closes fully when the brake pedal is released.
70	70	**Air Brake Reservoirs.** Examine the reservoirs to see that they are in good condition and securely mounted. Open drain cocks and drain off condensation.
71	71	**Transmission.** Examine all transmission seals and gaskets for evidence of leaks, and see that transmission is in good condition and securely mounted.
71		TIGHTEN. Tighten all mounting bolts securely.
73	73	**Rear Propeller Shafts.** Inspect propeller shaft for alinement, flange bolts for looseness, universal joints

89

10-TON 6 x 4 TRUCK (MACK MODEL NR)

MAINTENANCE		
6000 Mile	1000 Mile	
75	75	and slip joints for excessive wear and grease leaks. Tighten all bolts and mountings, if necessary. **Rear Axle.** Inspect axles for alinement, leaking grease seals, or clogged vents. Test pinion shaft for excessive end play. SERVE. Remove vent and clean thoroughly. Tighten all mounting bolts and cap screws as necessary.
76	76	**Rear Air Brake.** Examine the chambers and hose connections for air leaks. Make sure slack adjusters are securely mounted and that push rod seals are in place. Tighten all mountings and connections as necessary.
77	77	**Rear Springs.** Inspect springs for sag, broken or shifted leaves, loose or missing U-bolts, eyebolts and rebound clips. Tighten all mounting and assembly bolts and nuts.
78	78	**Cab and Body Mountings.** Examine vehicle for loose mounting bolts and deteriorated or missing cushions. Inspect body and hold-down bolts for looseness, and sills for cracked or damaged condition. CAUTION: *If necessary to replace cushion or tighten cab hold-down bolts, loosen steering column bracket before tightening cab bolts.*
80	80	**Frame.** Examine frame and cross members for cracks, misalinement, and loose or missing rivets. If frame appears to be out of alinement refer to higher authority.
81	81	**Wiring Looms and Grommets.** With all switches in "OFF" position, ammeter must read zero. Examine looms and wiring for damaged or frayed condition, and missing grommets. Tighten all loose connections.
82	82	**Fuel Tanks, Fittings and Lines.** Inspect fuel tank, lines, and connections for evidence of leaks or chafing. Tighten loose line clamps and connections as necessary.
83	83	**Brake Lines.** Inspect brake lines for good condition and see that clamps and connections are secure. Examine flexible hose for evidence of leaks, chafing, or deterioration. Tighten all hold-down clamps and connections as necessary.
84	84	**Exhaust Pipes and Muffler.** Examine exhaust pipe and muffler for loose mountings, connections, and leaks, and inspect tail pipe for possible obstructions. Tighten all mountings and connections as necessary.
85	85	**Vehicle Lubrication.** If due, lubricate in accordance with Lubrication Guide (par. 19), and current lubrica-

TM 9-818
27

ORGANIZATION PREVENTIVE MAINTENANCE SERVICE

MAINTENANCE	
6000 Mile	1000 Mile
86	86
91	91
91	
92	92
93	93
94	94
95	95
96	96
98	98
99	99
100	100
101	101

tion directives, using only clean lubricant, and omitting items that have had special lubrication during service. Replace damaged or missing fittings, vents, flexible lines, or plugs.

LOWER VEHICLE TO GROUND

Toe-in and Turning Stops. With front wheels in straight-ahead position, check the toe-in limit of $\frac{1}{8}$ to $\frac{3}{16}$ inch. Turn wheel full travel in both directions. Turning stops must hold wheel from rubbing against vehicle.

Lamps. With all light switches in "ON" position, including stop light, inspect lights to see that they are clean, securely mounted, in good condition, are lighted, and that they go out when switches are turned "OFF."

ADJUST. Adjust headlights for proper aim.

Safety Reflectors. Inspect reflectors for cracked or broken glass, or loose mountings. Wipe glass clean.

Front Bumper. Inspect bumper for loose mountings or damaged condition. Tighten mountings as necessary.

Hood (Hinges and Fasteners). Inspect to see that hood is properly alined, hinges operate freely, and latches lock securely. Tighten mountings as necessary.

Front Fenders and Running Boards. Examine fenders and running boards for good condition and secure mounting bolts. See that bracket scuff plates are present and secure. Tighten all mountings as necessary.

Cab Body (Doors, Hardware, Glass, Top and Frame, Curtains and Fasteners, Seats, Upholstery and Trim, Floor Boards, and Map Compartment). Inspect these units to see that they are in serviceable condition and securely mounted. Open and close doors, making sure they are in alinement and locked when closed.

Circuit Breaker and Fuse Block. Inspect circuit breakers and fuse block for loose mountings and electrical connections. Make sure fuses are not burned out, and that they fit tight in clips. Tighten all loose mountings and connections as necessary.

Splash Guards. Inspect splash guards for loose mounting brackets or damaged condition.

Body (Floor, Bows, Tops, Troop Seats, and Tail Gate). Inspect these items to see that they are in serviceable condition and securely mounted. Examine tarpaulins for tears, missing or damaged ropes or fasteners. Lubricate troop seat and tail gate hinges.

Pintle Hook. Inspect to see that pintle hook is in good condition and securely mounted and that latch operates

91

10-TON 6 x 4 TRUCK (MACK MODEL NR)

MAINTENANCE	
6000 Mile	1000 Mile
103	103
104	104
131	131
132	132
133	133
134	134
135	135
141	141
142	142

freely. Make sure pintle pin is securely attached to chain. Lubricate and free-up as necessary.

Paint and Markings. Inspect paint of entire vehicle for good condition. Make sure there are no bright spots that might cause glare or reflection, and that all identification plates and markings are legible.

Radio Bonding. Inspect all bonding and connections to see that they are in good condition and securely mounted. Clean and tighten as necessary.

TOOLS AND EQUIPMENT

Tools. Inspect all tools for good condition, proper mounting, and serviceability, using the vehicle stowage list.

Fire Extinguisher. Inspect fire extinguisher to see that it is in position and securely mounted. Shake to determine contents, and replace or refill as necessary.

Decontaminator. Inspect decontaminator to see that it is in proper position and securely mounted. Shake to determine contents and replace or recharge as necessary. Contents must be renewed every 90 days, as it deteriorates. Refer to tag for date of last recharge.

First Aid Kit. Inspect first aid kit and see that it is in position and securely mounted. Check contents with list on inside cover. Tighten mountings and refill supply as necessary.

Publications and Form No. 26. All vehicle publications and Standard Accident Form No. 26 must be present, legible, and properly stowed.

Modifications. Examine the vehicle to make certain that all Field Service Modification Work Orders pertaining to the vehicle have been completed and entered on W.D. A.G.O. Form 478. Enter any replacement of major unit assembly made at time of this service.

Final Road Test. Review items 2 to 15 inclusive, paying particular attention to those units on which work has been performed, to make certain they have been restored to proper operating condition. Correct any deficiencies found during the final road test.

TM 9-818
28

Section XI

ORGANIZATION TOOLS AND EQUIPMENT

                                                                                            Paragraph
Organization tools and equipment......................... 28

28. ORGANIZATION TOOLS AND EQUIPMENT.

a. In addition to the tools listed in paragraphs 20 through 22, the following are available to the using arms for the maintenance of this vehicle:

b. Standard Tool Sets. The tool sets available to individuals (specialists) and organizations, dependent upon the allocation in the Tables of Equipment, are listed in SNL N-19. The components of these tool sets are also listed and illustrated.

c. Special Tool Set. Special tools available to second echelon maintenance organizations are listed in the pertinent Organization Spare Parts and Equipment Lists, and in SNL G-19.

TM 9-818
29-30

10-TON 6 x 4 TRUCK (MACK MODEL NR)

Section XII

TROUBLE SHOOTING

	Paragraph
General	29
Engine	30
Clutch	31
Fuel	32
Exhaust	33
Cooling	34
Starting and generating systems	35
Transmission	36
Propeller shaft	37
Front axle	38
Rear axle	39
Brake system	40
Wheels	41
Springs and shock absorbers	42
Steering gear	43
Frame	44
Battery and lighting system	45

29. GENERAL.

a. Following are trouble shooting charts and diagnosis procedures for determining the cause of troubles that may develop in the vehicle. A separate chart is provided for each major unit or system. Wherever necessary in the trouble shooting chart, references are made to paragraphs containing further information to assist in diagnosing the particular trouble, or to amplify the suggested remedy.

30. ENGINE.

a. **Trouble Shooting Chart.**

(1) ENGINE WILL NOT TURN.

Possible Cause	Possible Remedy
Engine seizure due to internal damage.	Test (b (1) below) and/or notify higher authority.
Starting system inoperative.	Determined by diagnosis (par. 35 b (1) and (2)).
Incorrect oil viscosity.	Drain and refill (par. 19 b (4)).
Transmission in gear.	Move shifter lever to neutral position.

(2) ENGINE TURNS, BUT WILL NOT START.

Possible Cause	Possible Remedy
Slow cranking speed.	Determined by diagnosis (par. 35 b (1)).

TROUBLE SHOOTING

Possible Cause	Possible Remedy
No fuel in tank.	Fill tank, prime and start.
Inoperative fuel system.	Determined by diagnosis (par. 32 a (6)).
Water in fuel.	Drain fuel system and tanks, clean, refill, prime, and start.
Manifold heaters not operating (cold weather only).	Check for loose connections, broken wires and defective heater element ribbon, tighten and renew.
Poor compression.	Notify higher authority (except when due to leaky head gasket).
Air cleaner plugged, not allowing sufficient air to pass through.	Clean air cleaner.
Air in fuel system.	Bleed system (par. 68).

(3) ENGINE DOES NOT DEVELOP FULL POWER.

Possible Cause	Possible Remedy
Compression poor or improper valve timing.	Notify higher authority (except when due to leaky cylinder head gasket).
Faulty fuel system.	Determined by diagnosis (par. 32).
Air cleaner dirty.	Clean air cleaner (par. 74 b).
Engine overheating.	Determined by diagnosis (par. 34 a (2)).
Poor quality lubricating oil.	Change oil.
Improper valve adjustment.	Adjust clearance (par. 49 c).

(4) ENGINE STOPS SUDDENLY.

Possible Cause	Possible Remedy
No fuel.	Fill tank, prime and start.
Fuel lines air or gas bound.	Check for loose fitting, tighten and prime and start.
Fuel filters plugged.	Clean element or replace cartridge (depending on which type filter is affected) (par. 71).
Obstruction in, or broken main fuel line.	Locate and correct.
Water in fuel.	Drain tank and entire fuel system; clean, refill, prime and start.
Fuel supply pump not functioning properly.	Renew pump.
Piston or bearing seizure due to lack of lubrication.	Notify higher authority.

10-TON 6 x 4 TRUCK (MACK MODEL NR)

(5) Engine Missing Intermittently on All Cylinders.

Possible Cause	Possible Remedy
Improper fuel or fuel with poor burning qualities.	Drain tank and entire fuel system; clean, refill with fuel as per specifications, prime and start.
Water in fuel.	Follow above procedure.
Sticky injection nozzles, valve stems or injection-pump delivery valves.	Replace faulty nozzles with complete new ones and/or injection pump assembly.
Air in fuel system.	Bleed system (par. 68).
Low engine compression.	Notify higher authority.
Plugged fuel tank vents.	Open vents or replace cap.
Plugged air cleaner.	Clean air cleaner.
Improper valve adjustment.	Adjust clearance (par. 49 c).

(6) Engine Missing on One or More Cylinders.

Air or gas in injection pump or lines.	Determined by diagnosis (par. 32).
Injection pump delivery valve stuck in body or leaking.	Renew injection pump (par. 64).
Exhaust or inlet valve stuck.	Free with kerosene or alcohol.
Leaky exhaust or inlet valve.	Notify higher authority.
Broken exhaust or inlet valve spring.	Renew spring.
Improper tappet clearance between exhaust or inlet valve and rocker arm.	Reset to proper clearance (par. 49 c (3)).
Poor compression due to worn or stuck rings or badly worn cylinder sleeves.	Notify higher authority.
Broken high-pressure fuel line.	Replace with new line.

(7) Engine Overheats.

Coolant level low.	Refill radiator.
Air cleaner dirty.	Clean.
Cooling system inoperative.	Determined by diagnosis (par. 34 a (2)).
Injection timing incorrect—late.	Notify higher authority.
Valve timing incorrect.	Notify higher authority.
Oil pressure low.	Determined by diagnosis (b (3) below).
Brakes drag.	Adjust (par. 115 a).
Oil level low or poor grade oil.	Fill to correct level or replace.
Fan broken or bent.	Renew.

TROUBLE SHOOTING

(8) EXCESSIVE OIL CONSUMPTION.

Possible Cause	Possible Remedy
Piston rings broken, worn, stuck or incorrectly fitted.	Notify higher authority.
Cylinder bore out-of-round, tapered or scored.	Notify higher authority.
Main or connecting rod bearings loose, incorrectly fitted or too much end play.	Notify higher authority.
Engine overheating.	Determined by diagnosis (a (7) above).
Oil too light or of poor grade.	Drain and refill.
Oil pressure too high.	Notify higher authority.
Oil level too high.	Maintain correct level.
External leaks at gaskets or seals.	Renew or tighten as necessary.

(9) ENGINE KNOCKS OR PINGS.

Injection timing too early.	Notify higher authority.
Excess fuel injected.	Notify higher authority.
Excessive valve tappet clearance.	Adjust clearance (par. 49 c (3)).
Pistons, piston rings, or piston pins badly worn.	Advise higher authority.
Poor grade fuel.	Drain and refill with fuel as per specifications.

(10) ENGINE SMOKES.

Faulty combustion or excess oil in combustion chambers.	Determined by diagnosis (par. 32 a (7), (8), and (9)).

b. Diagnosis.

(1) SEIZURE. After making sure transmission is in neutral, and that crankcase oil is of proper viscosity for prevailing temperature, attempt to turn over engine with hand crank. If it cannot be turned, seizure is indicated. Notify higher authority.

(2) ENGINE OVERHEATS. Determine whether air passages through radiator are clear. Make sure fan belt is not slipping, that there is no loss of water through leaky hose, pump, radiator or engine connections and that suction hose is not collapsed.

(3) LOW OIL PRESSURE. If oil pressure is very low, and no leak exists in oil lines to filter or gage, advise higher authority.

(4) ENGINE MISSING. To determine which cylinder or cylinders are missing, loosen the nuts connecting high-pressure lines to injection nozzles, one at a time. If the engine speed drops and the exhaust loses its rhythm, that indicates the cylinder has been functioning. If the engine speed remains the same and the exhaust sounds the same, that indicates the cylinder has been missing. Usually, when

10-TON 6 x 4 TRUCK (MACK MODEL NR)

testing for a missing cylinder, if the injection pump or lines are air- or gas-bound, that condition will clear up, since loosening the nuts permits the air or gas to escape.

31. CLUTCH.
a. Trouble Shooting Chart.
(1) CLUTCH SLIPS.

Possible Cause	Possible Remedy
Adjustment incorrect.	Readjust pedal (par. 61 a).
Clutch disk lining worn out.	Replace disk.
One or more pressure springs cracked or set.	Notify higher authority.
Pressure plate warped.	Notify higher authority.

(2) CLUTCH GRABS.

Possible Cause	Possible Remedy
Oil or grease on disk, pressure plate or flywheel.	Clean thoroughly with suitable solvent.
Broken disk facings or pressure plate.	Renew disk assembly or notify higher authority.
Binding of release mechanism.	Adjust and renew necessary parts.

(3) CLUTCH DRAGGING.

Possible Cause	Possible Remedy
Excessive free movement of pedal.	Readjust correctly (par. 61 a).
Loose or broken disk facings.	Renew disk assembly.
Excessive friction in pilot bearing.	Renew.
Bent driven disk.	Renew.

(4) NOISE.

Possible Cause	Possible Remedy
Squealing (caused by clutch slipping).	Readjust pedal (par 61 a).
Grating or clicking when clutch pedal is depressed.	Renew release bearing if damaged.
Grating or squeaking caused by dry pilot bearing.	Notify higher authority.
Grease on disk, flywheel or pressure plate.	Clean thoroughly with suitable solvent.
Loose disk facing.	Renew disk assembly.
Loose engine mounting.	Inspect and renew all broken or damaged parts.
Broken disk facing or pressure plate.	Advise higher authority.

b. Diagnosis.
(1) TEST FOR OPERATION. Idle the engine at approximately 800 revolutions per minute. Push the clutch pedal to fully released posi-

TROUBLE SHOOTING

tion and allow time for the clutch to stop. Shift the transmission into first or reverse gear. If the shift cannot be made without a severe clash of the gears, or if, after engagement of the gears, there is a jumping or creeping movement of the truck with the clutch still fully released, the clutch is at fault.

32. FUEL.

a. Trouble Shooting Chart.

(1) ENGINE WILL NOT START, OR HARD STARTING.

Possible Cause	Possible Remedy
No fuel in tank.	Fill tank, prime and start.
No fuel in injection pump.	Prime by operating handle of pump.
Air in system.	Bleed system (par. 68).
Too low cranking speed.	Determined by diagnosis (par. 35).
Manifold heaters not operating (cold weather only).	Check for loose connections, broken wires and defective heater element ribbon, tighten and renew.
Fuel too heavy to flow through pipes properly.	Change to fuel as per specifications.
Water in fuel.	Drain tank and entire fuel system; clean, refill, prime and start.
Poor compression.	Notify higher authority (except when due to leaky cylinder head gasket).

(2) ENGINE STOPS SUDDENLY.

No fuel.	Fill tank, prime and start.
Injection pump or lines air- or gas-bound.	Check for loose fitting, tighten, prime and start.
Fuel filters plugged.	Clean element or replace cartridge (depending on which type filter is affected) (par. 71).
Obstruction in, or broken main fuel line.	Locate and correct.
Water in fuel.	Drain tank and entire fuel system; clean, refill, prime and start.
Fuel supply pump not functioning.	Replace pump (par. 64 b (5) and e (9)).

10-TON 6 x 4 TRUCK (MACK MODEL NR)

(3) ENGINE MISSING ERRATICALLY OR INTERMITTENTLY ON ALL CYLINDERS.

Possible Cause	Possible Remedy
Improper fuel or fuel with poor burning qualities.	Drain tank and entire fuel system; clean, refill with proper fuel, prime and start.
Water in fuel.	See preceding item.
Sticky injection nozzles, valve stems or injection pump delivery valves.	Renew nozzle assembly or injection pump as determined by diagnosis (pars. 64 and 67).
Air in fuel system.	Bleed system (par. 68).
Plugged air cleaner.	Clean air cleaner.
Poor compression due to leaky valves, worn rings or cylinder sleeves.	Notify higher authority.

(4) ENGINE MISSING ON ONE OR MORE CYLINDERS.

Possible Cause	Possible Remedy
Air or gas in injection pump or lines.	Determined by diagnosis (par. 32 b (1)).
Injection pump delivery valve stuck in body or leaking.	Renew injection pump.
Exhaust or inlet valve stuck.	Free up with alcohol or kerosene.
Leaky exhaust or inlet valve.	Notify higher authority.
Broken exhaust or inlet valve spring.	Replace with new part.
Improper clearance between exhaust or inlet valve and rocker arm.	Reset to proper clearance (par. 49 c (3)).
Poor compression.	Notify higher authority (except when due to leaky cylinder head gasket).
Broken injection nozzle lines.	Replace with new parts.

(5) EXCESSIVE FUEL CONSUMPTION.

Possible Cause	Possible Remedy
Air cleaner loaded.	Clean as directed.
Fuel leaks.	Check all lines and connections.
Sticky controls.	Free throttle so that it returns.
Long idling periods.	Shut off engine when parking.
Engine too hot.	Determined by diagnosis (par. 34 a (2)).
Dragging brakes.	Check and adjust (par. 111 a).
Low tire pressure.	Inflate to correct pressure (par. 147).
Truck overloaded.	Check weight.

TROUBLE SHOOTING

(6) FUEL DOES NOT REACH CYLINDERS.

Possible Cause	Possible Remedy
Injection nozzle not operating or leaky.	Renew (par. 67 b and d).
Emergency shut-off valve closed.	Open valve.
Either of two engine-mounted fuel filters loaded or plugged (pressure gage will indicate this condition).	Renew cartridge (par. 71).
Chassis-mounted fuel filter dirty.	Clean element (par. 71).
Fuel tank caps plugged.	Clear vent or replace cap.
Air leaks in suction line (between fuel supply pump and fuel tanks).	Must be made airtight.
Fuel leak in lines or connections between injection pump and nozzles.	Tighten and seal connections.
Water or dirt in any part of system.	Drain and clean entire system.
Air or gas in any part of system.	Bleed system (par. 68).
Injection pump does not operate properly.	Renew (par. 64).
Fuel supply pump does not operate properly.	Renew (par. 64 b and 64 e).

(7) EXHAUST SMOKE EXCESSIVE (WHITE SMOKE).

One or more cylinders not firing.	Test (par. 30 b (4)).
Engine too cold.	Check when engine reaches normal operating temperature.
Poor compression.	Notify higher authority.
Stuck or leaky valves.	Check (a (4) above).
Leaking head gasket.	Renew (par. 48 a and b).
Improper grade of fuel.	Renew (a (3) above).
Water or air in fuel system.	Drain or bleed system.

(8) EXHAUST SMOKE EXCESSIVE (BROWN AND BLACK SMOKE).

Dirty or leaking injection nozzle.	Renew (par. 67 b and d).
Leaking valves.	Notify higher authority.
Stuck rings.	Notify higher authority.
Worn rings, pistons and sleeves.	Notify higher authority.
Dirty and obstructed air cleaner.	Clean.
Improper grade of fuel.	Renew (a (3) above).
Rarefied atmosphere at high altitudes.	Notify higher authority.
Timing of injection too early (indicated by fuel knock and power loss).	Notify higher authority.

10-TON 6 x 4 TRUCK (MACK MODEL NR)

Possible Cause	Possible Remedy
Timing of injection too late (no fuel knock, but power loss).	Notify higher authority.
Excessive fuel feed.	Notify higher authority.
Injection pump out of calibration.	Notify higher authority.

(9) EXHAUST SMOKE (BLUE SMOKE).

Possible Cause	Possible Remedy
Excessive engine oil in combustion chamber (b (2) below).	Determined by diagnosis.
Carboned or stuck rings.	Notify higher authority.
Worn rings, pistons or cylinder sleeves.	Notify higher authority.
Improper engine oil.	Renew.
Same causes which produce black or brown smoke until engine reaches normal heat.	Determined by diagnosis ((8) above).

b. **Diagnosis.**

(1) ENGINE SMOKE. If blue smoke is evident when engine starts and does not disappear eventually, but increases with engine speed and decreases with load, it is an indication that excessive oil is passing the piston rings. Also it may be caused by the same reasons as black or brown smoke, the blue color being due to the underload and speed conditions specified under which the combustion chambers have not attained normal temperature.

33. EXHAUST.

a. **Trouble Shooting Chart.**

(1) EXCESSIVE NOISE.

Possible Cause	Possible Remedy
Burned-out muffler.	Renew (par. 75 b (3) and c (3)).
Leaky connections.	Tighten.

(2) LOSS OF POWER.

Possible Cause	Possible Remedy
Increased back pressure due to muffler tube collapsing.	Renew muffler (par. 75 c (3)).

(3) FUMES IN CAB.

Possible Cause	Possible Remedy
Loose gaskets at exhaust pipe to manifold connection.	Replace ring gaskets (par. 53 a (1) and b (2)).

(4) EXHAUST SMOKES EXCESSIVELY.

Possible Cause	Possible Remedy
Improper combustion, or oil pumping.	Determined by diagnosis (par. 32 a (7), (8) and (9) and b (2)).

TROUBLE SHOOTING

34. COOLING.
  a. Trouble Shooting Chart.
    (1) LOSS OF COOLANT.

Possible Cause	Possible Remedy
Loose hose clamps.	Tighten clamps.
Deteriorated hose.	Replace hose.
Water pump leakage.	Renew pump.
Leaks in radiator core and tanks.	Renew radiator.
Overheating.	Determined by diagnosis (par. 34 a (2)).

  (2) OVERHEATING.

Possible Cause	Possible Remedy
Coolant level low.	Refill radiator.
Obstruction preventing free air flow through radiator core.	Remove obstruction.
Fan belts slipping.	Tighten or renew if necessary.
Radiator tubes clogged.	Reverse-flush radiator.
Thermostat sticking.	Clean or renew thermostat (par. 85 a and b).
Suction hose collapsed.	Renew hose and wire insert.
Water pump failure.	Renew pump (par. 82 a and b).

  (3) INABILITY TO WARM UP.

Possible Cause	Possible Remedy
Thermostat defective.	Replace thermostat (par. 85 a and b).

35. STARTING AND GENERATING SYSTEMS.
  a. Trouble Shooting Chart.
    (1) CRANKING MOTOR SPINS BUT WILL NOT TURN ENGINE.

Possible Cause	Possible Remedy
Automatic drive pinion damaged or stuck.	Renew cranking motor if drive is damaged. If stuck with gummy oil, clean.

    (2) CRANKING MOTOR WILL NOT OPERATE.

Possible Cause	Possible Remedy
Open circuit between cranking motor and battery.	Examine circuit for loose or corroded connections.
Batteries run down.	Charge or replace batteries.
Switches inoperative.	Renew switches (pars. 89 and 90).
Faulty cranking motor.	Renew motor (par. 91).

    (3) SLOW CRANKING SPEED.

Possible Cause	Possible Remedy
Batteries run down.	Charge or replace.
Corroded or loose battery connections.	Clean or tighten.
Cranking motor faulty.	Renew motor.
Engine oil too heavy.	Change to correct grade of oil.

10-TON 6 x 4 TRUCK (MACK MODEL NR)

(4) CRANKING MOTOR DRIVE ENGAGES WITH FLYWHEEL GEAR BUT WILL NOT TURN ENGINE.

Possible Cause	Possible Remedy
Transmission in speed.	Shift to neutral.
Batteries run down.	Charge or replace battery.
Faulty cranking motor.	Determine by diagnosis (b (1) below).
Corroded or loose battery connections.	Clean and tighten.

(5) BATTERIES RUN DOWN.

Generator output low.	Determined by diagnosis (par. 35 b (5)).
Faulty voltage regulator.	Replace regulator (par. 88 a and b).
Loose connection in generator regulator or battery circuit.	Check and correct.
Faulty batteries.	Renew.
Switches left on when not in use.	Turn switches off.
Grounded or shorted circuits.	Check and eliminate.
Corroded battery terminals.	Clean and coat with vaseline.
Loose or dirty ground connections.	Clean and tighten.

(6) NOISY GENERATOR.

Loose mounting.	Adjust.
Drive pulley loose.	Tighten.
Internal noise.	Replace generator (par. 87 a and b).

(7) GENERATOR DOES NOT CHARGE.

Broken or loose drive belts.	Adjust or replace belts (par. 83).
Voltage regulator faulty.	Renew regulator (par. 88 a and b).
Generator inoperative.	Renew generator (par. 87 a and b).

(8) GENERATOR OUTPUT EXCESSIVE.

Faulty generator.	Test generator (b (4) below).
Voltage regulator faulty.	Test voltage regulator (b (4) below).

(9) GENERATOR OUTPUT LOW OR UNSTEADY.

Loose drive belt.	Adjust (par. 83 a).
Faulty generator.	Renew generator (par. 87 a and b).
Loose connections in battery generator, or voltage regulator circuits.	Check and correct.
Faulty voltage regulator.	Test (b (4) below) and/or renew regulator (par. 88 a and b).

TROUBLE SHOOTING

b. **Diagnosis.**

(1) CRANKING MOTOR TEST. To determine quickly the approximate location of trouble when motor cranks slowly, or not at all, turn on lights and press starter button. If the lights go out, there is a defective connection between the cranking motor and battery. Clean and tighten connections and replace any defective cables. If lights dim it indicates that the added burden of cranking causes the voltage to drop off. Check battery with hydrometer. Unless battery is low, check for heavy engine oil or engine seizure due to tight bearings or pistons. If lights stay bright and cranking motor does not turn, examine circuit and magnetic switch. If batteries, switches, cables and connection are not faulty and engine turns freely, replace the cranking motor.

(2) MAGNETIC SWITCH. Using a test lamp and prod, place one prod on one of the insulated coil terminals and the other on the switch case. If the lamp lights, the coil is grounded, and should be renewed.

(3) GENERATOR REGULATOR SYSTEM. If generator output appears too low or too high, it is necessary to determine whether regulator or generator is at fault. Before making any test, however, it is essential that all wiring and connections be checked for loose or dirty contacts, especially at the ammeter and battery terminals and ground connections.

(4) GENERATOR OUTPUT EXCESSIVE. Opening field circuit, by disconnecting lead from field terminal of regulator or generator with the generator operating at a medium speed, will determine which unit is at fault. If output drops off, regulator is at fault and should be renewed. If output remains high, generator is at fault and should be renewed.

(5) GENERATOR NOT PRODUCING FULL OUTPUT. Eliminate regulator from system, by momentarily connecting a jumper lead from the regulator armature to the field terminal, with all electrical accessories turned off. If generator output comes up, then regulator is indicated as the source of trouble and should be renewed. CAUTION: *Under the above conditions a good generator can produce a very high output, consequently, extreme care must be taken to avoid operating the generator for more than a second or two—just long enough to see if it can produce a high output.*

36. TRANSMISSION.

a. **Trouble Shooting Chart.**

(1) NOISE.

Possible Cause	Possible Remedy
Lack of lubrication.	Refill to proper level with correct grade of lubricant (par. 19).
Worn gears.	Notify higher authority.
Loose gear fit.	Notify higher authority.

10-TON 6 x 4 TRUCK (MACK MODEL NR)

Possible Cause	Possible Remedy
Worn or loose bearings on shafts or in gears.	Notify higher authority.
Transmission misalined with clutch housing.	Notify higher authority.
Clutch housing loose on engine.	Tighten clutch housing mounting bolts.
Transmission loose on clutch housing.	Tighten transmission mounting bolts.
Rear section loose on front section.	Tighten mounting bolts.
Propeller shaft misalined or out of balance.	Check loose universal joints for worn needle bearings. Renew if necessary. Check propeller shaft for alinement. Check universal flanges for loose bolts, tighten and renew when necessary.

(2) SLIPPING OUT OF GEAR.

Possible Cause	Possible Remedy
Weak shifter shaft ball springs.	Clean balls and springs and renew if necessary.
Clashing of gears, causing tapered wear of gear teeth.	Notify higher authority.
Loose fit of gears on splines or shaft due to wear.	Notify higher authority.
Partial engagement of gears, causing tapered wear of teeth.	Notify higher authority.
Worn bearings.	Notify higher authority.

(3) HARD SHIFTING.

Possible Cause	Possible Remedy
Clutch drags or does not release.	Adjust or renew worn parts (par. 61).
Shifter shaft scored.	Notify higher authority.

(4) LOSS OF LUBRICANT.

Possible Cause	Possible Remedy
Defective gaskets.	Notify higher authority.
Worn or damaged oil seal.	Notify higher authority.
Excessive lubricant.	Drain to level of filler plug.

37. PROPELLER SHAFT.

a. Trouble Shooting Chart.

(1) EXCESSIVE VIBRATION.

Possible Cause	Possible Remedy
Shaft sprung from contact with obstruction.	Renew shaft (par. 95 a, b or c).

(2) NOISE IN UNIVERSAL JOINT.

Possible Cause	Possible Remedy
Backlash due to worn journal, bearings or yoke.	Renew shaft (par. 95 a, b or c).

TROUBLE SHOOTING

(3) LOSS OF LUBRICANT.

Possible Cause	Possible Remedy
Faulty oil seals or loss of lubrication fitting.	Renew shaft (par. 95 a, b or c).

(4) NOISE IN CENTER BEARING.

Possible Cause	Possible Remedy
Pounding due to excessive wear of bearing or housing bore.	Renew bearing (par. 96).

38. FRONT AXLE.

a. Trouble Shooting Chart.

(1) HARD STEERING.

Possible Cause	Possible Remedy
Steering linkage too tight.	If due to faulty end connections replace drag link and/or tie rod (par. 146). Check for worn lever stud ball.
Steering post sprung.	Loosen steering post bracket and shim into alinement.
Lack of lubrication.	Lubricate drag link ends, tie rod ends and steering gear.

(2) WANDER.

Possible Cause	Possible Remedy
Loose steering linkage.	If due to faulty end connections, replace drag link and/or tie rod (par. 146). Check for loose steering arm or levers.
Insufficient caster due to overload on front axle.	Correct load distribution or advise higher authority.
Toe-out of front wheels.	Check and reset to $\frac{1}{8}$ to $\frac{3}{16}$-inch toe-in (par. 98).

(3) SHIMMY.

Possible Cause	Possible Remedy
Tire and wheel out of balance.	Notify higher authority.
Loose steering linkage.	Check and adjust.
Worn king pins.	Notify higher authority.

(4) NEGATIVE CAMBER.

Possible Cause	Possible Remedy
Overloaded axle.	Check loads and redistribute.
Bent axle or spindle.	Notify higher authority.

(5) IMPROPER TOE-IN.

Possible Cause	Possible Remedy
Bent tie rod.	Replace.
Bent cross-steering lever.	Notify higher authority.

(6) WHEELS DO NOT STRAIGHTEN AFTER TURNING.

Possible Cause	Possible Remedy
Insufficient caster.	Notify higher authority.
Tight steering parts.	Adjust (a (1) above).
Unequal caster or bent axle.	Notify higher authority.

10-TON 6 x 4 TRUCK (MACK MODEL NR)

39. REAR AXLE.

a. Trouble Shooting Chart.

(1) EXCESSIVE NOISE.

Possible Cause	Possible Remedy
Loose torque rod brackets, bolts and rivets.	Tighten.
Loose torque rod ball studs.	Tighten.
Trunnion clamp bolts loose.	Tighten.
End play of spring seats on trunnion tube.	Adjust.
Worn trunnion bushings (maximum clearance 0.025 in.).	Renew spring assembly.
Axle tilted.	Check torque rods.

(2) AXLE NOISE.

Possible Cause	Possible Remedy
Continuous hum.	Readjust wheel bearing. If this does not correct, notify higher authority.
Coasting hum.	Notify higher authority.
Pulling hum.	Notify higher authority.

(3) BACKLASH.

Possible Cause	Possible Remedy
Loose axle shaft flange.	Tighten nuts.
Excessive pinion and bevel gear clearance.	Notify higher authority.
Excessive axle shaft spline clearance.	Notify higher authority.

40. BRAKE SYSTEM.

a. Trouble Shooting Chart.

(1) INEFFECTIVE BRAKES.

Possible Cause	Possible Remedy
Grease on lining.	Replace shoe (pars. 111 b and c, or 112 b and c).
Lining worn to rivet heads.	Replace shoe (pars. 111 b and c, or 112 b and c).
Drum scored or out-of-round.	Replace drums (pars. 133 and 137).
Excessive clearance between shoes and drum.	Adjust slack adjuster (par. 111 a).
Partial lining contact with drum.	Repair or renew spring shoes or bent anchor pins.

(2) DRAGGING BRAKES.

Possible Cause	Possible Remedy
Shoes too close to drum.	Increase stroke of slack adjuster (par. 111 a).
Drum out-of-round.	Renew drum (par. 133 or 137).

TROUBLE SHOOTING

Possible Cause	Possible Remedy
Loose or worn wheel bearings.	Adjust or renew (par. 132 or 136).
Weak retracting springs.	Renew springs (par. 111 a or 112 b).
Seized or binding linkage.	Free up and lubricate.

(3) GRABBING BRAKES.

Grit on lining.	Clean lining or renew shoes.
Loose lining.	Tighten rivets or renew shoe.
Spring clips loose or broken.	Tighten or renew.
Drums scored or cracked.	Replace drums (par. 133 or 137).

(4) NOISY BRAKES.

Grit on lining.	Clean lining or renew shoe (par. 111 or 112).
Drums scored.	Renew drums (par. 133 or 137).
Loose lining rivets.	Tighten rivets or renew shoe (par. 111 or 112).
Shoes, anchor pins or plates distorted.	Replace damaged part (par. 111 or 112).

(5) INEFFECTIVE AIR SYSTEM.

Low brake line pressure.	Notify higher authority.
Brake chamber diaphragm leaking.	Renew brake chamber (par. 115 or 116).

(6) SLOW PRESSURE BUILD-UP IN RESERVOIRS.

Leaking application or brake valve.	Renew valve (par. 118).
Leaking compressor discharge valve.	Renew compressor head assembly (par. 113 d and e).
Leaking lines or connections.	Tighten or renew as necessary.
No clearance on unloader valve.	Renew compressor head assembly (par. 113 d and e).
Clogged air cleaner.	Clean (par. 113 a (1)).
Worn piston and rings, carbon in discharge line.	Replace compressor (par. 117).

(7) QUICK LOSS OF RESERVOIR PRESSURE WHEN ENGINE IS STOPPED.

Worn and leaking compressor discharge valves.	Renew compressor head assembly (par. 113 d and e).
Tubing or connections leaking.	Renew tubing or tighten fittings.
Leaking valves.	Renew compressor head assembly (par. 113 d and e).
Leaking governor.	Renew.

10-TON 6 x 4 TRUCK (MACK MODEL NR)

(8) COMPRESSOR NOT UNLOADING.

Possible Cause	Possible Remedy
Broken unloader diaphragm.	Renew compressor head assembly (par. 113 d and e).
Too much clearance on unloader valves.	Renew compressor head assembly (par. 113 d and e).
Obstruction in line from governor to unloader.	Clean or renew line.
Governor not operating.	Replace governor (par. 114).

(9) SLOW BRAKE APPLICATION.

Low brake line pressure (brake application valve to brake chambers).	Notify higher authority.
Brake chamber push rod travel excessive.	Adjust slack adjuster (par. 111 a).
Obstruction in line.	Clean or renew tubing or hose.
Leaking brake chamber diaphragm.	Renew brake chamber (par. 114 or 116).
Leaking brake application valve diaphragm.	Renew valve assembly (par. 118).

(10) SLOW BRAKE RELEASE.

Binding cam or camshafts.	Lubricate and free up.
Brake chamber push rod travel excessive.	Adjust slack adjuster (par. 111 a).
Obstruction in tubing or hose.	Clean or renew part.

41. WHEELS.

a. Trouble Shooting Chart.

(1) SHIMMY.

Possible Cause	Possible Remedy
Tires worn unevenly or improperly inflated.	Match tires or inflate to recommended pressure (par. 147 b).
Tire and wheel out of balance.	Notify higher authority.
Bent rim or wheel.	Replace damaged part (par. 148 a and b or 130 a and b, or 134 a and b).
Broken spring leaves.	Replace spring (par. 139 a and b or 141 a and b).
Loose or broken spring clamp bolts.	Tighten or replace.
Misalined camber, caster or toe-in.	Adjust toe-in (par. 98). If camber or caster correction is necessary, notify higher authority.
Wheel bearings loose.	Adjust (par. 130 b (8)).

TROUBLE SHOOTING

(2) WANDER.

Possible Cause	Possible Remedy
Uneven tire pressure.	Inflate to recommended pressure (par. 147 b).
Wheel bearing tight.	Adjust (par. 130 b (8)).

(3) EXCESSIVE TIRE WEAR.

Possible Cause	Possible Remedy
Improperly inflated tires.	Inflate to recommended pressure (par. 147 b).
Tire and wheel out of balance.	Notify higher authority.
Wheel misalined.	Correct (a (1) above).
Bent rim or wheel.	Replace damaged part (pars. 130, 134, 148 and 149).
Grabby brakes.	Adjust brakes (par. 111 a).

(4) WHEEL BEARING ADJUSTMENT.

Possible Cause	Possible Remedy
Overheated bearings.	Indicates tightness—adjust (par. 132 or 136).
Loose bearings.	Causes wheel pounding and shimmy—adjust bearing (par. 132 or 136).

42. SPRINGS AND SHOCK ABSORBERS.

a. Trouble Shooting Chart.

(1) AXLE STRIKES FRAME (FRONT).

Possible Cause	Possible Remedy
Broken leaves.	Renew springs (par. 139 or 141).
Sagged spring.	Renew spring (par. 139 or 141).

(2) SPRING BREAKAGE AT CENTER BOLT.

Possible Cause	Possible Remedy
Loose clips.	Tighten as per specifications.

(3) EXCESSIVE NOISE.

Possible Cause	Possible Remedy
No oil in trunnion.	Lubricate.
Spring clamp studs loose.	Tighten.
Loose shock absorber parts.	Tighten.

43. STEERING GEAR.

a. Trouble Shooting Chart.

(1) HARD STEERING.

Possible Cause	Possible Remedy
Steering linkage tight.	Readjust tie rod and drag link ends.
Steering column sprung.	Loosen steering column bracket and shim into correct alinement.

TM 9-818
43-45

10-TON 6 x 4 TRUCK (MACK MODEL NR)

Possible Cause	Possible Remedy
Steering column bracket clamp too tight.	Loosen.
Lack of lubrication.	Lubricate steering gear and tie rod and drag link ends.
Soft tires.	Inflate properly (par. 147 **b**).

(2) EXCESSIVE TIRE WEAR.

Bent axle.	Notify higher authority.

(3) WANDER.

Loose steering linkage.	If due to faulty end connections replace drag link and/or tie rod (par. 146). Check for loose steering arm or levers.
Insufficient caster—too much load on front axle.	Redistribute load to put more weight on rear axle or notify higher authority.
Front wheels toe out.	Reset to $\frac{1}{8}$ to $\frac{3}{16}$-in. toe-in.

(4) SHIMMY.

Tire and wheel out of balance.	Notify higher authority.
Loose steering linkage.	Check and adjust.
Worn king pin.	Notify higher authority.
Steering gear loose on frame.	Tighten.

44. FRAME.

a. Trouble Shooting Chart.

(1) EXCESSIVE TIRE WEAR.

Possible Cause	Possible Remedy
Sprung frame.	Notify higher authority.
Rear and front wheels do not track.	Notify higher authority.
Relative motion between side rails due to loose rivets.	Notify higher authority.

45. BATTERY AND LIGHTING SYSTEM.

a. Trouble Shooting Chart.

(1) BATTERIES RUN DOWN.

Possible Cause	Possible Remedy
Generator not charging properly.	Determined by diagnosis (par. 35 d (3)).
Cells shorted.	Replace battery.
Excessive resistance in cables.	Clean and tighten terminals.
Battery cells dry.	Fill to proper level.
Cranking motor faulty (causing excessive drain on battery).	Determined by diagnosis (par. 35 d (1)).
Engine starts hard.	Determined by diagnosis (par. 30 a (2)).

TM 9-818
45

TROUBLE SHOOTING

(2) LAMPS DO NOT LIGHT.

Possible Cause	Possible Remedy
Burned-out lamps.	Replace.
Broken headlight lens.	Replace sealed beam unit.
Broken wire or loose connection.	Locate and correct.
Defective light switch.	Replace.
Shorted wires.	Repair.
Poor ground.	Correct.

(3) FREQUENT BULB FAILURE.

Generator charging at excessive voltage.	Determined by diagnosis (par. 35 d (4)).
Voltage regulator faulty.	Determined by diagnosis (par. 35 d (4) and (5)).

(4) DIM LIGHTS.

Poor ground.	Repair.
Loose or dirty terminals.	Tighten or clean.
Discharged battery.	Recharge or replace.

10-TON 6 x 4 TRUCK (MACK MODEL NR)

Section XIII
ENGINE—DATA, MAINTENANCE AND ADJUSTMENT IN VEHICLE

	Paragraph
Description and tabulated data	46
Engine tune-up	47
Cylinder head and gasket replacement	48
Adjust valves	49
Valve rocker arm assembly	50
Push rods	51
Intake manifold	52
Exhaust manifold	53
Manifold air heater elements	54
Oil pan assembly	55
Oil filter	56
Crankcase breather	57

46. DESCRIPTION AND TABULATED DATA.

a. Description. The engine is of the four-cycle, valve-in-head, Diesel type, having six vertical cylinders. These are cast in one block integral with the upper crankcase and have renewable, dry-type cylinder sleeves. There are two cylinder heads, each serving for three cylinders in which there are energy cells opposite the fuel injection nozzles. The injection pump is of the multiple type, that is, an individual pump element serves each of the six nozzles. The pump is flange-mounted, and the coupling member of its built-in drive is a centrifugally actuated device which automatically advances and retards the time of injection. A centrifugal, gear-driven governor is incorporated in the injection-pump control. All lubrication is of the pressure type. The air compressor is flange-mounted, has a built-in drive, and is lubricated and water-cooled from the engine. The engine has a vibration damper and is rubber-mounted.

b. Data.

Make	Mack
Model	ED
Type	Diesel, overhead-valve, water-cooled
Number of cylinders	6
Bore and stroke	$4\frac{3}{8}$ x $5\frac{3}{4}$ in.
Piston displacement	518.64 cu in.
Compression ratio	14.57 to 1
Brake horsepower	131 at 2000 rpm
Governed speed:	
Full load	2000 rpm
No load	2110 rpm

ENGINE—DATA, MAINTENANCE AND ADJUSTMENT IN VEHICLE

Idling speed	500 rpm
Maximum torque	382 lb-ft
Speed at maximum torque	1200 rpm
Rotation of crankshaft (facing front of engine)	clockwise
Firing order	1-5-3-6-2-4

47. ENGINE TUNE-UP.

a. Engine tune-up is a procedure for checking engine and accessory equipment to determine if they are within original specifications.

RA PD 310537

Figure 41—Engine Installed—Right Side

At the same time a series of corrective operations should be performed to restore peak operating efficiency. It will be noted that a cylinder compression test is not specified in procedure given below. Although such a test might be helpful, it has been found impractical to obtain accurate readings for a Diesel engine while it is mounted in the chassis.

b. **Tune-up.** Before performing operations listed in following table, both the engine crankcase and injection-pump sump should be drained and refilled with fresh oil (par. 19 b (4) and (13)); oil filter

TM 9-818
10-TON 6 x 4 TRUCK (MACK MODEL NR)

serviced (par. 56 a) and crankcase breather air cleaner serviced (par. 57 b). Proceed with the following operations in order shown:

   (1) RADIATOR. Check for leaks, fluid level and clear air passages.

   (2) RADIATOR HOSE. Check for leaks, deteriorated hose and loose connections.

   (3) BATTERY. Service (par. 167).

   (4) BATTERY CABLES. Inspect, clean and tighten terminals.

   (5) CRANKING MOTOR. Test operation.

   (6) AIR CLEANER. Service.

RA PD 310538

Figure 42—Engine Installed—Left Side

   (7) CYLINDER HEAD BOLTS. Tighten (par. 48 b (2)).

   (8) VALVE CLEARANCE. Check and adjust (par. 49 a (3)).

   (9) FUEL FILTERS. Service (par. 19 c (7)).

   (10) INJECTION PUMP. Timing with engine (par. 64 c and d).

   (11) ENERGY CELLS. Remove, inspect and clean (par. 73 b).

   (12) FUEL LINES. Check for oil or air leaks. Tighten (par. 66 b).

   (13) INJECTION NOZZLES. Inspect for leaks (par. 67 b and d).

   (14) FAN BELT. Adjust (par. 83 a (1)).

   (15) GENERATOR. Test operation.

   (16) OIL PUMP. Check pressure reading on gage (par. 17 b (3) (a)).

TM 9-818
48

ENGINE—DATA, MAINTENANCE AND ADJUSTMENT IN VEHICLE

48. CYLINDER HEAD AND GASKET REPLACEMENT.

a. **Removal.**

(1) DRAIN SYSTEM. Open drain plug on lower radiator pipe and drain cooling system.

(2) REMOVE VALVE COVERS. Follow procedure outlined in paragraph 49 b (2) to remove valve covers.

(3) REMOVE ROCKER ARMS. Follow procedure outlined in paragraph 50 a (2).

(4) REMOVE PUSH RODS. Follow procedure outlined in paragraph 51 a (2).

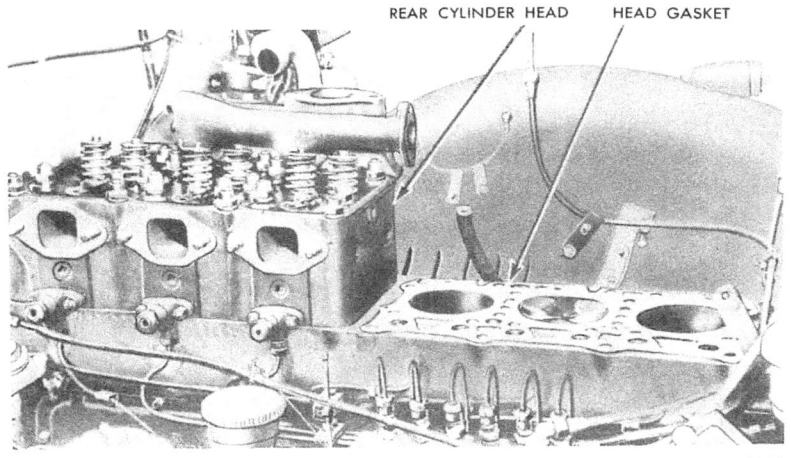

Figure 43—Front Head Removed

(5) REMOVE EXHAUST MANIFOLD. Follow procedure outlined in paragraph 53 a.

(6) DETACH FUEL LINES. Follow procedure outlined in paragraph 64 b (4), to remove fuel return lines and high-pressure lines from the nozzles.

(7) REMOVE AIR CLEANER. Follow procedure outlined in paragraph 74 c.

(8) REMOVE FUEL FILTERS. Follow procedures outlined in paragraph 71 b (3) and c (3), to remove No. 2 and No. 3 fuel oil filters excepting that it will not be necessary to disconnect the line between them as they may be removed from the engine as a unit with the mounting bracket plate by removing four cap screws and lock washers.

(9) REMOVE CENTER INTAKE MANIFOLD. Follow procedure outlined in paragraph 52 a (6).

TM 9-818
48

10-TON 6 x 4 TRUCK (MACK MODEL NR)

(10) DETACH HEATER CABLES. Disconnect the manifold heater cables at the terminals.

(11) REMOVE HOSE CONNECTIONS. Follow procedure outlined in paragraph 85 a and 80 e (1) to remove thermostat and bypass elbow hose.

(12) DETACH WATER MANIFOLD. Remove two bolts, nuts and lock washers and the gasket at the center flanges on the water manifold.

(13) REMOVE WATER TEMPERATURE GAGE ELEMENT. Remove the gage element and adapter from the end of the water manifold.

(14) REMOVE CYLINDER HEADS. Remove the 20 nuts and washers on each cylinder head. Remove 20 cylinder head studs by unscrewing

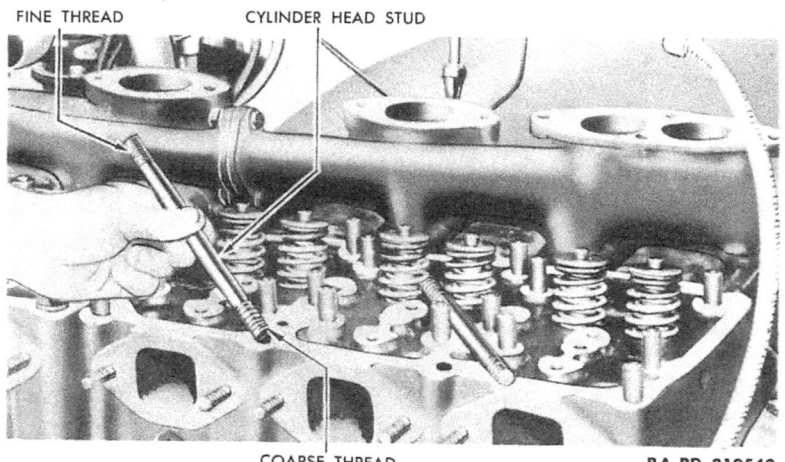

Figure 44—Inserting Cylinder Head Studs

from cylinder block. If any stud is so tight that it cannot be turned with the fingers, use two nuts jammed together on the stud to remove it. Lift heads from engine block and remove.

(15) REMOVE GASKETS. Lift gaskets off cylinder block, prying loose if necessary.

(16) CARBON REMOVAL. Loosen carbon from cylinder head, block, and tops of pistons with a stiff wire brush and carbon scraper. Thoroughly clean loose carbon from cylinder walls, tops of pistons, cylinder and block surfaces.

b. **Installation.**

(1) INSTALL GASKETS. Thoroughly clean the upper surface of the engine block and the lower surfaces of the cylinder heads, removing carbon and any particles of the gaskets or other foreign matter, and place new gaskets on cylinder block, with the holes accurately located by the dowels.

TM 9-818
48

ENGINE—DATA, MAINTENANCE AND ADJUSTMENT IN VEHICLE

(2) INSTALL CYLINDER HEADS. Carefully avoiding damage to the newly placed gaskets, lower cylinder heads onto engine block and gasket, locating them accurately on dowels so that studs may be inserted easily. Place gasket between water manifold flanges. Attach them with two bolts, nuts and lock washers, and insert 20 studs in each head, with coarse threads at the bottom, and screw into cylinder block fingertight. Place nuts and washers on studs. It is important that these stud nuts be tightened in proper sequence, as shown by the diagram (fig. 46). Start with nut No. 1 and progress numerically. Do not tighten nuts fully first time around. Use torque indicating

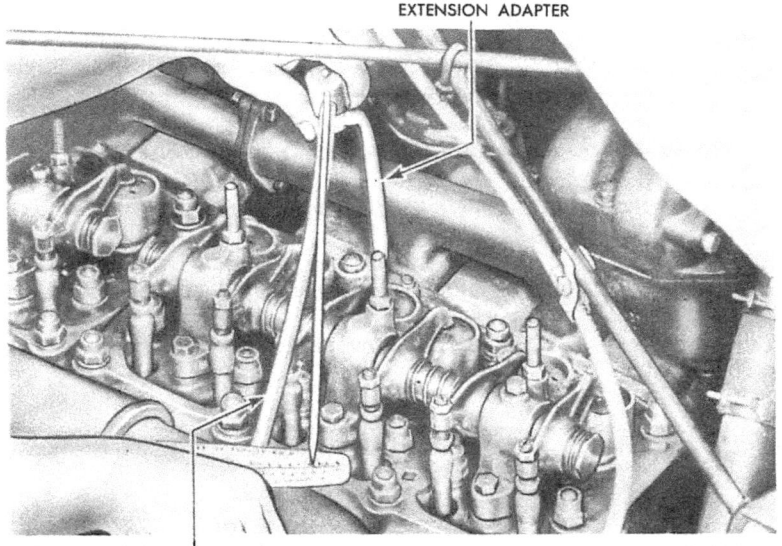

Figure 45—Tightening Cylinder Head Nut

wrench (41-W-3630) and extension adapter (41-W-2965-5) to obtain proper tightening of nuts. Go over nuts several times so that wrench indicates 125 to 130 foot-pounds as nuts are finally tightened. Tighten water manifold flange nuts.

(3) ATTACH FUEL LINES. Follow procedure outlined in paragraph 67 d (4) and (5) to install fuel return lines and high-pressure lines at nozzles, but do not tighten pressure lines until directed.

(4) INSTALL CENTER INTAKE MANIFOLD. Follow procedure outlined in paragraph 52 b (1) to install center manifold.

(5) INSTALL WATER TEMPERATURE GAGE ELEMENT. Install gage element and adapter in rear end of water manifold.

TM 9-818
48

10-TON 6 x 4 TRUCK (MACK MODEL NR)

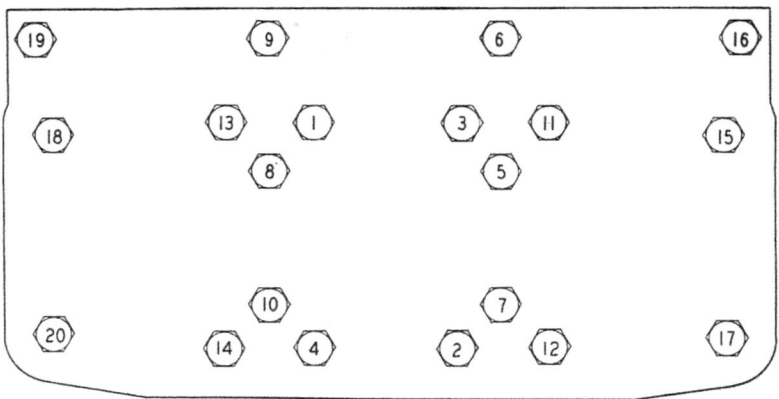

Figure 46—Sequence for Tightening Cylinder Head Nuts

(6) CONNECT HEATER CABLES. Connect manifold heater cables at terminals at cylinder intakes.

(7) INSTALL FUEL FILTERS. Install fuel filters and bracket assembly with four cap screws and lock washers, and connect filters to lines (par. 71 **b** (4) and **c** (4)).

(8) INSTALL AIR CLEANER. Follow procedure outlined in paragraph 74 **d** (1).

Figure 47—Adjusting Valve Clearance

ENGINE—DATA, MAINTENANCE AND ADJUSTMENT IN VEHICLE

(9) INSTALL HOSE CONNECTIONS. Install thermostat housing and bypass elbow hose (pars. 85 b and 80 e (2)).

(10) TIGHTEN FUEL LINES. Follow procedure outlined in paragraph 67 d (4) to tighten fuel pressure lines at nozzles, and bleed fuel system following procedure outlined in paragraph 68.

(11) INSTALL PUSH RODS. Follow procedure outlined in paragraph 51 b (1).

(12) INSTALL ROCKER ARM ASSEMBLY. Follow procedure outlined in paragraph 50 b.

(13) ADJUST VALVES. Follow procedure outlined in paragraph 49 b (3), then install valve covers.

(14) INSTALL EXHAUST MANIFOLD. Follow procedure outlined in paragraph 53 b (1).

(15) FILL COOLING SYSTEM. Fill cooling system with water, and check for leaks.

49. ADJUST VALVES.

a. **Remove Breather Pipe.** Loosen hose clamp on valve cover breather pipe and remove hose. Remove four cap screws and flat washers on breather cross tube flange and remove breather cross tube and gaskets.

b. **Remove Valve Covers.** Remove three acorn nuts and washers from each valve cover and remove covers and gaskets. Push rear breather pipe away if necessary to remove rear cover.

c. **Adjust Valves.** Use screwdriver with box wrench in the manner shown in figure 47 to adjust valves. Insert feeler gage (41-G-400) as shown. The clearances with engine cold are 0.012 to 0.014 inch for inlet valve and 0.017 to 0.019 inch for exhaust. At hot idling speed inlet valve clearance is 0.010 to 0.012 inch and exhaust valve clearance is 0.015 to 0.017 inch. Cold settings are those determined at room temperatures (approximately 70°F) and are given only for preliminary checking. Valves should be adjusted at idling heat when water and oil have reached steady temperatures. A cold engine will require about one-half hour of idle running to heat up. An engine of a truck just in from operations should be allowed to idle about 15 minutes.

d. **Install Valve Covers.** Place gaskets on cylinder heads and install covers with three acorn nuts and washers on each cover. Do not tighten the acorn nuts until the breather cross tube is in place.

e. **Install Breather Tubes.** Place breather cross tube on covers with outlet to left-hand side of truck and with gaskets in place. Attach by four cap screws and flat washers. Attach breather pipe to cross tube with rubber hose, and tighten hose clamps.

50. VALVE ROCKER ARM ASSEMBLY.

a. **Removal.**

(1) REMOVE COVERS. Remove valve covers as outlined in paragraph 49 b.

TM 9-818
50-51
10-TON 6 x 4 TRUCK (MACK MODEL NR)

(2) REMOVE ROCKER ARM ASSEMBLY. Remove three center studs from each cylinder head. Remove three cap screws and flat washers, and remove rocker arm assembly, lifting carefully to avoid dropping out valve thrust cups.

b. **Installation.**

(1) INSTALL ROCKER ARM ASSEMBLY. After checking cups for wear or damage and replacing any defective cups, place assembly on cylinder head, in same position as before removal, and install and tighten three cap screws and washers on right-hand side, and three studs on center of each cylinder head.

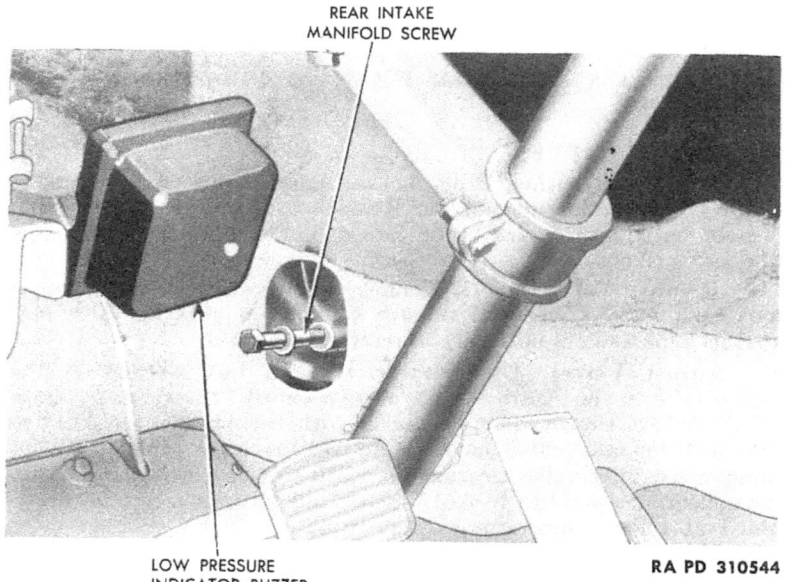

Figure 48—Rear Intake Manifold Screw

(2) INSTALL COVERS. Install valve covers and complete engine assembly as outlined on paragraph 49 d.

51. PUSH RODS.

a. **Removal.**

(1) REMOVE ROCKER ARM ASSEMBLY. Follow procedure outlined in paragraph 49 b, and paragraph 50 a (2) to remove valve covers and rocker arm assembly.

(2) REMOVE PUSH RODS. Pull out push rods.

b. **Installation.**

(1) INSTALL PUSH RODS. Insert push rods through head openings with the ball end downward.

ENGINE—DATA, MAINTENANCE AND ADJUSTMENT IN VEHICLE

(2) INSTALL ROCKER ARM ASSEMBLY. Follow procedure outlined in paragraph 50 b (1), and paragraph 49 c, d, and e to complete engine assembly.

52. INTAKE MANIFOLD.

a. Removal.

(1) REMOVE FILTERS. Remove No. 2 and No. 3 fuel filters as outlined in paragraph 71 b (3) and c (3).

(2) REMOVE AIR CLEANER. Remove air cleaner assembly as outlined in paragraph 74 c.

(3) REMOVE AIR CLEANER BRACKET. Remove four cap screws and lock washers, and remove air cleaner bracket. The upper inside cap screw has fuel line clamp attached to it. Remove clamp.

(4) REMOVE INTAKE HOSE. Loosen hose clamp on intake hose and remove.

(5) REMOVE FILTER MOUNTING BRACKET. Remove four cap screws and lock washers, and remove fuel filter mounting bracket. The bottom rear cap screw holds fuel line bracket which should be removed.

(6) REMOVE CENTER MANIFOLD SECTION. Remove four cap screws and lock washers attaching center manifold section, and remove the manifold. Remove plug and gasket from manifold.

(7) REMOVE FORWARD MANIFOLD SECTION. Remove six cap screws and lock washers attaching forward manifold section, and remove manifold.

(8) REMOVE REAR MANIFOLD SECTION. Inside cab remove nuts and lock washers attaching low-pressure indicator buzzer to dash sheet. Loosen two dash mat buttons at lower left of mat, and lift mat up to clear hole in dash sheet. Through this opening remove rear cap screw and lock washer on rear manifold section. Remove the other five cap screws from engine compartment, and remove rear manifold section.

b. Installation.

(1) INSTALL MANIFOLD. Install plug and gasket in center manifold section. After making sure that manifold flanges are clean, line up intake heater assembly. Place manifold in position, and start six cap screws and lock washers to attach front and rear manifold sections. Do not tighten. Install center manifold section on forward and rear sections, with gaskets in place, and install four cap screws and lock washers. Tighten these four cap screws, and then tighten the 12 cap screws on forward and rear manifold sections, the rearmost screw being placed through hole in dash sheet.

(2) INSTALL CAB FITTINGS. Place dash mat in position, and install the two mat buttons. Install buzzer with nuts and lock washers.

(3) INSTALL FILTER BRACKETS. Attach fuel filter bracket with four cap screws and lock washers, with fuel line clamp bracket on lower rear screw. Tighten cap screws.

(4) INSTALL AIR CLEANER BRACKET. Place air cleaner bracket

TM 9-818
52-53

10-TON 6 x 4 TRUCK (MACK MODEL NR)

in position and install four cap screws, placing fuel line clamp on rear cap screw.

(5) ATTACH INTAKE HOSE. Place intake hose on center manifold and tighten clamp.

(6) INSTALL AIR CLEANER. Follow procedure outlined in paragraph 74 d.

(7) INSTALL FUEL FILTERS. Follow procedure outlined in paragraph 71 b (4) and c (4).

53. EXHAUST MANIFOLD.

a. Removal.

(1) DETACH EXHAUST PIPE. Free nuts from locks on exhaust manifold flange with hammer and chisel, and remove the four nuts. Pull exhaust pipe down to free it from manifold, and remove two gaskets. Disconnect accelerator spring.

Figure 49—Exhaust Pipe to Manifold Connection

(2) DETACH MANIFOLD. Remove the 12 flange nuts attaching exhaust manifold to engine block. Also remove 11 lock washers and bracket on front stud. Pull back spring clamp on No. 2 stud.

(3) REMOVE MANIFOLD. Pull manifold and gaskets off studs.

b. Installation.

(1) INSTALL MANIFOLD. Clean flange faces and engine block faces, removing particles of old gaskets from flanges. Install new gaskets, and place new manifold as one piece on studs. Install 12 nuts with 11 lock washers, the bracket on the forward stud, and the accelerator spring clamp on No. 2 stud. If manifold has been only partially removed, it can be installed as three separate pieces instead of one. Place rear, center, and front pieces in that order. Tighten nuts evenly and alternately. Connect accelerator spring.

ENGINE—DATA, MAINTENANCE AND ADJUSTMENT IN VEHICLE

(2) ATTACH EXHAUST PIPE. Attach exhaust pipe to manifold flange with parts assembled as shown in figure 49. Tighten nuts and bend locks into place. Attach accelerator spring.

54. MANIFOLD AIR HEATER ELEMENTS.
  a. Removal.

(1) REMOVE MANIFOLD. Remove intake manifold sections as outlined in paragraph 52 a.

(2) DETACH CABLES. Remove nuts and lock washers and detach cables.

(3) REMOVE HEATER. Lift heater assembly out of engine compartment and remove jumpers by removing nuts and washers.

  b. Installation.

(1) INSTALL JUMPERS. Assemble three heater units with two jumpers, and install nuts and washers. Place on intake ports, but do not tighten nuts until manifold mounting cap screws are in place.

(2) ATTACH CABLES. Attach the two cables from the magnetic switches to terminals at front and rear ends of heater assemblies, and the two cables from batteries to heater assembly terminals center, and tighten nuts and washers.

(3) INSTALL MANIFOLD. Follow procedure outlined in paragraph 52 b to install rear, front, and center manifold sections in that order.

55. OIL PAN ASSEMBLY.
  a. Replace Oil Pan Sump. Drain oil from pan by removing plug and gasket in sump; then remove 10 cap screws and lock washers from sump flange, and remove sump and gasket. Wash sump screen thoroughly with dry-cleaning solvent and reinstall, reversing removal procedure. Install drain plug and gasket, and refill oil pan to required level.

  b. Replace Oil Pan Gasket.

(1) REMOVAL.

(a) *Drain Oil.* Remove sump plug and drain oil from pan.

(b) *Remove Flywheel Housing Front Cover.* Remove four cap screws and lock washers, and remove flywheel housing front cover and felt seal.

(c) *Remove Oil Pan.* Remove 26 $\frac{3}{8}$-inch cap screws and lock washers from oil pan flange. Remove the two $\frac{1}{2}$-inch cap screws and lock washers at front of pan and the two $\frac{1}{2}$-inch set screws at rear of pan. Remove oil pan.

(2) INSTALLATION.

(a) *Install Oil Pan.* Place new gasket on oil pan flange, and position under cylinder block. Attach pan by installing 26 $\frac{3}{8}$-inch cap screws with lock washers, the two $\frac{1}{2}$-inch cap screws and lock washers at front, and the two $\frac{1}{2}$-inch set screws at rear.

(b) *Install Flywheel Housing Front Cover.* Place flywheel front cover and felt, and install the four cap screws with lock washers.

10-TON 6 x 4 TRUCK (MACK MODEL NR)

(c) Refill with Oil. Install drain plug and gasket in sump, and refill with oil as specified in paragraph 19 **b** (4) and figure 31.

56. OIL FILTER.

a. Cartridge Replacement. Unscrew the wing nut, remove the cover, and pull up on wire to withdraw the bag. Install new bag after draining sediment out of the casing, and install cover. Apply wing nut and tighten.

b. Removal.

(1) REMOVE AIR CLEANER SUMP. Unclamp the two snap fastenings holding air cleaner sump, and remove sump.

(2) REMOVE FILTER. Remove the four cap screws and lock washers. Remove filter and gaskets.

c. Installation.

(1) INSTALL FILTER. Place gasket on filter mounting pad, and install filter with four cap screws and lock washers.

(2) INSTALL SUMP. Place air cleaner sump on bottom of air cleaner, and install by clamping spring clamps to the sump.

57. CRANKCASE BREATHER.

a. Removal. Turn wing nut on top of breather air cleaner counterclockwise to remove both cover and cleaner. Loosen lock nut and cap screw, and unscrew cleaner mounting bracket.

b. Service Breather Assembly. Remove cover and sump as noted above, and wash screen thoroughly in dry-cleaning solvent. Wash sump in dry-cleaning solvent. Blow each dry and reinstall in cleaner.

c. Installation. Turn filter mounting bracket onto crankcase breather pipe. Install sump, and tighten locking nut and cap screw. Fill sump to bead with engine oil and install screen and cover, turning wing nut clockwise to fasten.

TM 9-818
58

Section XIV

ENGINE—REMOVAL AND INSTALLATION

	Paragraph
Removal	58
Installation	59

58. REMOVAL.

  a. **Remove Hood.** Loosen and remove two flathead bolts, nuts, and lock washers at radiator shell. Loosen two bolts on cowl, using recessed screwdriver and wrench. Raising hood at front end, pull out of rear clamp and remove.

Figure 50—Generator Regulator and Wiring on Dash Sheet

  b. **Relieve Pressure in Air Tanks.** Open pet cock on air reservoir.

  c. **Drain Water.** Open pet cock on lower radiator pipe and drain water.

  d. **Roll Back Cab Top.** Follow procedure outlined in paragraph 161 e (3).

  e. **Disconnect Batteries.** Follow procedure outlined in paragraph 167 c (1) (a) and (b) to disconnect batteries.

127

10-TON 6 x 4 TRUCK (MACK MODEL NR)

f. Remove Toe- and Floorboards. Follow procedure outlined in paragraph 159 a, to remove toe- and floorboards, accelerator pedal, brake pedal, etc.

g. Disconnect Cab Wiring. From junction block on dash sheet, pull out the eight connections farthest to left. Remove nuts and lock washers from six connections at right of junction block (fig. 50). Disconnect plug on single No. 14 black wire by pulling wire out of plug. Disconnect two cables, one on left-hand side of each magnetic switch for intake heaters, by removing nuts and lock washers.

h. Disconnect Heater Relay Switch Cables. Disconnect the four heater relay switch cables on left side of engine, at first, third, fourth, and sixth cylinders.

i. Remove Water Temperature Gage Element. Remove temperature gage element from rear of water manifold and remove adapter.

j. Disconnect Hand Throttle. Remove hand throttle wire from accelerator cross shaft bell crank, remove stop, and pull out wire.

k. Remove Generator Cables. Follow procedure outlined in paragraph 87 a (2) to disconnect shielded cables from generator. Disconnect other ends from regulator.

l. Disconnect Fuel Pressure Gage Line. Remove fuel pressure gage line from clamps at top of air cleaner and No. 3 fuel filter by removing cap screw at air cleaner and nut and lock washer at filter. Remove line from top of No. 3 fuel filter. Remove strap clamp attaching line to radiator tie rod.

m. Disconnect Tachometer Cable. Remove cable from two clamps on upper radiator tie rods. Remove nut, bolt, and lock washer at first cylinder exhaust manifold bolt bracket. Remove cable from adapter on right-hand side of engine. Remove cable from engine.

n. Disconnect Engine Stop Control. Loosen two set screws, one on governor lever and one on clamp on governor. Remove nut, bolt and lock washer from clamp bracket on right-hand fender. Remove cable from engine.

o. Disconnect Engine Emergency Stop Control. Loosen set screw in stop on end of wire and remove stop. Remove wire from clamping brackets on right-hand fender by removing nut, bolt and lock washer from each bracket. Remove cable from engine.

p. Disconnect Oil Pressure Gage Line. Disconnect oil pressure line from engine at right-hand side of engine. Remove clamp from top rear screw of valve push rod inspection cover by removing cap screw and lock washer. Remove line from engine.

q. Disconnect Air Pressure Governor Line. Remove air governor line at top of air compressor and at low-pressure indicator.

r. Disconnect Air Supply Valve Line. Disconnect line from air supply fitting to low-pressure indicator at indicator switch. Remove clamp, and push line forward.

TM 9-818
58

ENGINE—REMOVAL AND INSTALLATION

s. **Remove Tie Rods.** Loosen and remove nuts from front and rear of upper tie rods. Turn stop nuts on rear ends as far forward as possible, and remove rods.

t. **Remove Steering Wheel.** Follow procedure outlined in paragraph 143 a to remove horn button, horn wire and steering wheel.

u. **Remove Brake Lever.** Remove cotter and clevis pins from clevises on each end of brake rod, one inside of cab at hand lever and the other at bell crank on clutch pedal bracket. Remove two bolts and two cap screws from underneath chassis, and remove lever assembly.

v. **Disconnect Speedometer Cable.** Disconnect speedometer cable at rear of transmission, and remove cap screw on left side of shifter cover. Pull cable away from transmission.

w. **Disconnect Steering Post.** Remove two bolts, nuts, and lock washers from bracket attachment to cab frame.

RA PD 310548

Figure 51—Cab Raising Sling

x. **Remove Cab Mounting Bolts.** (Cab removal authorized only to facilitate the removal of the engine.) Remove two bolts, nuts, and lock washers from bracket on each side at front of cab, and also clamp for battery cables. Remove palnut, nut and spring from cab rocker assembly.

TM 9-818

10-TON 6 x 4 TRUCK (MACK MODEL NR)

y. **Remove Cab.** Place slings for cab removal around upper hinges on each door, around grab handle on left-hand side, and around either grab handle or pioneer tool kit bracket on right-hand side as illustrated. With slings attached to hoist, raise and remove cab, placing it on blocks or other support so as not to damage the aprons.

z. **Disconnect Front Propeller Shaft.** Remove eight bolts, nuts and lock washers from forward propeller shaft flange and support propeller shaft on cab cross member.

RA PD 310549

Figure 52—Removing Cab

aa. **Remove Cutch Lever.** Remove nut, bolt and lock washer from clamp on end of clutch shaft, and use hammer to remove lever from shaft.

bb. **Remove Fan and Belts.** Remove six cap screws and lock washers from fan hub. Remove fan assembly to left-hand side, and remove fan belts (pars. 81 a and 83 b (1)).

cc. **Disconnect Generator Bond Strap.** Remove screw and two toothed lock washers from generator bond strap and remove strap.

dd. **Disconnect Engine-to-Frame Bond Straps.** Remove cap screw and two toothed lock washers from bond straps on each side at front of engine and detach bond straps from engine.

TM 9-818

ENGINE—REMOVAL AND INSTALLATION

ee. **Remove Radiator Connections.** Remove water pump inlet pipe and hose connecting upper tank to water pump outlet, following procedure outlined in paragraph 80 b (1).

ff. **Disconnect Brake Application Valve Air Line.** For clearance in removing engine, disconnect the air line (which extends to the relay valve) at the front of brake application valve.

gg. **Detach Exhaust Pipe at Manifold.** Follow procedure outlined in paragraph 53 a (1).

RA PD 310550

Figure 53—Engine and Transmission Ready for Removal

hh. **Disconnect Fuel Return Line.** Use two wrenches to disconnect flexible fuel return line from rigid tubing, disconnect latter from return valve and swing it upward to clear engine.

ii. **Disconnect Fuel Supply Line.** Use two wrenches to disconnect fuel supply line at rigid copper tubing.

jj. **Disconnect Main Discharge Air Line.** Using two wrenches, disconnect the air compressor flexible line from the rigid copper tubing.

kk. **Disconnect Cranking Motor.** Disconnect two No. 00 cables at cranking motor by removing nuts and washers from terminals.

ll. **Adjust Slings for Engine Removal.** Place a sling (41-S-3832-8) around transmission at front of auxiliary case and another at front of engine around crankshaft pulley, and attach slings to hoist.

TM 9-818
58

10-TON 6 x 4 TRUCK (MACK MODEL NR)

RA PD 310551

Figure 54—Removing Engine and Transmission

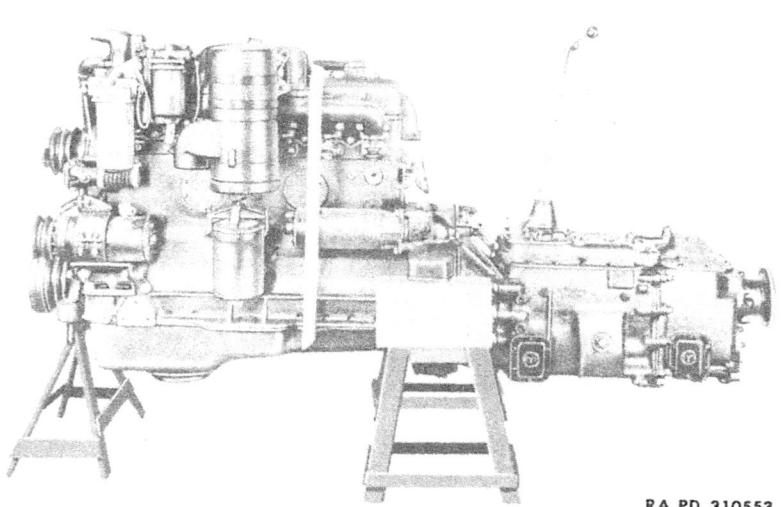

RA PD 310553

Figure 55—Engine and Transmission Removed

132

TM 9-818
58-59

ENGINE—REMOVAL AND INSTALLATION

mm. **Remove Engine Support Bolts.** Remove two palnuts and nuts from bolts in front engine supports, using one wrench to turn and another to hold. Remove bolts. Remove four nuts, lock washers and flat washers from rear engine support bolts, using two wrenches, and remove bolts.

nn. **Remove Engine and Transmission.** Raise hoist to remove engine and transmission as a unit. Remove by turning transmission end outward to the right, when the oil pan has cleared frame and exhaust pipe, unless there is enough room vertically to remove by lifting over right fender. Place on blocks or other support, with transmission clear for removal.

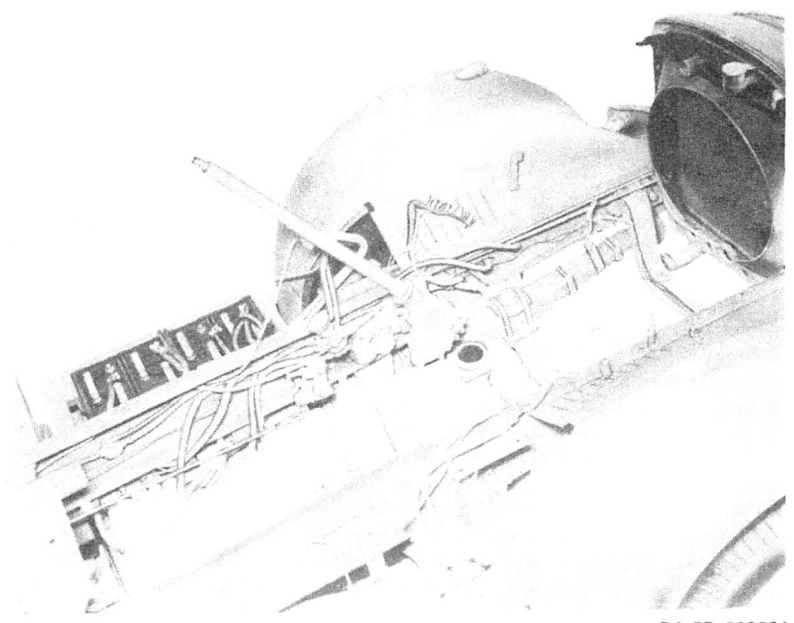

RA PD 310554

Figure 56—Chassis with Engine and Transmission Removal

oo. **Remove Transmission.** Remove the 12 cap screws and lock washers from transmission bell housing. Place sling from hoist around transmission. Remove the two nuts and lock washers from pilot studs. Remove transmission with pry-bar to aid in separating transmission from engine bell housing.

59. INSTALLATION.

a. **Install Transmission.** Place sling around transmission and raise, guiding onto pilot studs on engine bell housing with spline shaft through clutch disk. Place lock washers and nuts on pilot

TM 9-818
10-TON 6 x 4 TRUCK (MACK MODEL NR)

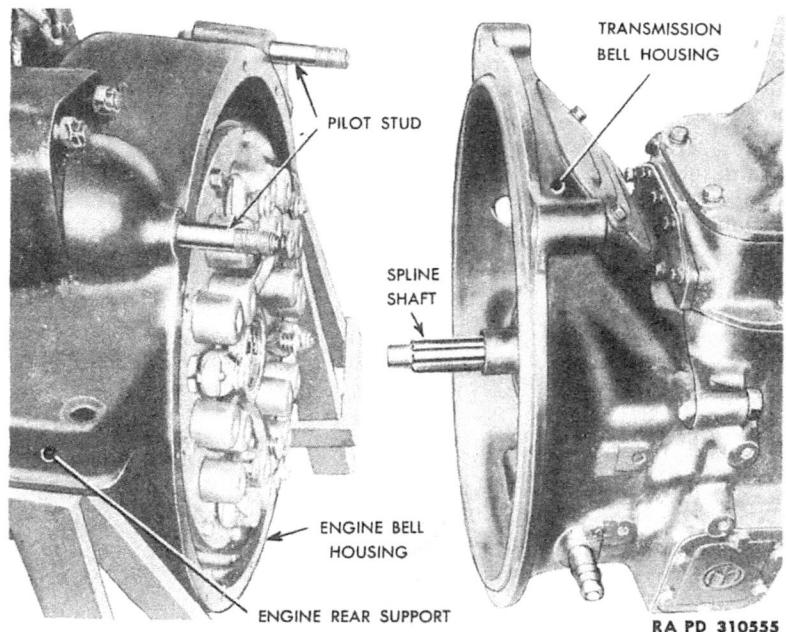

Figure 57 — Engine and Transmission Separated

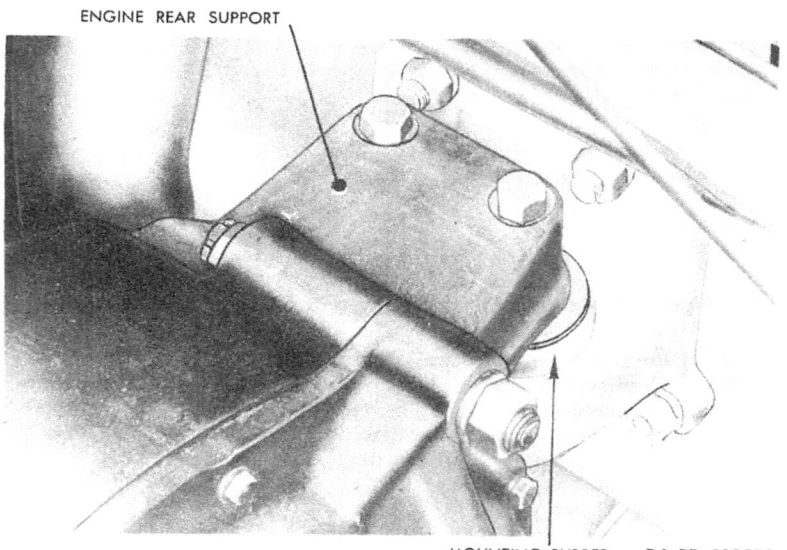

Engine 58 — Rear Engine Mounting Bolts

TM 9-818
59

ENGINE—REMOVAL AND INSTALLATION

studs. Place 12 cap screws and lock washers in bell housing and tighten. Also tighten pilot stud nuts. Remove sling. Place slings about transmission and engine.

b. **Install Engine and Transmission.** Raise engine assembly over chassis and install from right-hand side. Lower onto mounting brackets. Be sure large flat washers are in place on rear mounting rubbers. Line up rear mounting holes with drift pin and install bolts. Line up front holes and install bolts. Install nuts and palnuts on front bolts, using two wrenches. Install large, flat washers, lock washers and units on rear bolts, using two wrenches.

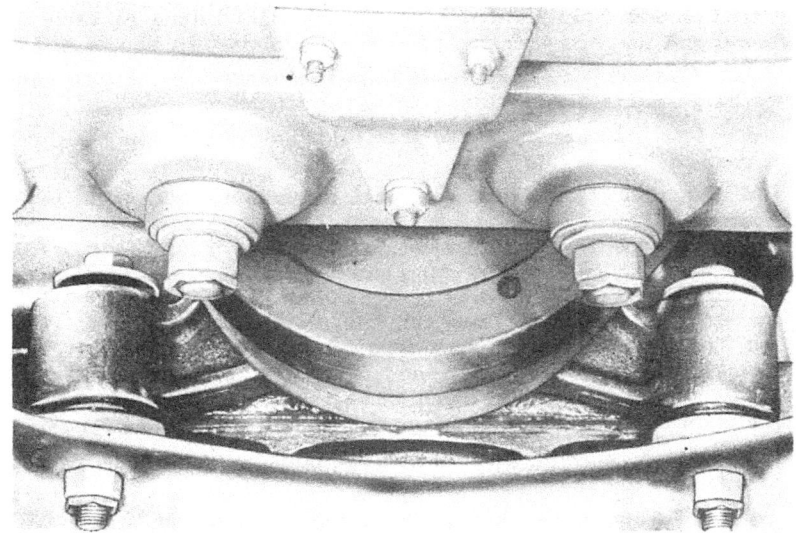

RA PD 310557

Figure 59—Front Engine Mounting Bolts

c. **Connect Propeller Shaft.** Attach propeller shaft flange to transmission flange by installing eight bolts, lock washers and nut bolt heads to front, and tighten.

d. **Install Clutch Lever.** Turn clutch lever shaft toward front and mount lever on shaft, tapping it into place with hammer. Adjust clutch pedal assembly as outlined in paragraph 61 a.

e. **Connect Brake Application Valve Air Line.** Attach air line to front of valve.

f. **Connect Cranking Motor Cables.** Attach the two No. 00 cables to cranking motor, following wiring diagram for proper connections. Install lock washer and nut and tighten.

g. **Install Fan Belts and Fan.** Follow procedure outlined in paragraph 81 b and 83 b and adjust belts.

10-TON 6 x 4 TRUCK (MACK MODEL NR)

h. *Attach Generator Bond Strap.* Attach bond strap to generator with toothed lock washer above and below strap. Install cross-recessed head screw.

i. *Install Water Connections.* Install water pipe and upper hose as outlined in paragraph 80 b (2).

j. *Connect Frame Bond Straps.* Connect bond straps from frame to engine at front on each side with cap screws, placing toothed lock washer above and below each strap.

k. *Connect Fuel Feed Line.* Attach fuel feed line to flexible tube, using two wrenches.

l. *Connect Fuel Return Line.* Attach rigid tubing to valve on frame, and use two wrenches to attach flexible return line to it.

m. *Connect Air Compressor Main Discharge Line.* Attach main discharge line to flexible tubing, using two wrenches.

n. *Attach Exhaust Pipe to Manifold.* Follow procedure outlined in paragraph 53 b (2).

o. *Install Cab.* Place slings on cab in manner described in paragraph 58 y, and raise cab over chassis. Lower cab nearly into position and place emergency brake lever box (with rubber edging) on bracket at left-hand end of cab support cross member. Insert brake rod into box through opening in front, and leave it suspended until brake lever is installed. Drop cab into place on mounting brackets, and install two bolts in each forward bracket, attaching battery cable clamp to forward left-hand bolt, with antisqueak in place between bracket and cab. Place lock washers and nuts and tighten. Place spring, nut and palnut on rear mounting stud and tighten.

p. *Attach Steering Post.* Install two bolts, nuts and lock washers in bracket attachment to cab frame.

q. *Install Speedometer Cable.* Install speedometer cable in rear end of transmission and tighten. Attach clamp to cap screw on left-hand side of transmission and tighten.

r. *Install Parking Brake Lever.* Place brake lever assembly on cross member bracket, install two cap screws and lock washers from beneath, and two bolts and lock washers from above, and tighten. Install brake lever rod and place clevis pins in the clevis at the lower end of brake lever assembly, and at bell crank at forward end of rod. Install cotter pins in clevis pins.

s. *Install Toe- and Floorboards.* Follow procedure outlined in paragraph 159 b.

t. *Install Steering Wheel.* Follow procedure outlined in paragraph 143 b (1).

u. *Connect Cab Wiring.* Follow wiring diagram to connect all wires and cables at junction block and heater switches on left-hand of dash sheet.

v. *Install Water Temperature Gage Element.* Install adapter in rear of water manifold. Install gage element in adapter.

TM 9-818
59

ENGINE—REMOVAL AND INSTALLATION

w. **Connect Hand Throttle.** Place wire from hand throttle control in hole in accelerator cross shaft bell crank and, with control in as far as possible, install stop and tighten screw.

x. **Connect Generator Cables.** Install two clamps attaching generator cables to left-hand side rail. Install shielded cable ends in terminal shields, the heavier or No. 8 cable to the connection on top of generator, and the No. 14 cable to the lower connection. Place lugs over terminal screws and install and tighten lock washers and nuts. Install two cable shield nuts on shields. Place condenser on upper radio shield and the cap on lower shield, tightening with pliers. Connect other ends of cables to regulator.

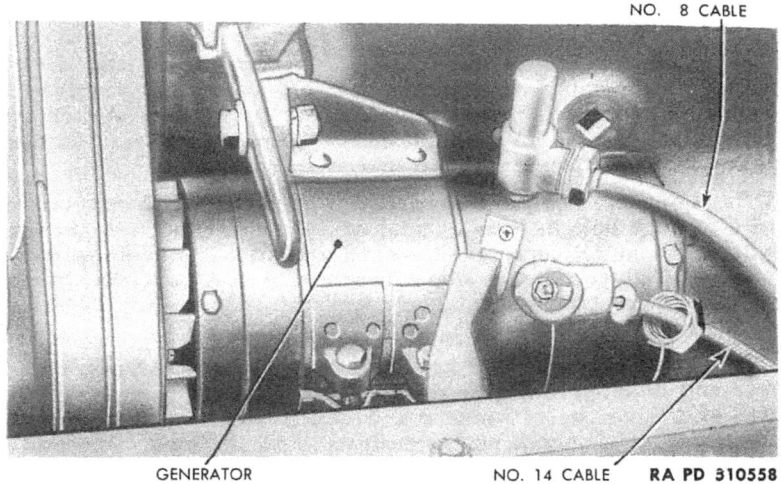

Figure 60—Generator Connections

y. **Install Radiator Tie Rods.** Install radiator tie rods in dash sheet and in forward brackets, placing offset rod from right of dash to left of radiator with offset toward front, and other rod from left of dash to right of radiator, each with curved end toward rear. Install lock washers and nuts to rear ends of rods inside cab and tighten. Install lock washers and nuts to front ends, but do not tighten until hood is installed.

z. **Connect Fuel Pressure Gage Line.** Install fuel pressure gage line in port on top of third fuel filter mounting. Install two clamps, one on air cleaner mounting bracket with cap screw and lock washer, and one on third fuel filter mounting bracket with nut and lock washer, and tighten both. Place strap clamp around line and radiator tie rod.

aa. **Connect Tachometer Cable.** Place tachometer cable end in adapter on engine toward front on right-hand side and tighten. At-

10-TON 6 x 4 TRUCK (MACK MODEL NR)

tach cable clamp to first cylinder exhaust manifold bolt bracket with bolt, nut and lock washer, and tighten. Attach cable to tie rod in two places with clamps, and tighten bolts.

bb. **Install Air Supply Valve Line.** Bend line into place and attach to low-pressure indicator.

cc. **Install Air Pressure Governor Line.** Connect governor air line at governor and at top of compressor. Attach clamp attaching governor line and air supply valve line to dash with screw, lock washer and nut and tighten.

dd. **Connect Oil Pressure Gage Line.** Insert oil pressure gage line fitting into port in engine and tighten. Clamp line to cap screw at upper rear of rear valve push rod inspection cover with lock washer on screw.

ee. **Connect Engine Stop Control.** Insert engine stop control wire through clamp on governor and into hole in governor lever. Tighten set screw in lever, and clamping screw with wrench and screwdriver. Clamp to rearmost fender bracket with bolt, nut and lock washer, tightening with screwdriver and wrench.

ff. **Connect Engine Emergency Stop Control.** Insert control wire through hole in lever on shut-off valve and place stop on wire, with valve in full open position and control on dash in as far as possible. Tighten screw in stop with screwdriver. Attach cable to each of two fender brackets with bolt, nut and lock washer, tightening with screwdriver and wrench.

gg. **Install Hood.** Place hood on top of cowl and radiator, sliding rear end of hinge rod into clamp. Place clamping piece over front end of rod on top of radiator and attach with two flathead bolts, tightening with wrench and screwdriver. Line up hood, if necessary, by adjusting tie rod nuts to vary the distance needed. Tighten the rod nuts. Check top radiator hose for tightness if rods have been readjusted.

hh. **Connect Batteries.** Follow procedure outlined in par. 167.

ii. **Adjust Canvas Top.** Roll cab top back into place over bows and adjust stay ropes.

jj. **Inspection and Adjustment After Installation.** Before attempting to start engine, make the following inspection and adjustment.

(1) Fill crankcase with correct lubricant (par. 19 b (4)).
(2) Close drain cock in lower water pipe and fill cooling system.
(3) Inspect cooling system for leaks.
(4) Bleed fuel system (par. 68 a).
(5) Inspect fuel lines for leaks.
(6) Check air cleaners for correct lubricant level (par 21 b (3)).
(7) Check level of transmission lubricant (par. 21 b (6)).
(8) Check clutch pedal free travel (par. 61 a).

ENGINE—REMOVAL AND INSTALLATION

(9) Check and adjust fan belt tension (par. 83 a).

(10) Close air reservoir valve opened to drain air reservoirs.

kk. Start engine, following instructions in par. 5 a and then perform the following inspections and adjustments.

(1) Immediately after the engine has started, note oil pressure gage. If no pressure is indicated or is abnormally low, stop engine, locate and correct cause.

(2) Check fuel pressure.

(3) Check air pressure and air system for leaks.

(4) Check ammeters.

(5) Inspect engine for oil leaks.

(6) Inspect cooling system for leaks.

(7) Accomplish complete engine check and tune-up (par. 47 **b**).

10-TON 6 x 4 TRUCK (MACK MODEL NR)

Section XV

CLUTCH

	Paragraph
Description and tabulated data	60
Clutch	61
Clutch controls and linkage	62

60. DESCRIPTION AND TABULATED DATA.

a. Description. Located between the engine and the transmission, the clutch is a friction device for transmitting the power from the engine to the transmission, and thence to the propeller shaft. The clutch is of the single-plate, dry-disk type, is fully enclosed by the bell housing and flywheel housing, and is controlled by the clutch pedal. This produces a push against the release (ball-thrust) bearing, through a cam-and-roller arrangement. The three major elements of the clutch are the driven (friction) disk, the pressure plate and, in assembly with this, the cover which carries the multiple springs and operating parts. Drive of the disk is between the pressure plate and the flywheel face. The pilot bearing for stabilizing the end of the shaft on which the disk hub is splined is in the bore of the flywheel. The clutch itself is nonadjustable; it is necessary to adjust only the pedal position. The clutch is secured in assembly with the flywheel by the cover. No attempt should be made to detach this cover, without first installing the proper spring retention (screw) tools to keep the pressure springs compressed.

b. Data.

```
Make .................................................... Mack
Model ................................................... CL-28
Type ................. Single-plate, dry-disk, pusher type
Disk diameter, outside ................................. 15 in.
Disk facing diameter, inside ........................... 8 in.
Total area ........................................... 252.8 sq in.
Number of spring elements (two springs each) ........... 18
Release bearing:
    make and number ......... M. R. C. Gurney 214-C T Q
    Type ............................... Shielded ball thrust
Pilot bearing:
    make and number ................. S. K. F. 6305-Z-C
                    (0.004-0.002 in. radial clearance)
Pedal adjustment:
    Pedal pad free travel .......................... 2 in. max
                    (At 1 in. adjust release lever)
```

TM 9-818
61

CLUTCH

61. CLUTCH.

a. **Adjustment.** When the facings wear enough to decrease the free travel of the clutch pedal to about 1 inch, adjust the clutch lever as follows: Back off cam setting screw and cam locking screw, and reset so that release lever on outside of clutch has a $3/8$-inch clearance

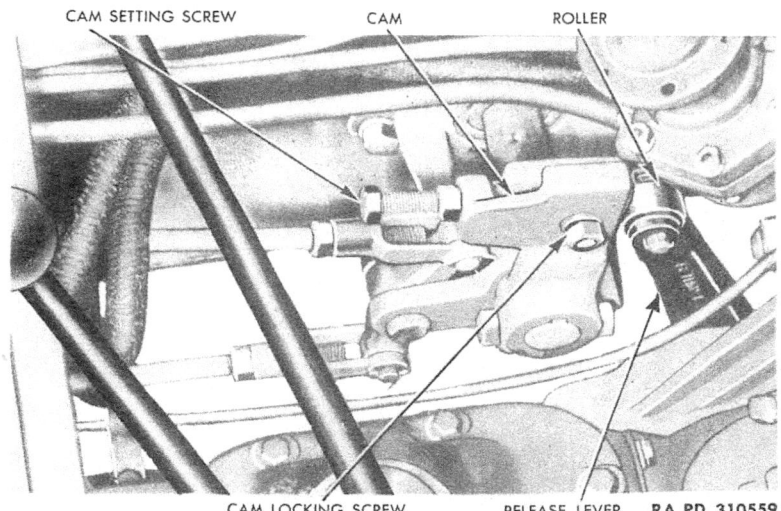

Figure 61—Clutch Control Linkage

between roller and cam (fig. 61). This will provide a $3/16$-inch clearance between internal thrust washer and clutch release bearing, and an approximate 2-inch free clutch pedal travel. When all adjustment is used up on cam, the release lever should be removed from its serrated shaft, and reinstalled so that additional adjustments may be made. If clutch pedal mechanism has been disassembled and must be readjusted, first make proper clutch adjustments as above. Then adjust pedal stop screw (underneath the cam-setting screw in back of bell crank) so that pedal does not strike floorboards. Do not attempt to adjust the free travel of pedal by adjustment of the pedal stop screw.

b. **Removal** (fig. 62).

(1) REMOVE TRANSMISSION. Follow procedure outlined in par. 93 a to remove transmission.

(2) REMOVE CAP SCREWS. Remove the 10 clutch-cover cap screws and lock washers.

(3) REMOVE CLUTCH ASSEMBLY. Remove clutch assembly from flywheel, and lift out clutch disk. NOTE: *If only facing disk is to be replaced, mark cover and flywheel to insure reinstallation in previous position.*

141

TM 9-818
61-62

10-TON 6 x 4 TRUCK (MACK MODEL NR)

c. Installation (figs. 63 and 64).

(1) INSTALL CLUTCH ASSEMBLY. Using alining tool (41-T-3085), place clutch disk on tool and place tool in pilot bearing. Place clutch assembly in flywheel, with marks (make in accordance with step 6 (3) above) properly coinciding.

(2) PLACE CAP SCREWS. Attach assembly to flywheel with 10 cap screws and lock washers. Tighten screws alternately until all are tight. Remove alining tool.

(3) INSTALL TRANSMISSION. Follow procedure outlined in paragraph 93 b to install transmission.

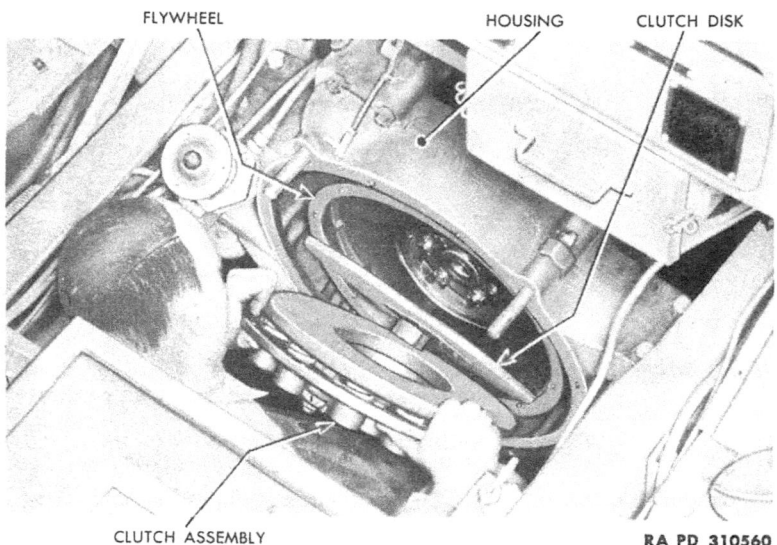

Figure 62—Removing Clutch

62. CLUTCH CONTROLS AND LINKAGE.

a. Removal.

(1) REMOVE TOE- AND FLOORBOARDS. Follow procedure outlined in par. 159 a (1) through (5).

(2) REMOVE CLUTCH RELEASE LEVER. Remove bolt, nut and lock washer from clutch release lever and remove lever.

(3) DISCONNECT HAND BRAKE LINKAGE. Remove two cotter and clevis pins from hand brake bell crank lever on clutch shaft bracket and detach hand brake linkage.

(4) REMOVE CLUTCH RETRACTOR SPRING. Remove two cap screws attaching battery cover to left-hand apron and remove cover. Remove clutch retractor spring.

(5) REMOVE CLUTCH PEDAL AND KEY. Remove bolt, nut and lock washer from clutch pedal and pry pedal off shaft. Remove key from shaft.

CLUTCH

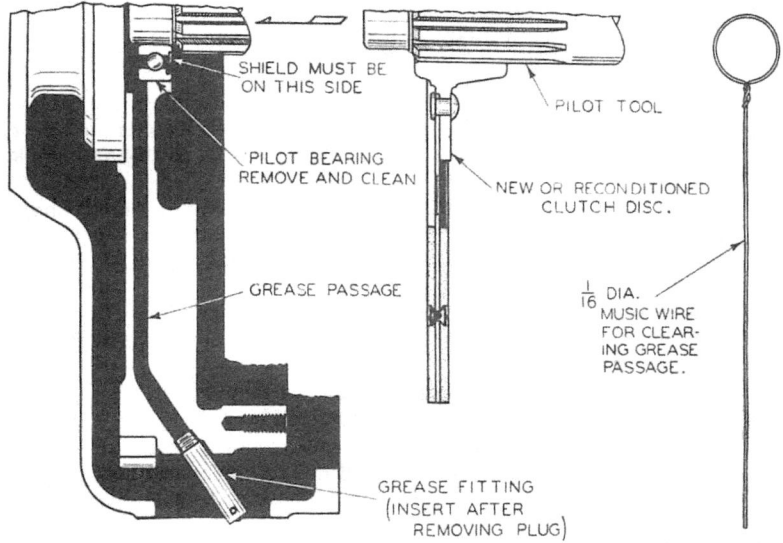

Figure 63—Clutch Alining Tool

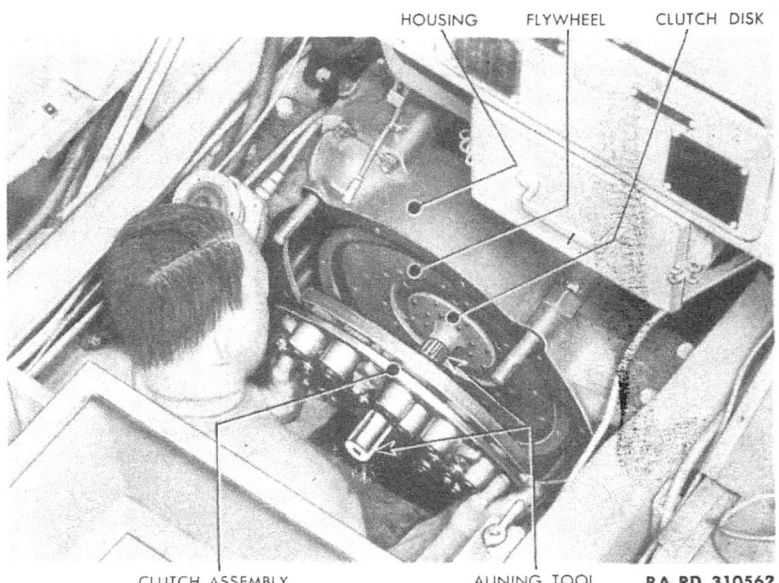

Figure 64—Installing Clutch

10-TON 6 x 4 TRUCK (MACK MODEL NR)

(6) REMOVE CLUTCH SHAFT MOUNTING BRACKET. Remove three bolts, nuts and lock washers. Use two wrenches to remove the other bolt and lock washer, and remove clutch shaft bracket.

b. Installation.

(1) INSTALL CLUTCH SHAFT BRACKET. Place clutch shaft bracket in position on frame and install bolt, nut and lock washer in lower rear hole with two wrenches. Install three bolts, nuts and lock washers.

(2) INSTALL CLUTCH PEDAL. Place key in keyway on clutch shaft and tap into place with hammer. Place clutch pedal on shaft with key in keyway and tap onto shaft. Install bolt, nut and lock washer and tighten.

(3) INSTALL CLUTCH RETRACTOR SPRING. With pliers hook the clutch retractor spring over the lug on the frame and the ear on the pedal.

(4) INSTALL BATTERY COVER. Place battery cover in position over battery and attach to apron with two cap screws.

(5) INSTALL EMERGENCY BRAKE LINKAGE. Place rod from hand brake lever on upper bell crank arm and install clevis and cotter pins. Install rod to hand brake on lower bell crank arm with clevis and cotter pins.

(6) INSTALL CLUTCH RELEASE LEVER. Place clutch release lever on shaft, tap into place, and tighten locking bolt, nut and washer.

(7) INSTALL TOE- AND FLOORBOARD. Follow procedure outlined in par. 159 **b** (5) through (9).

TM 9-818
63

Section XVI

FUEL SYSTEM

	Paragraph
Description and tabulated data	63
Injection pump	64
Fuel supply pump	65
High-pressure fuel lines	66
Injection nozzles	67
Bleed fuel system	68
Synchrovance	69
Governor	70
Fuel filters	71
Fuel tanks	72
Energy cells	73

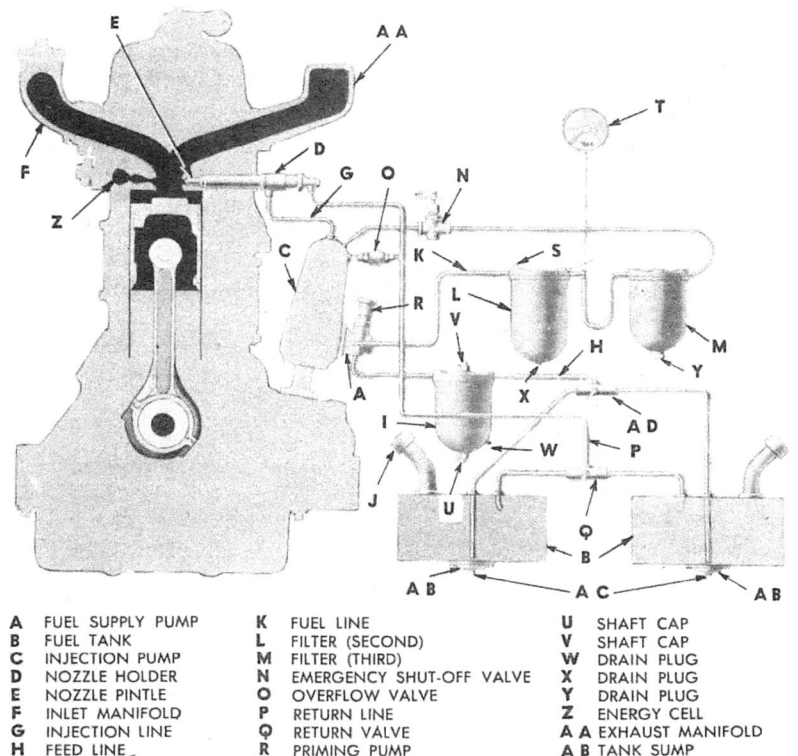

A	FUEL SUPPLY PUMP	K	FUEL LINE	U	SHAFT CAP
B	FUEL TANK	L	FILTER (SECOND)	V	SHAFT CAP
C	INJECTION PUMP	M	FILTER (THIRD)	W	DRAIN PLUG
D	NOZZLE HOLDER	N	EMERGENCY SHUT-OFF VALVE	X	DRAIN PLUG
E	NOZZLE PINTLE	O	OVERFLOW VALVE	Y	DRAIN PLUG
F	INLET MANIFOLD	P	RETURN LINE	Z	ENERGY CELL
G	INJECTION LINE	Q	RETURN VALVE	AA	EXHAUST MANIFOLD
H	FEED LINE	R	PRIMING PUMP	AB	TANK SUMP
I	FILTER (FIRST)	S	VENT PLUG	AC	DRAIN PLUG
J	FILLER CAP	T	PRESSURE GAGE	AD	FEED VALVE

RA PD 310563

Figure 65—Fuel System

145

10-TON 6 x 4 TRUCK (MACK MODEL NR)

63. DESCRIPTION AND TABULATED DATA.

   a. **Description.** There are two divisions of the arrangement by which fuel is served to the engine: the supply system, consisting of the fuel tanks, tube lines, the No. 1 filter, the fuel supply pump, which is on the injection pump, and Nos. 2 and 3 filters; and the injection system which comprises the injection pump, the six injection nozzles in the cylinder heads, and the connecting supply and return tube lines. There is also an air cleaner through which air is drawn directly into the intake manifold and then into the cylinders.

   b. **Data.**

   (1) INJECTION PUMP.
   Make .... American Bosch
   Type .... Multiple, APE (six outlets)

   (2) FUEL SUPPLY PUMP.
   Make .... American Bosch
   Type .... AFP/K (plunger; compensating, variable stroke)

   (3) INJECTION NOZZLES.
   Make .... American Bosch
   Type, nozzle .... Closed, pintle
   Nozzle holder .... AKB

   (4) ENERGY CELLS.
   Make .... Mack
   Type .... Lanova

   (5) SYNCHROVANCE.
   Make .... Mack
   Type .... Fly-ball

   (6) GOVERNOR.
   Make .... Pierce
   Model MA-1073

   (7) FUEL FILTERS.
   No. 1 filter
   Make .... Purolator
   Model G-132-J
   No. 2 filter
   Make .... Purolator
   Model N-704
   No. 3 filter
   Make .... Purolator
   Model DN-21-4

   (8) FUEL TANKS.
   Number 2
   Location .... On sides
   Capacity .... 75 gal ea.
   U. S. Army Specification 2-102B

   (9) FUEL OIL (DIESEL).

TM 9-818
64

FUEL SYSTEM

64. INJECTION PUMP.

a. Description and Tabulated Data (fig. 66).

(1) DESCRIPTION. Each of the six pump elements of the injection pump is connected to one of the six injection nozzles by a seamless steel tube. There are also return tubes for excess fuel. The purpose of the injection pump is to meter the fuel and deliver it at timed intervals to the injection nozzles, which atomize the fuel as it passes into the engine combustion chamber.

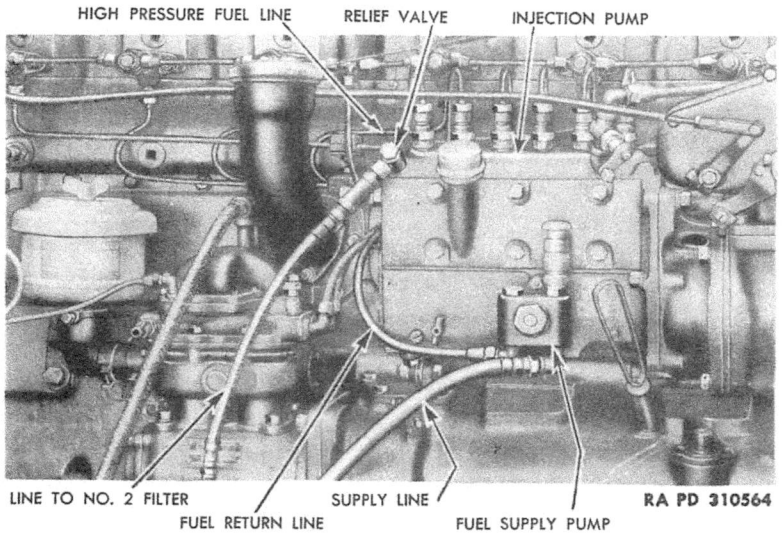

Figure 66—Injection Pump and Fuel Lines

(2) DATA.

Make .... American Bosch
Type .... Multiple, APE (six outlets)
Size .... "B", medium
Mounting .... Flanged, and outboard support
Drive .... Through gears and Synchrovance coupling, at half engine speed
Internal timing .... Indexed by port closings by plungers
External timing (to engine) .... 19° BTDC (before top dead center)—Synchrovance
Drive ratio .... One-half engine speed
Rotation .... Clockwise, viewed from drive end

b. Injection Pump Removal (fig. 67).

(1) REMOVE ENGINE EMERGENCY VALVE STOP. Use cross-recessed screwdriver to remove stop on emergency valve shut-off wire.

TM 9-818
64

10-TON 6 x 4 TRUCK (MACK MODEL NR)

(2) REMOVE CABLE CLAMP. Use cross-recessed screwdriver and open-end wrench to remove clamp from fender bracket to free emergency shut-off valve cable.

(3) REMOVE FILLER SPOUT. Remove three cap screws and lock washers attaching oil filler spout to engine and remove spout.

(4) FUEL LINES. Disconnect high pressure fuel lines at nozzles, using two wrenches. Disconnect flexible fuel return line at copper pipe, and remove tee from relief valve. Remove return lines from nozzle holder drain tube tee fittings, and from relief valve. Remove reducer and 45-degree elbow from relief valve tee, using box wrench,

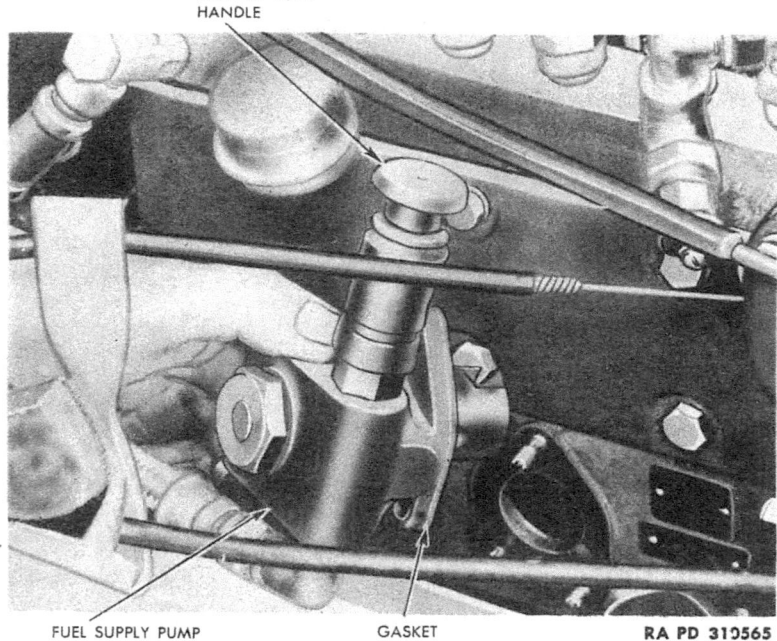

Figure 67 — Removing Fuel Supply Pump

with pipe wrench to hold tee. Remove tee from relief valve. Disconnect flexible supply line from rigid line, using two wrenches. Disconnect fuel line from fuel pump to No. 2 filter.

(5) REMOVE FUEL SUPPLY PUMP. Remove three nuts and toothed lock washers attaching fuel pump to injection pump, and remove fuel pump and flexible suction line.

(6) DISCONNECT ACCELERATOR SPRING. Use pliers to open the hook on end of accelerator spring, and remove.

(7) DISCONNECT GOVERNOR LEVER. Remove cotter and clevis pins from the governor lever with pliers.

TM 9-818
64

FUEL SYSTEM

(8) DISCONNECT FEED LINE. Disconnect the feed line at emergency shut-off valve elbow.

(9) REMOVE OIL LEVEL GAGE. Pull gage stick out of guide tube.

(10) DISCONNECT TACHOMETER CABLE. Remove tachometer cable from engine at adapter and remove clamp to free tachometer cable from bracket, using cross-recessed screwdriver and wrench.

(11) REMOVE GOVERNOR LINKAGE. Break governor seal and three cap screws and lock washers. Remove governor. Remove hairpin lock and clevis pin at governor yoke shaft lever. Remove governor tube shield and remove hairpin lock and clevis pin at end of pump rack. Remove linkage.

(12) DISCONNECT INJECTION LINES. Disconnect injection lines at pump using two wrenches.

(13) REMOVE RACK CAP. Break seal with pliers and use wrench to remove cap on end of pump rack.

(14) REMOVE FEED LINE. Loosen two bolts and nuts to free feed line. Disconnect feed line at No. 2 filter. Remove feed line between injection pump and nozzle lines.

(15) REMOVE INJECTION PUMP. Remove bolt, nut, and lock washer from rear support bracket with socket and open-end wrenches. Remove locking wire from mounting cap screws, and remove the four cap screws with flat washers and lock washers (two of each) from mounting flange, using box and open-end wrenches. Lift injection pump away from engine and remove gasket.

(16) REMOVE REAR BRACKET. Remove two cap screws and lock washers, and remove rear bracket.

(17) REMOVE EMERGENCY SHUT-OFF VALVE AND SHAFT. Remove emergency shut-off valve shaft, and use two pipe wrenches to remove the emergency valve and feed elbow.

(18) REMOVE FITTINGS. Remove 45-degree emergency shut-off valve elbow. Remove flexible tube from suction line on transfer pump, and remove elbow and street elbow.

c. **Adjustment (Timing of Pump to Engine)** (figs. 67 and 68).

(1) This adjustment pertains exclusively to the drive of the pump and "Adjustment" under the subject of "Injection Pump" is not to be taken as encouragement to disturb the mechanism of the pump itself.

(2) Since adjustment of timing by the provision at the timing gear requires some disassembly to expose the gears, for convenience, a second provision is made in the pump drive, where for access it is only necessary to detach the pump. Here, several step corrections may be accomplished by installing, adjacent to the Synchrovance, a different ring or disk (one of four special coupling rings), provided the timing deviation is such that the ring selected permits its elimination. These special rings are identified by a narrow groove cut in the edge surface; and they are marked $+2$, $+4$, -2 and -4. Those marked $+$ will advance the pump timing, and those marked $-$ will retard it. The numerals are representative of degrees of flywheel

TM 9-818
10-TON 6 x 4 TRUCK (MACK MODEL NR)

travel. Example: If when FPI (fuel pump injection) mark, at the Synchrovance coupling is alined with its pointer, the reading at the flywheel is 21 (that is 2 degrees ahead of the specified 19 degrees) use the ring marked —2. The ring must be installed with the face having the counterbore towards the pump; otherwise it will have no clearance, and will become bound in assembly.

(3) CAUTION FOR MAKING INITIAL AND FINAL CHECK OF TIMING. Due to the slight cumulative backlash inherent to any gear train, such as exists between the engine crankshaft and the injection pump, it is important to approach mark settings with correct rotation. Always approach the mark moving in the direction in which the engine turns when running; never the reverse. That is, if the desired mark position

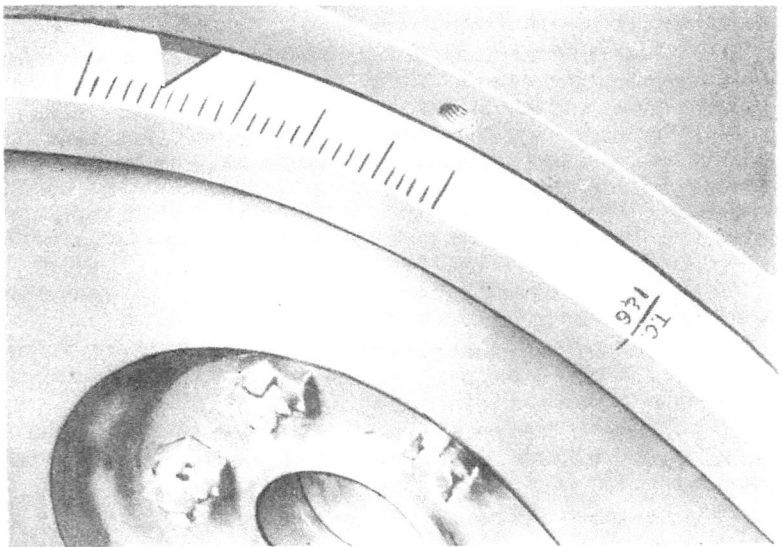

RA PD 310566

Figure 68—Flywheel and Timing Pointer

is passed, back up appreciably so that mark may again be approached in the correct direction, with backlash taken up in the running direction. Running rotation is clockwise when view is from the front toward the rear of the engine.

d. **Timing of Pump to Engine Procedure** (fig. 67).

(1) Turn over engine by hand until position of first cylinder from timing gear end approaches but does not reach top center on compression stroke (the exhaust valve of second cylinder, the third valve, should be opening). This can be seen with front valve cover removed.

(2) With an assistant to turn the engine at this point, observe through the inspection hole in flywheel housing, the marking, in

TM 9-818
64

FUEL SYSTEM

degrees, on the flywheel with relation to the pointer. Stop flywheel when the 19-degree mark is exactly opposite the pointer.

(3) Remove the small plug located in the injection pump adapter housing, and note the position of the line stamped FPI (fuel pump injection) on the Synchrovance-driven member, with respect to the

RA PD 310567

Figure 69 — Pointer and FPI (Fuel Pump Injection) Mark on Synchrovance

fixed pointer mounted on the housing. The pointer and the line must match for correct timing.

(4) If pump timing is not exact it may be corrected by installing one of four coupling rings provided for the purpose. These rings make corrections, however, only if the deviation from the 19-degree mark is approximately 2 or 4 degrees in either direction. Those

151

10-TON 6 x 4 TRUCK (MACK MODEL NR)

marked + will advance pump timing and those marked − will retard it. To install the ring:

(a) *Remove Injection Pump.* Follow procedure in paragraph 64 **b** (1) through (15).

(b) *Replace Coupling Ring.* Remove coupling ring from Synchrovance drive, and install new ring.

(c) *Install Injection Pump.* Follow procedure in paragraph 64 **e** (4) through (17).

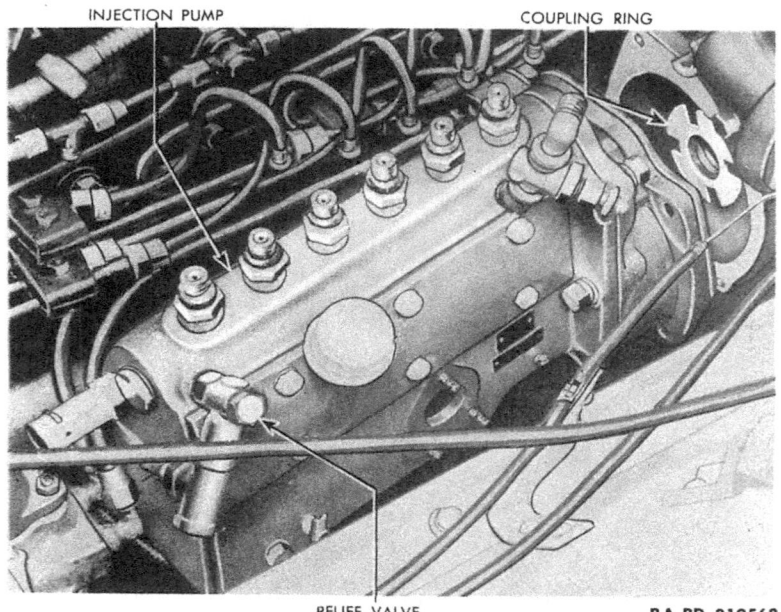

Figure 70—Installing Injection Pump

e. **Installation.**

(1) ATTACH BRACKET. Attach rear bracket on under-side of pump with two cap screws and lock washers.

(2) INSTALL EMERGENCY SHUT-OFF VALVE AND SHAFT. Install 45-degree elbow in emergency shut-off valve port at front end of pump, using pipe wrench, then use two pipe wrenches to install emergency shut-off valve. Install emergency shut-off valve shaft.

(3) INSTALL FITTINGS ON FUEL SUPPLY PUMP. Install the elbow in supply port and the street elbow in suction port on bottom of fuel supply pump, using pipe wrench. Install the flexible line on the suction elbow.

(4) INSTALL INJECTION PUMP (FIG. 70). Place the coupling ring on the Synchrovance shaft with counterbore hole toward injection

FUEL SYSTEM

pump, and place gasket between pump and flange on engine. Install pump on engine, using four cap screws, with lock washers on the two inner screws and flat washers on the outer screws. Tighten screws and place locking wire through them. Install spacer between pump and rear bracket, and install bolt, nut and lock washer, using socket and open-end wrenches.

(5) INSTALL FEED LINE. Attach feed line from No. 2 filter to the fuel pump at the filter.

(6) INSTALL RACK CAP. Install cap on end of injection pump rack.

(7) INSTALL GOVERNOR LINKAGE. Attach linkage between rack and governor yoke shaft with spring end toward rack and install clevis pin and hairpin lock in rack end. Install tube shield and attach governor yoke shaft lever with clevis pin and hairpin lock. Install governor cover and attach with three cap screws and lock washers.

(8) INSTALL FITTINGS. Install the tee connection at relief valve on rear end of pump. Install reducer in tee toward injector nozzles, and install 45-degree elbow in reducer. Attach nozzle leak off line return tee to relief valve.

(9) INSTALL FUEL SUPPLY PUMP. Place gasket on fuel pump mounting flange, and install pump on side of injection pump with three nuts and toothed lock washers.

(10) CONNECT FUEL LINES. Connect feed line to filter No. 2 and at elbow on transfer pump. Use two wrenches to attach flexible suction line from fuel pump to the rigid line.

(11) INSTALL OIL LEVEL GAGE. Place gage stick in guide tube.

(12) INSTALL FLEXIBLE FUEL RETURN LINE. Attach flexible return line to tee connection at relief valve and to rigid return line, using two wrenches, being careful not to twist or strain the flexible tubing.

(13) INSTALL OIL FILLER SPOUT. Place gasket on oil filler spout elbow, and install on engine with three cap screws and lock washers.

(14) LUBRICATION. Remove oil filler spout cap on injection pump, and open oil level cock on side. Add engine oil until it just flows from drain cock. Close cock and replace oil filler spout cap.

(15) BLEED SYSTEM. Follow procedure outlined in paragraph 68 a to bleed system of air after injection pump installation.

(16) CHECK FUEL LINE SYSTEM. All fuel lines must be securely clamped in place so that they do not rub against each other or other parts, and all connections must be tight.

(17) INSTALL SEALS. Install wires and seals on governor cover and on rack cap.

65. FUEL SUPPLY PUMP.

a. Description and Tabulated Data.

(1) DESCRIPTION. The fuel supply pump is a plunger-type pump for transferring the oil from the fuel tanks to the reservoir, or oil

TM 9-818
65-66

10-TON 6 x 4 TRUCK (MACK MODEL NR)

gallery, within the injection pump. It is secured to the injection pump, and is actuated by one of the cams of its camshaft. It maintains a return flow through an overflow valve, back to the fuel tank, and carries with it any air or gas that may have entered the supply system through leaks, or that may have been present in the fuel initially, and released by heat of agitation. A simple hand pump of the plunger type for priming and for bleeding lines is combined with the fuel pump.

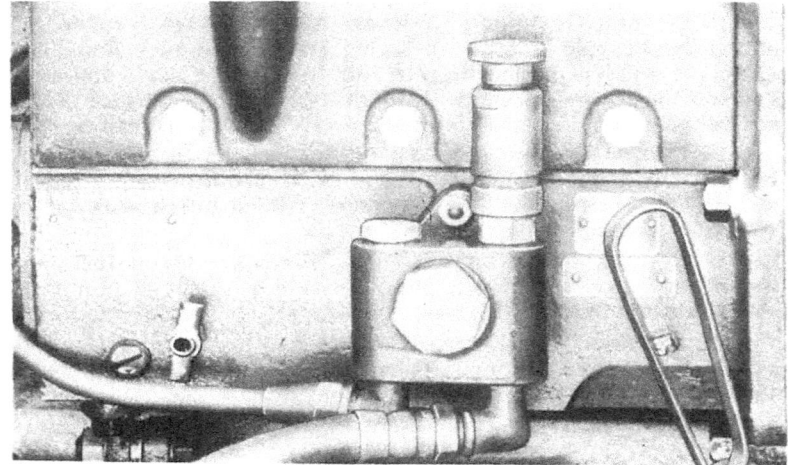

RA PD 310569

Figure 71—Fuel Supply Pump

(2) DATA.
Make ....American Bosch
Type ....AFP/K (Plunger; compensating, variable stroke)
Size ....Large
Location ....Side of injection pump
Mounting ....Flanged, stud fastened
Drive ....Injection pump camshaft
Type of Priming Pump ....Hand, plunger, vertical (over inlet valve)

b. **Removal.** Follow procedure outlined in par. 61 b (5).

c. **Installation.** Follow procedure outlined in par. 61 e (9).

66. HIGH-PRESSURE FUEL LINES.

a. **Removal.**

(1) DISCONNECT LINES. Use open-end wrench to disconnect lines from fittings at injector nozzles and at injector pump.

TM 9-818
66-67

FUEL SYSTEM

(2) REMOVE CLAMP. Loosen clamping bolts (cap screws on Nos. 1 and 2).

(3) REMOVE FUEL LINES. With care not to bend lines out of shape, remove lines from engine.

(4) SPECIAL NOTE. It is necessary to remove No. 1 line in order to remove No. 2 line, and to remove No. 6 line to remove lines Nos. 3, 4 and 5. It is also necessary to disconnect the accelerator spring to remove lines Nos. 3, 4, 5 and 6.

b. Installation.

(1) ATTACH LINES. Place fuel lines in position and connect fittings at nozzle and at pump.

(2) INSTALL CLAMPS. Place clamps on lines where removed and install ¼-inch clamping bolt, nut and lock washer (cap screws on lines Nos. 1 and 2).

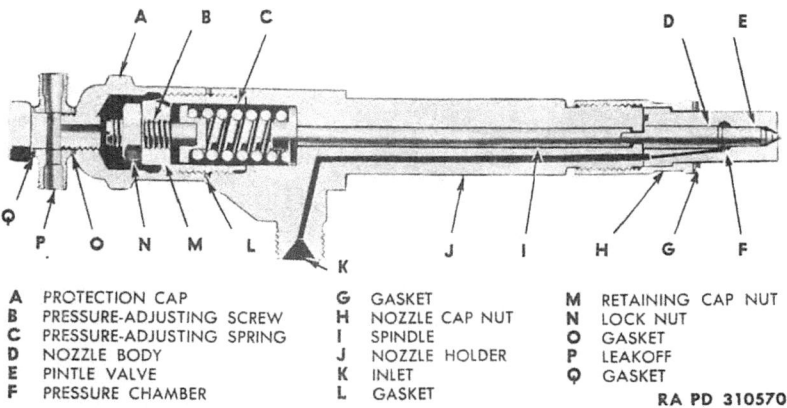

A	PROTECTION CAP	G	GASKET	M	RETAINING CAP NUT
B	PRESSURE-ADJUSTING SCREW	H	NOZZLE CAP NUT	N	LOCK NUT
C	PRESSURE-ADJUSTING SPRING	I	SPINDLE	O	GASKET
D	NOZZLE BODY	J	NOZZLE HOLDER	P	LEAKOFF
E	PINTLE VALVE	K	INLET	Q	GASKET
F	PRESSURE CHAMBER	L	GASKET		RA PD 310570

Figure 72—Cross-section of Injection Nozzle

67. INJECTION NOZZLES.

a. Description and Tabulated Data.

(1) DESCRIPTION (fig. 72). Injection of fuel into the engine is through six injection nozzles, one for each cylinder. Through hydraulic tube connections, these nozzles are connected to the individual pump elements of the injection pump. The nozzles are of the "closed," differential-needle (or pintle), hydraulically operated type. They are termed "closed" because the valve is spring-loaded, seated at the orifice, and opened by the hydraulic pressure of injection.

(2) DATA.
Make .... American Bosch
Type, nozzle .... AKB, closed, pintle
Location .... Cylinder heads, side

155

TM 9-818
67

10-TON 6 x 4 TRUCK (MACK MODEL NR)

b. Removal.

(1) DISCONNECT LINES. Disconnect the accelerator spring, and detach return line to relief valve from the tee. Detach return lines from nozzles, and remove lines and gaskets. Disconnect pressure lines from nozzles.

(2) REMOVE NOZZLES (fig. 74). Remove two nuts and lock washers at each nozzle. Remove nozzle with pry-bar as shown in figure 75. To remove No. 2 nozzle, it is necessary to disconnect the pressure line to the No. 2 cylinder at the injection pump, using an

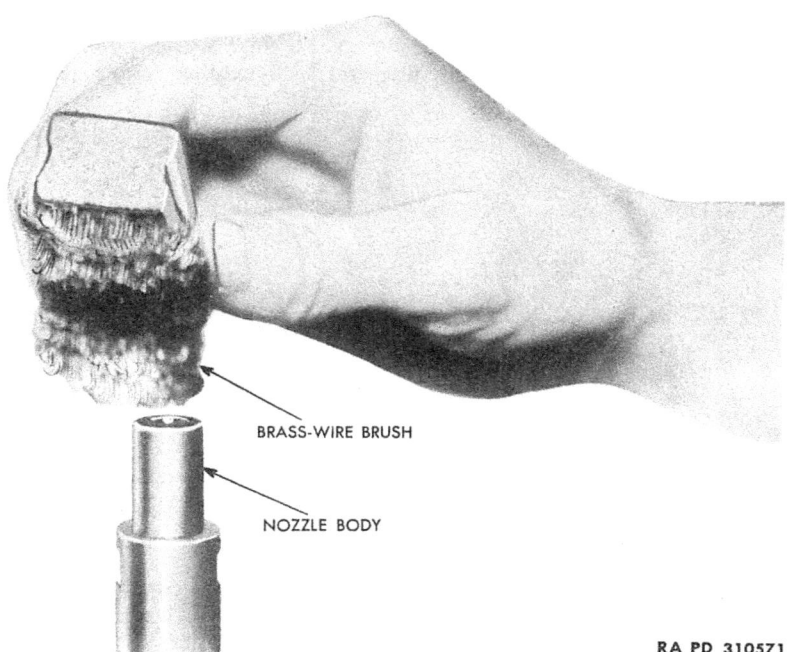

Figure 73—Cleaning Pintle End of Nozzle

open-end wrench to remove line, and another to hold fitting. To remove the No. 3 nozzle it will be necessary to disconnect the Nos. 5 and 6 pressure lines in the same manner. Attach these high-pressure lines to injection pump as soon as nozzles are removed, to prevent introduction of dirt or carbon into system.

c. Cleaning (fig. 73). That there must be no disassembly whatever, for cleaning purposes cannot be over-emphasized. The only cleaning permitted is without disassembly, and is confined to applying a *brass-wire* brush (41-T-3535-400) to the pintle end of the nozzle to remove carbon. *DO NOT USE A STEEL-WIRE BRUSH.* In order to guard against any possibility of damage to the vital parts of the

FUEL SYSTEM

nozzle, hold the unit in a vise, gripping the nozzle holder near the connections end. Never use any metal tools or implements to remove stubborn deposits; replace the nozzle unit. Brush cleaning is routine whenever nozzle units are removed for inspection.

d. Installation.

(1) CLEAN NOZZLE PORTS (fig. 75). Use clean rag to wipe off nozzle so as to remove carbon accumulation. Use reamer to remove carbon from nozzle port. Turn the engine over several times with cranking motor to blow any loose carbon out of cylinders.

(2) CONNECT PRESSURE LINES. For ease of installation of nozzle and pressure lines, connect pressure lines to nozzles before installing. Use extreme care not to damage nozzle ends.

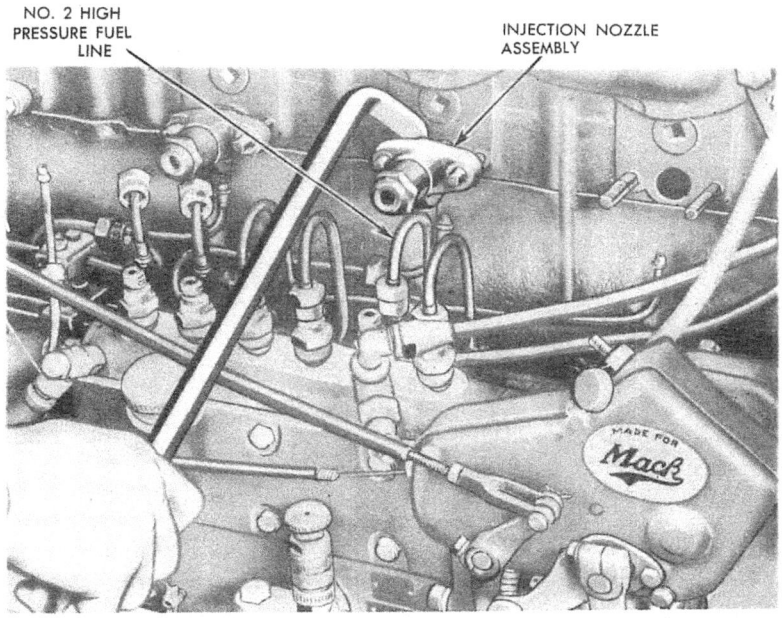

Figure 74—Removing Nozzle Assembly

(3) INSTALL NOZZLES. Place nozzles in ports, using new gasket for each, being careful not to damage the nozzle tips. To install the Nos. 2 and 3 nozzles it will again be necessary to disconnect the Nos. 2, 5 and 6 pressure lines at the injection pump as noted above. Place nuts and lock washers on nozzle studs, being sure to place nozzles with the inlet downward. Turn nuts fingertight and then tighten alternately and cautiously so no cocking or distortion of parts will occur. Tighten firmly, but not excessively.

TM 9-818
67-68
10-TON 6 x 4 TRUCK (MACK MODEL NR)

(4) CONNECT PRESSURE LINES. Reinstall the Nos. 2, 5 and 6 pressure lines at the injection pump and tighten pressure lines at nozzles.

(5) CONNECT RETURN LINE. Attach return lines to nozzle leak-off fittings with two gaskets in place, one under nut and one under fitting, and tighten. Connect return lines at tee on relief valve of injection pump and attach accelerator spring.

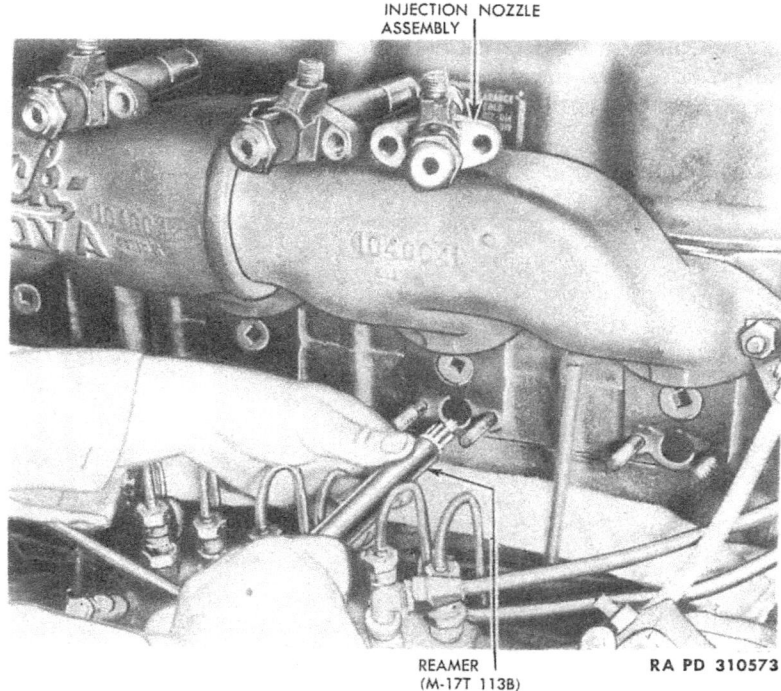

Figure 75—Removing Carbon from Nozzle Port

68. BLEED FUEL SYSTEM.

a. Whenever any fuel line between the discharge side of the transfer pump and the injection pump is removed, or even loosened, the system must be bled free of air. This is true also when either of the cartridges are renewed in the filters.

b. Usually it is not necessary to open the vents at the different points of the system, merely operate the priming pump handle until the pressure indicated by the gage is 15 pounds. Continue pumping a few moments longer to completely pump the air past the injection pump overflow valve.

FUEL SYSTEM

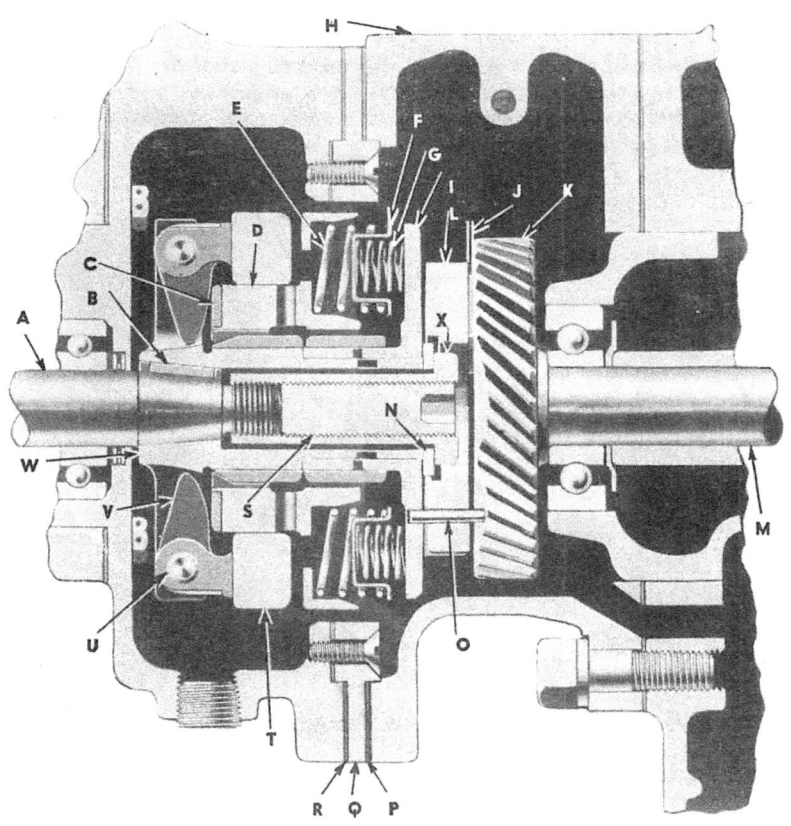

A	PUMP SHAFT	I	DRIVING MEMBER	Q	SPACER
B	KEY	J	ASSEMBLY CLEARANCE	R	GASKET
C	THRUST WASHER	K	GOVERNOR DRIVE GEAR	S	LOCK SCREW
D	SPLINED SLEEVE	L	COUPLING DISK	T	WEIGHT
E	SPRING	M	DRIVESHAFT	U	PIVOT SHAFT
F	SPRING SPACER THIMBLE	N	WASHER	W	DRIVEN MEMBER
G	SPRING	O	REGISTRATION PIN	X	SLEEVE NUT
H	HOUSING	P	GASKET	V	WEIGHT CAM

RA PD 310574

Figure 76—Cross-section of Synchrovance Drive

TM 9-818
68-69

10-TON 6 x 4 TRUCK (MACK MODEL NR)

c. Should it be further necessary to bleed system, proceed as follows: With engine dead, operate the priming pump handle opening air bleed or vent plug on filter No. 2 to allow escape of air. Continue pumping until solid stream, free from air bubbles, is discharged. Close vent and open vent on injection pump. Pump until gallery fills, and solid stream is discharged at this point. Close the connection, and the low-pressure lines should now be free of air.

d. The hand priming pump is also used to pump the fuel through the units to prime a dry system of a new engine, or one which has had the fuel system drained. It is not necessary to crank or turn the engine over by using the starter.

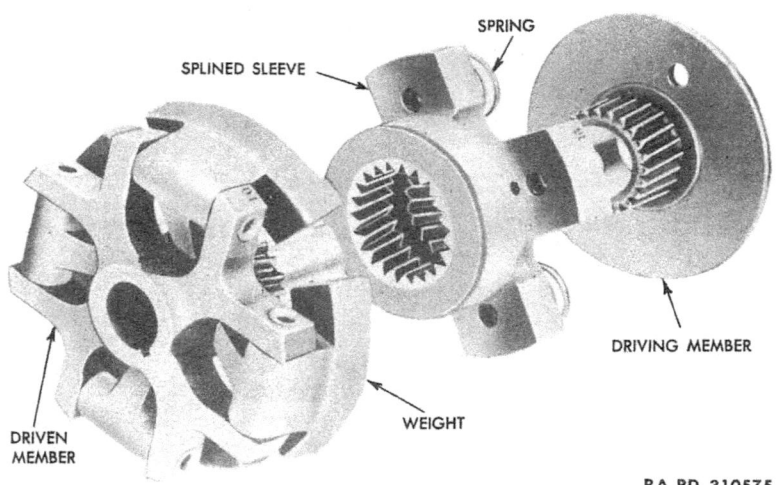

Figure 77 — Synchrovance Elements Disassembled

e. Tighten all fittings securely, and when finished with hand priming pump, lock in down position by screwing in on plunger knob. A loose fitting on the pressure side of a line will be evident by fuel leak. A loose fitting on suction side may not, but can cause more difficulty by admitting air. Lack of power can often be traced to loose fittings on suction line.

69. SYNCHROVANCE.

a. Description and Tabulated Data.

(1) DESCRIPTION (figs. 76 and 77). The name Synchrovance identifies the drive coupling unit which is enclosed within the injection pump mounting case. This is a centrifugal device which advances and retards the time of injection.

The assembly may be hand-tested for freedom of action, wear, or play, but complete replacement is the only corrective procedure

TM 9-818
69

FUEL SYSTEM

permitted. CAUTION: *Springs must not be altered or replaced because they came as calibrated sets, and replacement requires the facilities of a higher echelon.*

(2) DATA.
Make .... Mack
Type .... Fly-ball, action through helical elements
Location .... Injection pump drive

b. **Removal.**

(1) REMOVE INJECTION PUMP. Follow procedure outlined in paragraph 63 **b** (1) through (15).

(2) REMOVE LOCK SCREW. Using ¼-inch hexagon bar stock, remove lock screw from center of sleeve nut.

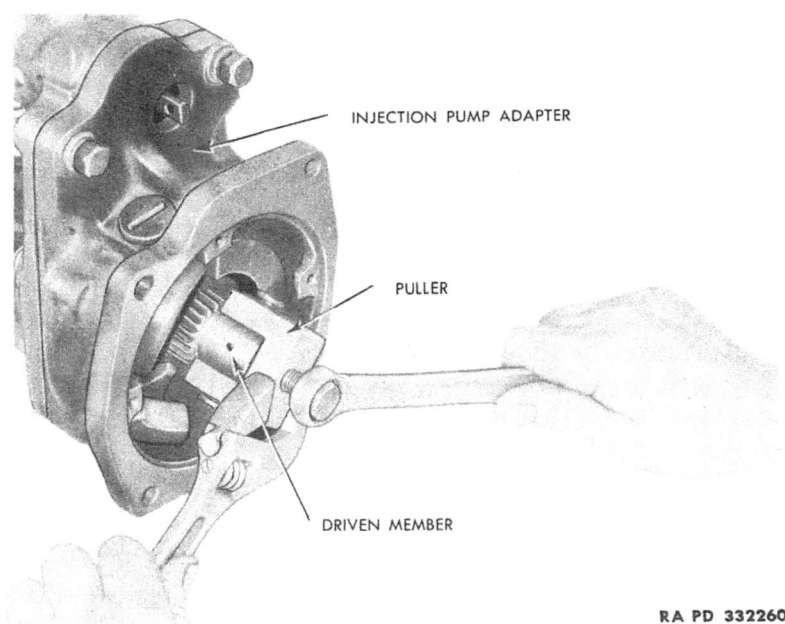

RA PD 332260

Figure 78—Removing Driven Member of Synchrovance, with Puller (41-P-2955-500)

(3) REMOVE SLEEVE NUT. Use socket wrench to unscrew sleeve nut, holding the Synchrovance assembly with an adjustable open-end wrench or a spanner-type wrench.

(4) REMOVE LOOSE PARTS. Remove the spline sleeve with spring cups and driving member, keeping them in order.

(5) REMOVE DRIVEN MEMBER (figs. 79 and 80). Using puller (41-P-2955-500), remove the driven member from the shaft by

TM 9-818

10-TON 6 x 4 TRUCK (MACK MODEL NR)

applying puller to shaft, with the pins in holes on opposite sides, and turning the puller screw with box wrench while holding the puller assembly with an adjustable open-end wrench.

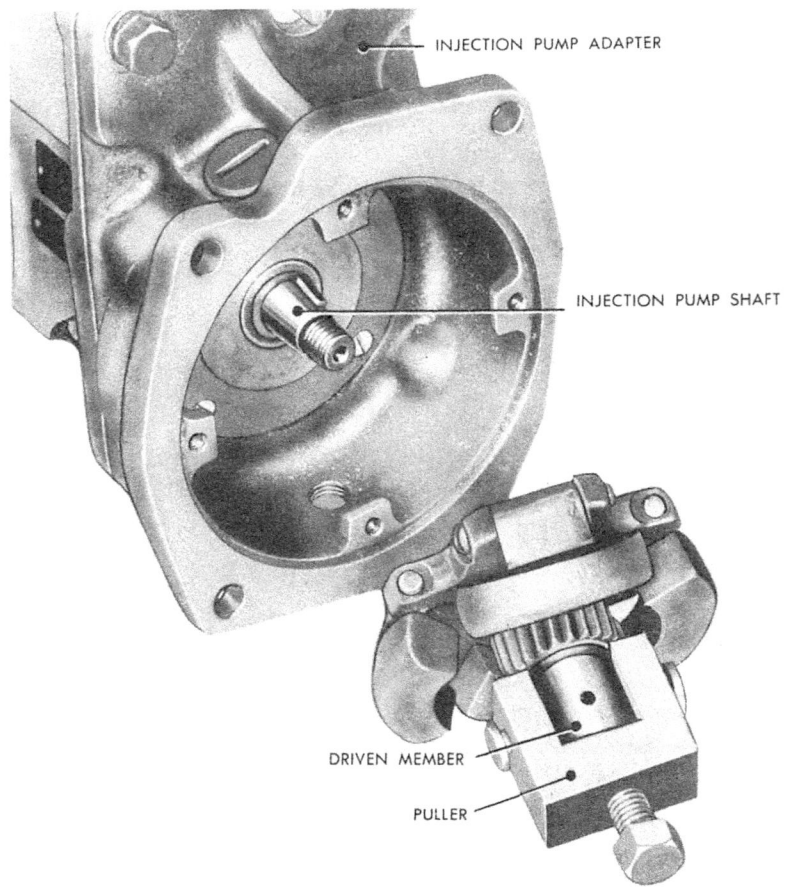

Figure 79—Driven Member Removed, Showing Puller (41-P-2955-500)

c. **Installation.**

(1) INSTALL DRIVEN MEMBER (fig. 80). With the pump and cap Synchrovance unit in the vertical position for ease of assembly, and with key in keyway on shaft of injection pump, place driven member on shaft, with weights to outside. Tap the assembly into place, using block of wood and hammer.

FUEL SYSTEM

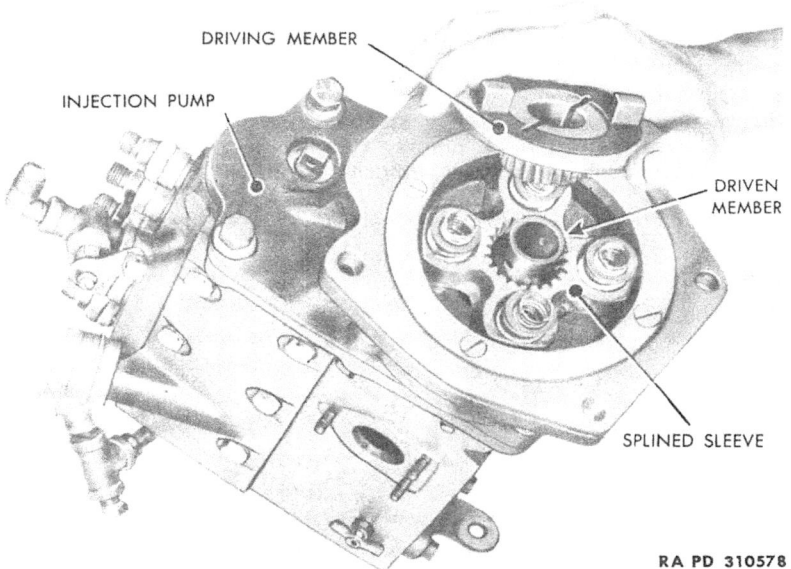

Figure 80 — Installing Driving Member of Synchrovance

Figure 81 — Governor and External Control Levers

10-TON 6 x 4 TRUCK (MACK MODEL NR)

(2) INSTALL LOOSE PARTS. Slide spline sleeve onto driven member, engaging splines, and put springs and spring cups into place. Install driving member by placing it on shaft, engaging its splines with those of spline sleeve.

(3) INSTALL SLEEVE NUT. Place washer on sleeve nut, and place sleeve nut in cap Synchrovance assembly, turning into place with socket wrench until tight, holding the assembly with an adjustable open-end wrench or spanner wrench. Check the clearance between

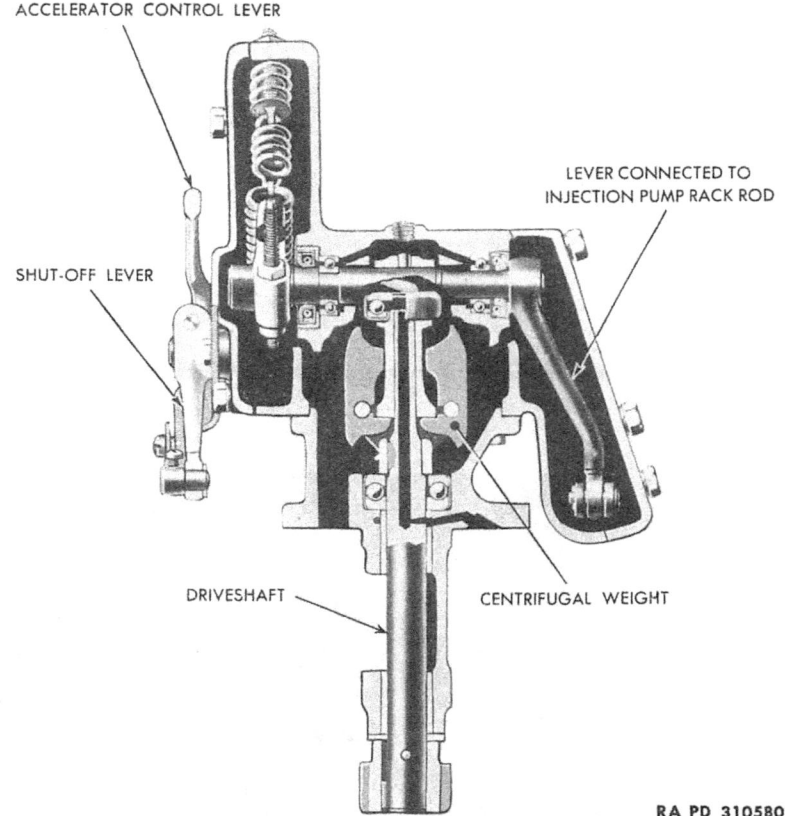

Figure 82—Cross-section of Governor

the washer and the driving member. This should be from 0.003 to 0.008-inch. If it is less, the end of the driving member should be ground off, and if it is more, a new washer should be installed.

(4) INSTALL LOCK SCREW. Again holding the unit with a wrench, and after making sure the parts move freely, insert the lock screw in the center of the sleeve nut and tighten with $1/4$-inch hex bar stock.

TM 9-818
69

FUEL SYSTEM

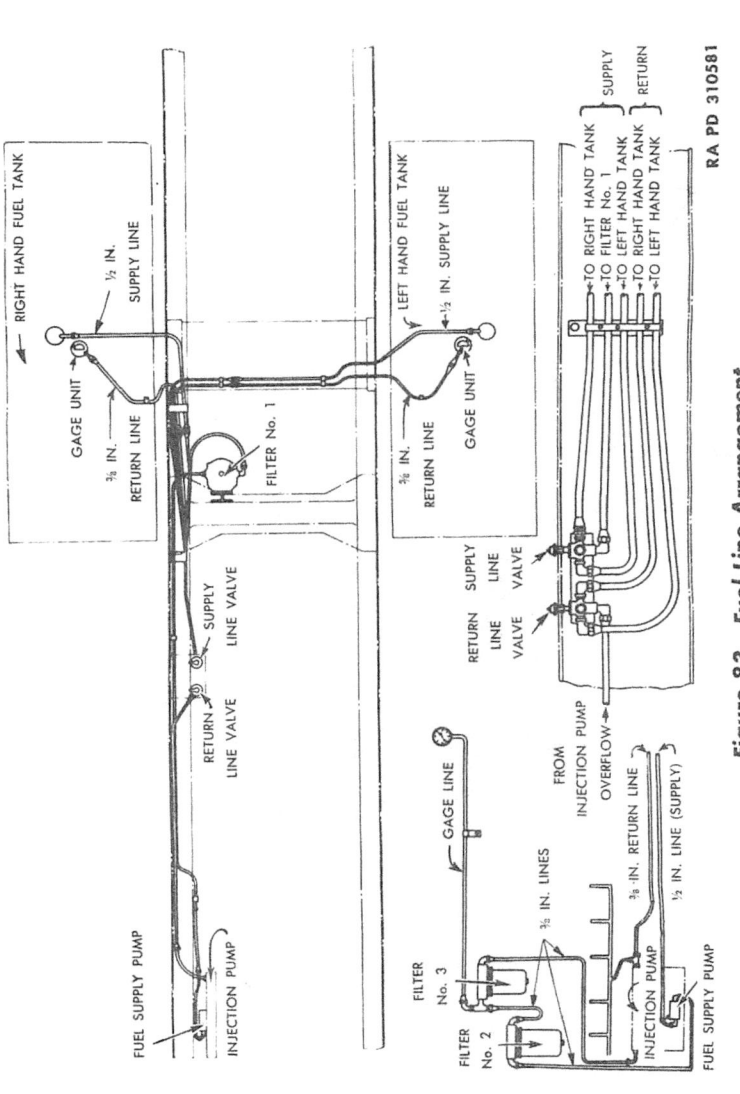

Figure 83—Fuel Line Arrangement

TM 9-818
69

10-TON 6 x 4 TRUCK (MACK MODEL NR)

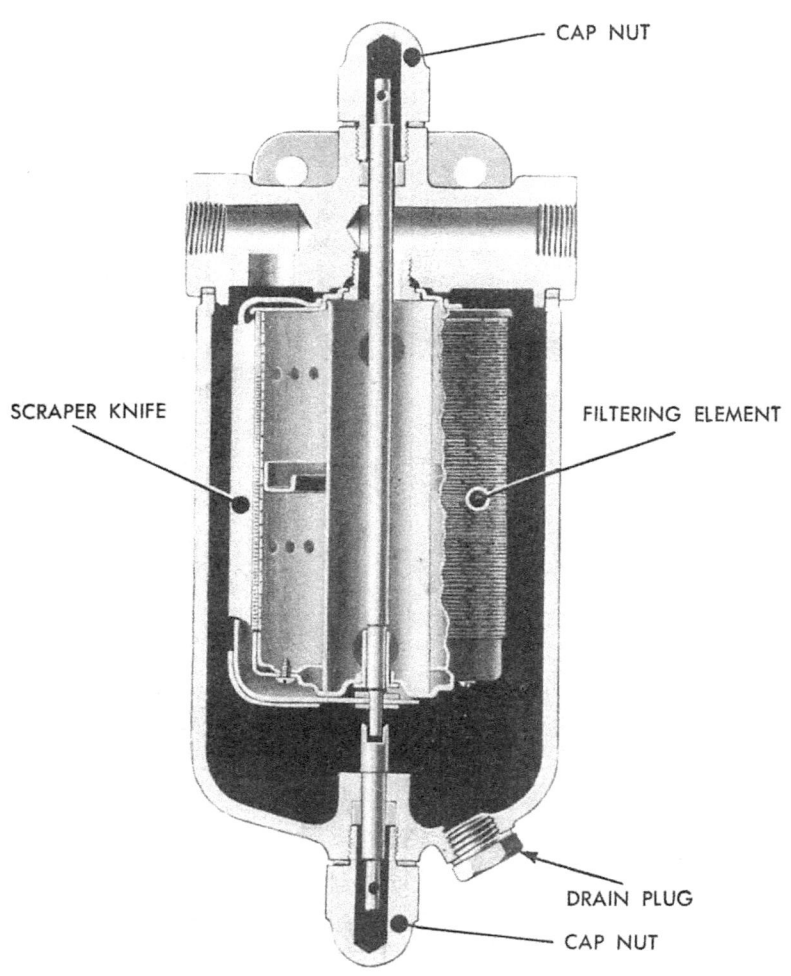

Figure 84—Cross-section of Filter No. 1

FUEL SYSTEM

70. GOVERNOR.

a. **Description and Tabulated Data.**

(1) DESCRIPTION (figs. 81 and 82). Engine speed is governed by a mechanical fly-ball type governor which is driven through a vertical shaft from a gear on the shaft of the injection-pump drive.

(2) DATA.

Make	Pierce
Model	MA-1073
Type	Centrifugal (fly-ball) vertical
Location	Over injection pump drive
Drive	Gear
Governed speed (no load)	2,100 rpm

71. FUEL FILTERS.

a. **No. 1 Filter.**

(1) DESCRIPTION AND TABULATED DATA.

(a) Description (figs. 83 and 84). Fuel filter No. 1 is in the fuel line between the fuel tanks and the fuel supply pump, and is attached to the chassis frame on the right side. It is cylindrically shaped, and has a permanent metal element with a scraper knife. The scraper knife shaft projects through both the upper and lower ends of the filter, and has its ends covered by screw cap nuts. There is a hole through each end for applying a tool (after removing the cap nut) to turn the scraper knife shaft which, plus drainage, is the cleaning operation. There is a drain plug at the bottom of the filter.

(b) Data.

Make	Purolator
Model	G-132-J
Type	Metal element, manually operated scraper
Location	Chassis frame, right side

(2) REMOVAL. Remove lines from both sides of fuel filter. Remove two mounting bolts, nuts, and lock washers and remove filter. Remove straight fitting from filter outlet port, and remove elbow from inlet port with adjustable open-end wrench.

(3) INSTALLATION. Install elbow in the inlet port, and install the straight fitting in the outlet port. Attach the filter to the cross member with two bolts, nuts, and lock washers. Attach the two lines to the filter fittings.

b. **No. 2 Filter.**

(1) DESCRIPTION AND TABULATED DATA.

(a) Description (figs. 83 and 85). In immediate series with No. 3, filter No. 2 is the first in the fuel line between the fuel (or transfer) pump and the injection pump and is carried on a bracket on ,the engine left side together with filter No. 3. Filter No. 2 is the one toward the front. It has a renewable, metal-cased, cartridge-type of element. There is a vent plug at the top, and a drain plug at the bottom.

10-TON 6 x 4 TRUCK (MACK MODEL NR)

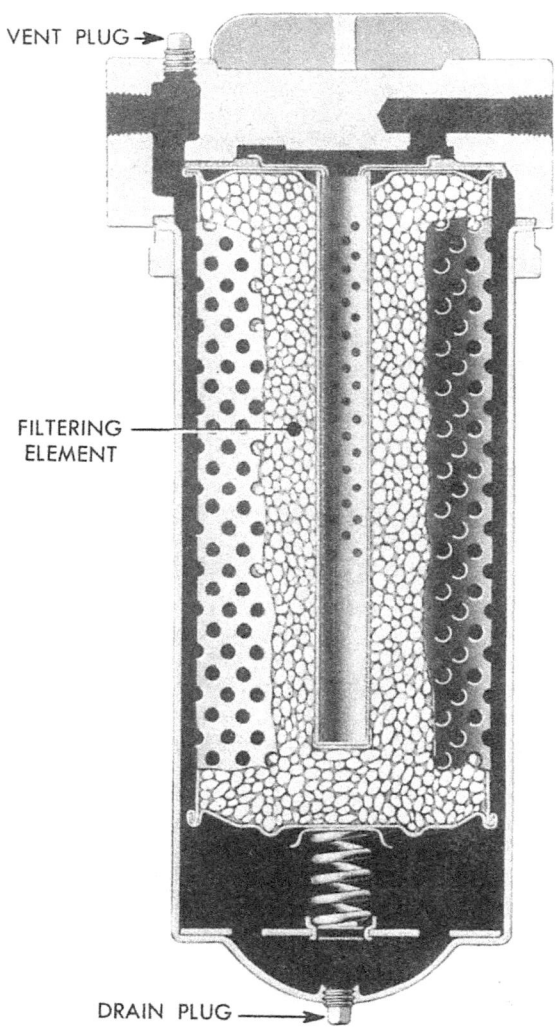

Figure 85—Cross-section of Filter No. 2

TM 9-818
71

FUEL SYSTEM

(b) Data.

Make .... Purolator
Model N-704
Type .... Renewable cartridge (metal-cased cartridge)
Location .... On engine, left side
Cartridge .... Purolator, N-704, 24661

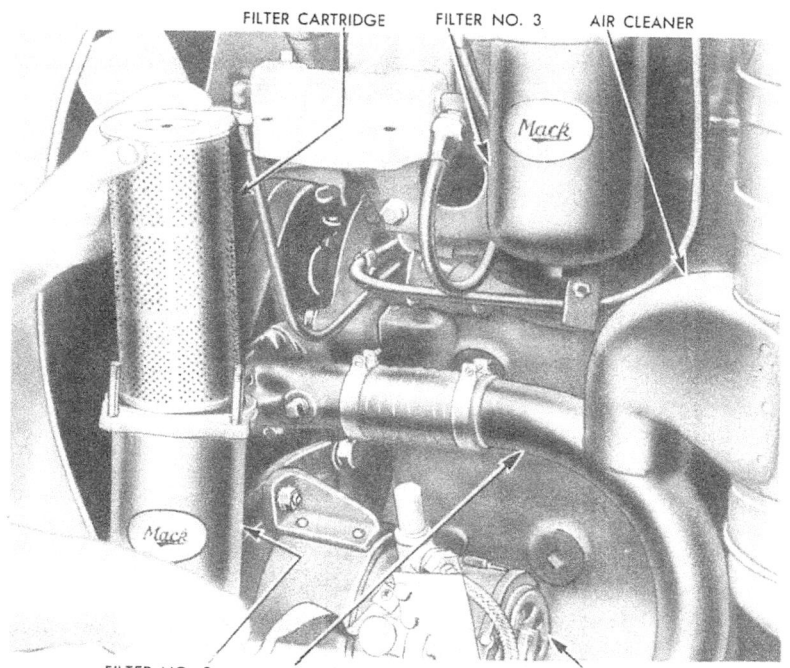

Figure 86—Removing Element from Filter No. 2

(2) REMOVAL. Remove the two fuel lines. Remove two mounting bolts, nuts, and lock washers and remove filter. Remove both elbows from filter.

(3) INSTALLATION. Install both elbows. Install filter, using two bolts, nuts, and lock washers to attach. Attach fuel line to each side of the filter.

c. **No. 3 Filter.**

(1) DESCRIPTION AND TABULATED DATA.

(a) Description (figs. 83 and 87). Final filtering before the fuel enters the injection pump is attained in Filter No. 3. This is in immediate series with No. 2 between the fuel supply pump and the injection pump, and is carried with and just to the rear of No. 2 on

169

TM 9-818
10-TON 6 x 4 TRUCK (MACK MODEL NR)

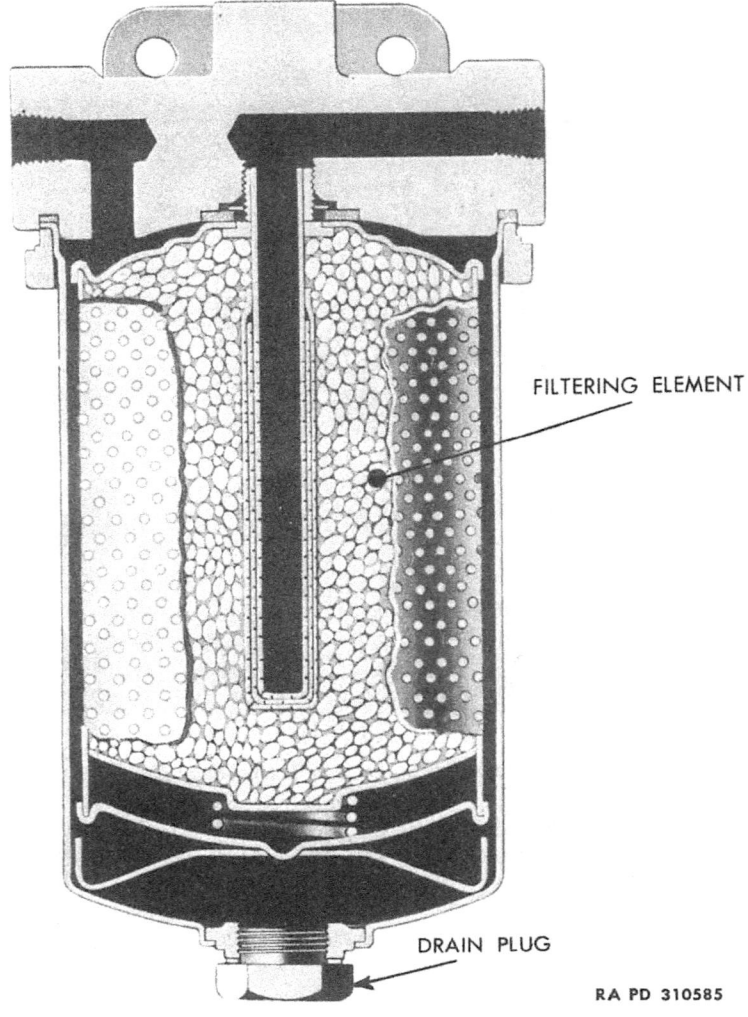

Figure 87—Cross-section of Filter No. 3

TM 9-818
71-72

FUEL SYSTEM

a bracket on the engine left side. Like the second filter, it has a renewable, metal-cased, cartridge-type of element; but its proportions and internal details are slightly different. It has a drain plug at the bottom.

(b) Data.

Make	Purolator
Model	DN-21-4
Type	Removable, cartridge (metal-cased cartridge)
Location	On engine, left side
Cartridge	Purolator, DN-21-D, 28865

(2) REMOVAL. Disconnect two fuel lines. Remove the pressure gage connection at the tee. Remove the two mounting bolts, nuts, and lock washers and the flexible line clamp on the rear bolt. Remove the tee and nipple from inlet port and the elbow from the outlet port.

(3) INSTALLATION. Install tee and nipple in inlet port and the elbow in the outlet port. Install filter with two bolts, nuts, and lock washers. Install the pressure gage line at the tee, and install the two fuel lines.

Figure 88—Fuel Tank Filler Spout

72. **FUEL TANKS.**

  a. **Description and Tabulated Data.**

  (1) DESCRIPTION (figs. 88 and 89). There are two 75-gallon fuel tanks, one on each side of the chassis frame immediately aft of the cab. They are located under the tire carrier and forward portion of the body. Attachment is by brackets directly to the chassis frame rails. A guard protects the bottom of the tank. Each tank has a capped filler spout with a wire screen and filter bag, and a drain sump with drain plug, fuel-gage element and fittings for a supply and a return line. The supply lines from the two tanks join at the feed valve and from this valve there is a single line to the No. 1 fuel filter. The return line which carries back the "leak-off" from the injection nozzles, and

TM 9-818
72

10-TON 6 x 4 TRUCK (MACK MODEL NR)

the flow from the overflow valve on the injection pump, becomes two lines at the return valve, one for each tank. The feed and return valves are set to employ one or the other of the tanks. The one fuel gage on the instrument panel shows the contents of either tank through operation of a two-way energizing switch.

(2) DATA.

Number of tanks ....Two
Total capacity ....150 gal (75 gal, ea)
Location ....One on each side of chassis
frame (longitudinal), aft of cab
Shape ....Basically rectangular

b. Filter Bag Removal. Remove filter bag and filler spout assembly by removing the eight cross-recessed head screws. Remove wire screen from filler spout and take out filter bag. Turn bag inside out and wash in fuel oil or dry-cleaning solvent to remove dirt. If there are holes or signs of deterioration, use a new bag. Reassemble by reversing disassembly instruction above.

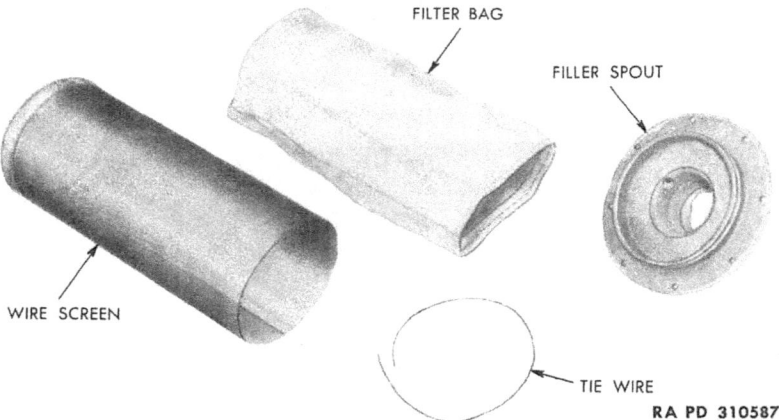

Figure 89—Fuel Tank Filler Spout Disassembled

c. Removal.

(1) DISCONNECT FUEL FEED LINE. With one open-end wrench to turn union and another to hold fitting, remove feed line connection at top of tank.

(2) DISCONNECT FUEL RETURN LINE. With one wrench to turn union and another to hold fitting, disconnect return line.

(3) DISCONNECT FUEL GAGE WIRE. Remove nut and gage wire from the terminal on top of tank. Remove wire-clamping strap from return line connection.

(4) UNFASTEN STRAPS. Remove lock nuts and nuts from tank strap T-bolts, and remove outrigger cross-angle iron. Pull straps out of the way and, wedging tank to raise from protecting frame, remove tank.

FUEL SYSTEM

(5) REMOVE FUEL SUPPLY STAND PIPE. Remove the five cross-recessed head screws from fuel supply stand pipe flange, and remove pipe and gasket.

(6) REMOVE GAGE TANK UNIT AND RETURN LINE FITTING. Remove five cross-recessed head screws from gage unit flange and return line fitting flange. Remove gage unit and gasket.

(7) REMOVE VENT PIPE. Remove vent pipe from filler spout. Remove elbow connection from tank. Remove gasket.

(8) REMOVE FILLER SPOUT. Remove eight cross-recessed head screws from filler spout mounting flange, and remove spout and gasket, taking care not to damage screen filter.

d. Installation.

(1) INSTALL FILLER SPOUT. Place filler spout in tank with gasket in place, and attach with eight cross-recessed head screws. Install breather elbow by screwing into tank with adjustable wrench. Install vent pipe by connecting fittings on filler spout.

(2) INSTALL GAGE AND FUEL RETURN LINE. Place gage and return line fitting in tank with gasket in place and attach with five cross-recessed head screws.

(3) INSTALL FUEL SUPPLY STAND PIPE. Place stand pipe fitting on tank with gasket in place and attach with five cross-recessed head screws.

(4) INSTALL TANK. Place tank on tank protector frame with tank mounting straps under tank. Elevate tank by use of bar or wooden block. Put antisqueak in place on straps and under outriggers, and place strap bolts through outriggers. Place outrigger cross piece on top of outrigger ends and insert strap bolts, applying nuts and lock nuts and tighten. Be sure that straps fit properly in place and hold tank snugly against outriggers, with back lower seams properly placed in recesses provided for seams on straps.

(5) INSTALL FUEL FEED LINE. With open-end wrench screw union into place on feed line fitting with wrench to hold fitting.

(6) INSTALL FUEL RETURN LINE AND GAGE WIRE FITTING. Attach return line to fitting, with gasket in place on tank, using one wrench to turn union and another to hold fitting. Attach gage wire by means of nut on terminal fitting, using adjustable open-end wrench to tighten. Place clamp strap around wire and return line connection.

73. ENERGY CELLS.

a. Description and Tabulated Data.

(1) DESCRIPTION. These cells, one in each cylinder, are located in the cylinder heads on the left side, directly opposite the injection nozzles, and increase turbulence and control combustion. The cell is composed of a body and a cover. There are no gaskets in the cell assembly; the tapered portion of the body has a lapped fit in the cylinder head and the body and cover joint is lapped.

10-TON 6 x 4 TRUCK (MACK MODEL NR)

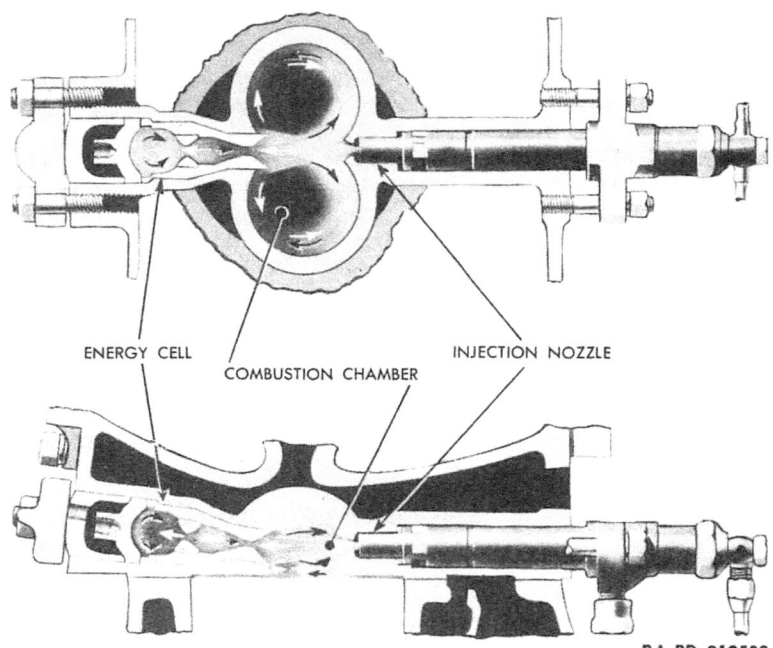

Figure 90—Combustion Chamber, Showing Injection Nozzle and Energy Cell

(2) DATA.
Make ....Mack
Type ....Lanova, removable
Location ....Cylinder head, left side

b. **Cleaning.** There is no attachment between the cell body and its cover, and they become separable when the cell is withdrawn from the cylinder head. If distortion has destroyed fits, or if the parts show burns, replace the cell, body, and cover. A knife will serve for removing carbon from the conical end cavity of the cell body. For cleaning the smaller orifice use a properly sized piece of hardwood (**never a drill or reamer, or any metal tool**). Also use carbon tetrachloride for cleaning of parts, including the spacer cap. Any cell whose cavities cannot be freed of carbon deposits must be replaced. Inspect clean parts for defects, and replace any doubtful cells with a mated body and cover.

c. **Removal.**

(1) REMOVE AIR CLEANER. Follow procedure outlined in paragraph 71 c (3).

(2) REMOVE CLAMPS. Remove the two nuts from each energy cell clamp and remove clamp. It will be necessary to remove cable

FUEL SYSTEM

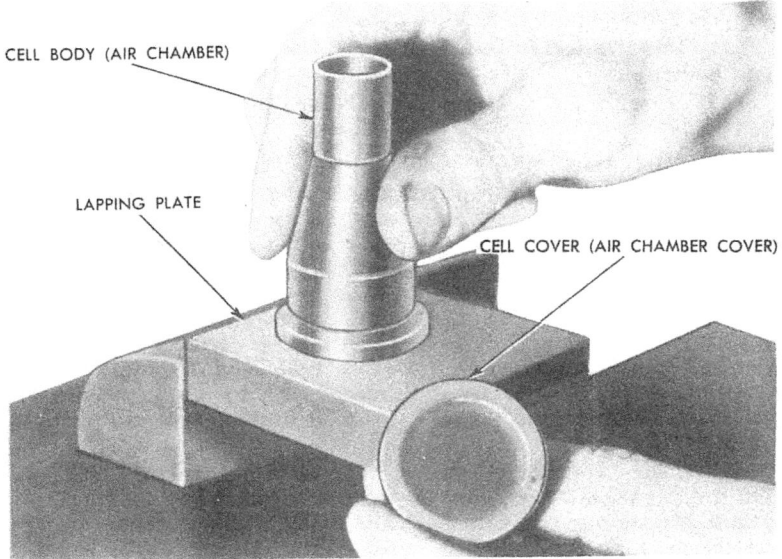

Figure 91 — Lapping Joint Surfaces of Energy Cell Body and Cover

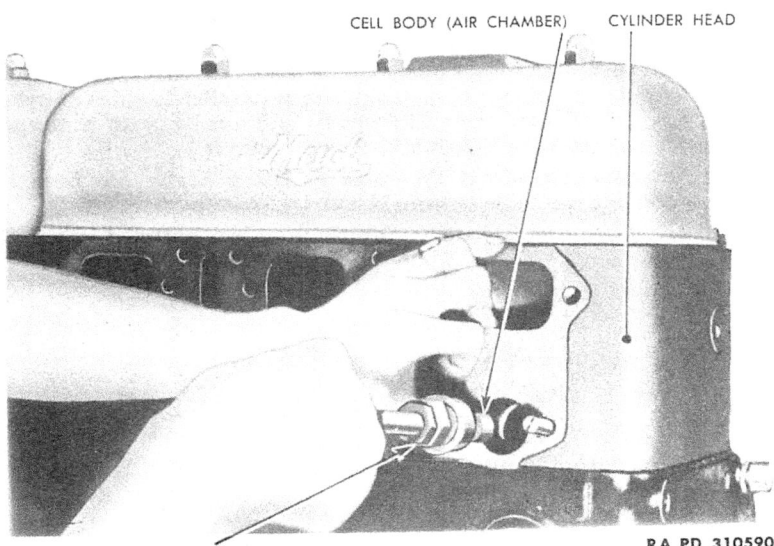

Figure 92 — Lapping Cell Body in Cylinder Head, with Puller Tool (41-T-3535-400)

10-TON 6 x 4 TRUCK (MACK MODEL NR)

from manifold heater on No. 6 cylinder to remove this cell, using wrench to disconnect cable at terminal.

(3) REMOVE CELL CAP AND COVER. Use puller to remove cell cap and cover.

(4) REMOVE CELL. Use puller (41-T-3535-400) to remove energy cell body.

d. Installation.

(1) LAP JOINT SURFACES OF CELLS (fig. 91). Lap the flat contacting surfaces of the cell body and cell cover so that they are perfectly flat and gastight. For this operation, use cast iron lapping plate and a very fine lapping compound. When such lapping does not result in achieving a perfect bearing between the parts, use new parts. Do not use any further means to restore the fit of parts which may have become distorted, or whose mating surfaces have suffered damage from the leakage of hot gases. Following lapping, be sure to carefully clean off all compound. Burned cells are usually attributable to overheating of the engine. Check operating temperature (par. 76 b.)

(2) INSTALL CELL BODY (fig. 77). With puller (41-T-3535-400) lap the energy cell body into the cylinder head with a medium valve grinding compound to insure a close fit. Note the numbers stamped on the cells, as they must be reinstalled in the same cylinders from which they were removed. Be sure to wipe off all valve grinding compound before reassembling and then insert the cell body in the cylinder head.

(3) INSTALL CELL COVER AND CAP. Place energy cell cover and cap over cell bodies in that order.

(4) INSTALL CLAMPS. Place clamps over cell covers and install with two nuts on each cell, keeping clamps parallel to sides of cylinder heads and tighten nuts alternately and evenly with a torque-indicating wrench (41-W-3630) 50-60 foot-pounds.

(5) INSTALL CABLE. If the cable to No. 6 heater has been removed, connect the cable terminal.

(6) INSTALL AIR CLEANER. Follow procedure outlined in par. 71 c (3).

TM 9-818
74

Section XVII

INTAKE AND EXHAUST SYSTEMS

	Paragraph
Intake system	74
Exhaust system	75

74. INTAKE SYSTEM.

a. Description and Tabulated Data.

(1) DESCRIPTION. The intake system consists of the intake manifold and an oil-bath air cleaner. The air cleaner is mounted on a bracket on the left side of the engine, and directly connected to the intake manifold (fig. 93).

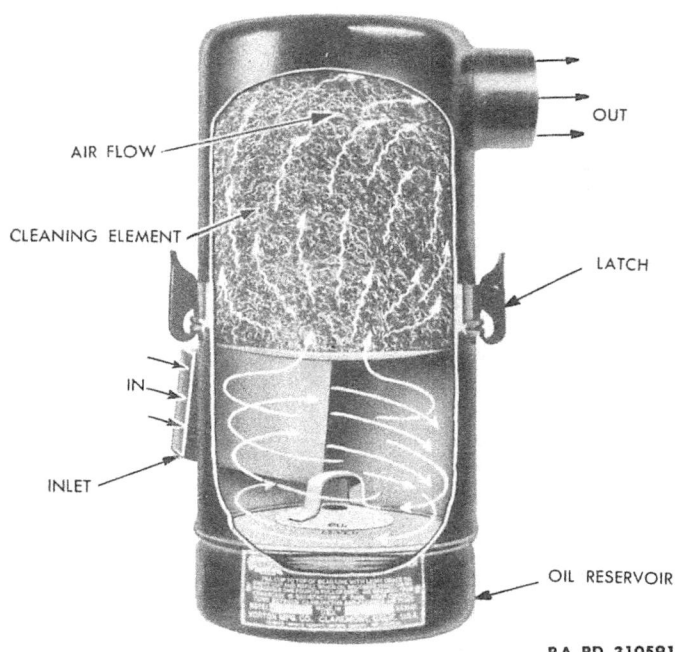

RA PD 310591

Figure 93—Air Cleaner

(2) DATA (AIR CLEANER).
```
Make ........................................ Vortox
Model ........................................ 3480-D
Type ........................................ Oil bath
Location ........................... Engine, left side
```

177

TM 9-818
74-75

10-TON 6 x 4 TRUCK (MACK MODEL NR)

b. **Removal.**

(1) REMOVE SUMP. Remove the oil sump by loosening the clamps.

(2) DISCONNECT HOSE. Disconnect hose, using a screwdriver to loosen the clamping bolt on the outlet hose, and disconnect from cleaner.

(3) REMOVE AIR CLEANER. Remove the two cross-recessed head screws and lock washers from the forward bracket connection. Open straps and remove cleaner.

c. **Installation.**

(1) INSTALL CLEANER. Place air cleaner in position in mounting straps and attach strap ends to bracket with cross-recessed head screws and lock washers.

(2) ATTACH HOSE. Place outlet hose over pipe on cleaner and tighten clamps.

(3) INSTALL SUMP. Fill the sump to oil level and install on cleaner, clamping with the two spring clamps.

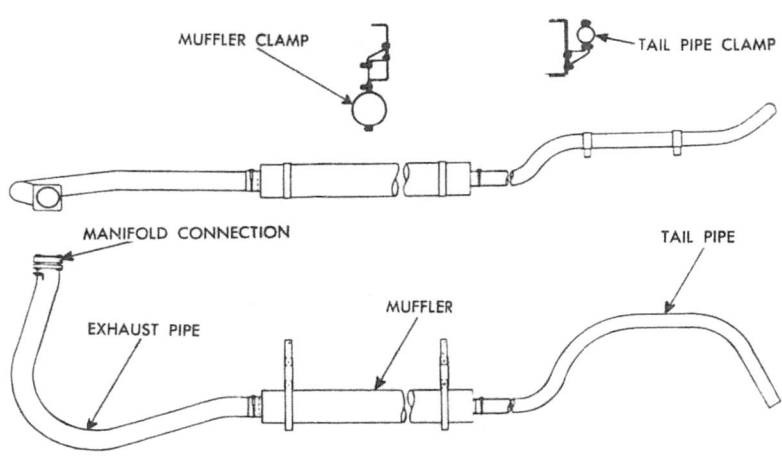

RA PD 310592

Figure 94—Exhaust System

75. EXHAUST SYSTEM.

a. **Description and Tabulated Data.**

(1) DESCRIPTION. The exhaust system consists of an exhaust pipe, muffler and tail pipe, and is supported from the chassis frame rail. The connection at the manifold is of the compression-type with ring gaskets, but the pipe can not slip out of position because one of the four flange bolts, longer than the other three, passes through a lug which is welded to the pipe. The muffler is of the "straight through" type (fig. 94).

178

INTAKE AND EXHAUST SYSTEMS

(2) Data (Muffler).
- Make .... Powell
- Type .... "Straight-through"
- Manufacturer's No. SM-473
- Length .... 39 in.
- Diameter .... 6 in.
- Inlet inside diameter .... $3\frac{1}{2}$ in.
- Outlet outside diameter .... $2\frac{5}{8}$ in.

b. Removal.

(1) Remove Forward Pipe. Detach exhaust pipe at manifold as outlined in paragraph 53 a (1). Loosen nut on muffler clamp. Tap pipe to loosen it from muffler and remove pipe as well as flange and two gaskets. Remove the long bolt from bracket on pipe and remove flange.

(2) Remove Tail Pipe. Remove upper palnut and nut from the two clamps on right-hand side rail, and loosen the lower nuts. Loosen clamp on rear of the muffler. Pull pipe to rear with a turn and remove.

(3) Remove Muffler. Remove the two bolts from the bottom clamps. Remove cotter pins and loosen castle nuts on top of clamps. Turn top clamps and remove muffler.

c. Installation.

(1) Install Forward Pipe. Place long bolt through bracket on exhaust pipe, with lock under head on bracket and install nut on bolt. Place pipe end into muffler and tighten $\frac{5}{16}$-inch clamping nut with $\frac{1}{2}$-inch nut. Install flange, gaskets and bolts with parts assembled as shown in figure 49, paragraph 53 b (2). There are two gaskets between the flanges.

(2) Install Tail Pipe. Place pipe in brackets, push it forward into the muffler and tighten nut on clamp. Attach pipe to clamps, with nut and lock washer on lower clamp fastenings, and nut and palnut on upper fastenings.

(3) Install Muffler. Place muffler against clamps and install four bolts. Place nuts on bottom bolts and tighten them, also tighten the castle nuts above them. Install cotter pins in upper bolts.

10-TON 6 x 4 TRUCK (MACK MODEL NR)

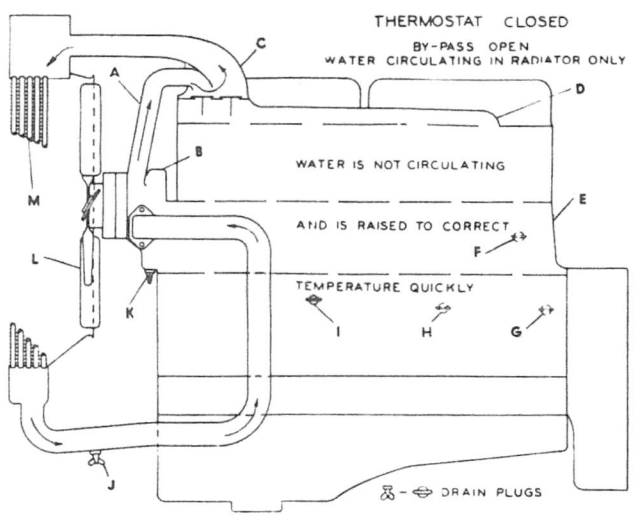

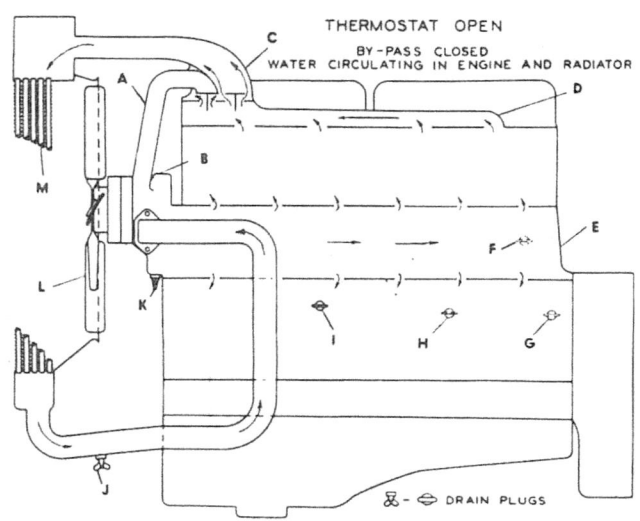

A BY-PASS TUBE
B WATER PUMP
C THERMOSTAT HOUSING
D WATER MANIFOLD
E WATER JACKET
F WATER-MANIFOLD DRAIN PLUG (R.H. SIDE)
G CYLINDER BLOCK DRAIN PLUG (R.H. SIDE)
H COMPRESSOR-HEAD DRAIN PLUG (R.H. SIDE)
I CYLINDER BLOCK DRAIN PLUG (L.H. SIDE)
J RADIATOR DRAIN VALVE
K WATER PUMP DRAIN
L FAN
M RADIATOR

RA PD 310593

Figure 95—Cooling System Circulation

Section XVIII

COOLING SYSTEM

	Paragraph
Description and tabulated data	76
Service cooling system	77
Radiator assembly removal	78
Radiator assembly installation	79
Cooling system connections	80
Fan assembly	81
Water pump assembly	82
Fan belts	83
Water manifold	84
Thermostats	85

76. DESCRIPTION AND TABULATED DATA.

a. Description (fig. 95). In arrangement the cooling system is conventional, having a frontal-type radiator, fan, engine-mounted water pump, thermostatic temperature control and the necessary connecting hose. Temperature control is of the "cold-circulation" type, in which, during warm-up, there is free circulation through the radiator (without any back pressure on the pump) and no circulation through the engine water jackets. Valving of the water flow from cold to hot circulation is by twin, bellows-type thermostats incorporated in the front end of the water manifold. The radiator has a vertical tube-and-fin core and a deep upper tank to allow for expansion of the coolant. The six-blade, propeller-type fan is driven by double V-belts from the crankshaft in three-point arrangement to drive also the generator. Its shaft is ball-bearing mounted, and projects into the pump to carry the pump impeller. The bearings are grease-lubricated. The water seal is packless consisting of a spring-loaded carbon washer which turns with the shaft and is sealed thereto with a rubber sleeve.

b. Data.

```
Cooling system capacity......................60 qt
Radiator:
    Make ....................................Modine
    Model ...................................AD3741
    Frontal area ............................736 sq in.
    Core depth ..............................3 in.
    Location ................................Front of chassis
    Mounting.................................Cross member, rubber
                                             insulators, tie rod bracing
Fan:
    Make ....................................Service Products
    Type ....................................Propeller
    Diameter ................................24 in.
    Number of blades.........................Six
```

TM 9-818
76-77

10-TON 6 x 4 TRUCK (MACK MODEL NR)

    Drive .... Double V-belts
    Arrangement .... With water pump
    Location .... Behind radiator
    Shroud .... Attached to radiator
Fan belt:
    Make .... Dayton
    Number .... Two (a double belt)
    Inside circumference .... $59\frac{17}{64}$ in.
    Width .... $1\frac{5}{64}$ in.
    Depth .... $1\frac{9}{32}$ in.
    Angle .... 40 degrees
    Arrangement Three-point (crankshaft, fan and generator)
Water pump:
    Make .... Mack
    Type .... Centrifugal impeller
    Drive .... Fan belt
    Arrangement .... With fan
    Location .... Front of engine block

Figure 96—Radiator Filler Cap

Thermostat:
    Make .... Fulton Sylphon
    Number .... Two (duplicates)
    Location .... Front end of water manifold
    Open fully .... 185°F
    Close fully .... 165°F

77. SERVICE COOLING SYSTEM.

    a. Draining.

    (1) DRAIN RADIATOR. Remove the filler cap (fig. 96) slowly to permit pressure to escape through vent in cap if radiator is hot. Open

TM 9-818
77-78

COOLING SYSTEM

radiator drain cock on lower radiator pipe. If cooling system contains antifreeze, drain in pan and save.

(2) DRAIN CYLINDER BLOCK. Open drain plugs shown on figure 97.

b. **Filling.**

(1) Close drain valve in radiator pipe and install drain plugs in cylinder block.

(2) Fill system with cooling solution (60 qt).

(3) Add rust inhibitor.

(4) Turn filler cap all the way to the right to seal the system.

c. **Flushing Cooling System.**

(1) Remove radiator hose upper and lower (par. 79 a (5)).

(2) Remove thermostats from cylinder head (par. 85 a).

(3) Install thermostat housing without thermostat.

(4) Flush radiator core and engine water jackets as follows:

(a) Place end of water hose in radiator filler opening and wrap cloth around connection to hold some of the water pressure. Turn on water and allow water to flow through radiator until water is clean as it leaves radiator.

(b) Place end of water hose in opening at the thermostat housing and allow water to flow through the engine water jackets until water is clean as it leaves water jackets.

(5) Remove thermostat housing and install the thermostat (par. 85 b (1)).

(6) Install upper and lower radiator hose (par. 79 b (9)).

(7) Refill cooling system (par. 79 b (12)).

d. **Antifreeze Solutions.** Proceed as outlined in paragraph 8 b.

78. RADIATOR ASSEMBLY REMOVAL.

a. **Drain System.** Drain the water from cooling system by opening drain valve on radiator outlet pipe.

b. **Remove Hood.** Remove forward hood hinge clamp by removing two cross-recessed head screws, with screwdriver and open-end wrench. Loosen the rear clamp using screwdriver and open-end wrench. Remove hood.

c. **Remove Fan Assembly.** Remove fan assembly as outlined in par. 81 a.

d. **Detach Radiator Tie Rods.** Remove nut and lock washer from each upper tie rod. Detach lower radiator tie-rod bolts, nuts, and lock washers, using two open-end wrenches.

e. **Loosen Hose.** Loosen the upper and lower radiator hose clamp bolts.

f. **Remove Bond Strap.** Remove bolt, three toothed lock washers and nut attaching bond strap to radiator shell.

g. **Remove Strap.** Remove lower strap from shell to tank by removing one nut and lock washer from stud on tank and two bolts, nuts, and lock washers from shell.

TM 9-818
78-79

h. **Remove Lower Shield.** Remove three cap screws and lock washers and remove shield.

i. **Remove Radiator Mounting Nuts.** Remove two palnuts and mounting stud nuts from radiator mounting studs. Remove washers and lower rubber insulators.

j. **Remove Radiator.** If it is possible to lift radiator out of cross member without removing brush guard, do so, if not, remove brush guard as outlined in paragraph 168 a (2).

k. **Remove Accessories.** If the upper insulators remain on mounting studs, remove them. Remove fan shroud by removing six cap screws and lock washers. Remove hood latch stops by removing two screws from each. Remove radiator cap.

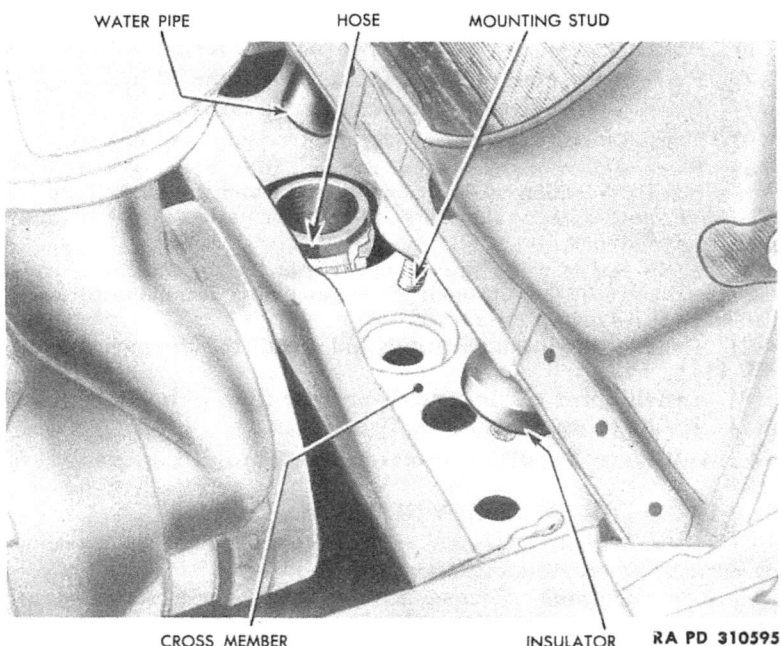

Figure 97—Installing Radiator

l. **Remove Shell.** Remove cross-recessed head screws from shell. Remove the 10 flathead screws and nuts under antisqueak on radiator shell. Remove shell and two spacers under top screws.

m. **Remove Hood Rest Bracket.** Use cross-recessed screwdriver and open-end wrench to remove two screws and lock washers from each hood rest bracket, and remove bracket.

79. RADIATOR ASSEMBLY INSTALLATION.

a. **Install Hood Rest Bracket.** Use cross-recessed screwdriver and wrench to attach hood rests on each side with two screws and lock washers in each.

COOLING SYSTEM

b. Install Hood Latch Stops. Attach hood latch stops with four cap screws and lock washers.

c. Install Shell. Install 10 flathead screws in nuts. Also place one bolt in hole at top of shell to secure hood clamp spacer in position.

d. Install Shroud. Place shroud on radiator, and attach with five screws and lock washers.

e. Install Radiator. Lift radiator into place with larger insulators either on mounting studs or placed in radiator support cross member with flat side down. Lower radiator into position on cross member with water pipe in hose. Insert tubing leading to pump in rubber hose on pump elbow (par. 97).

f. Install Tie Rods. Place upper tie rods through holes in radiator bracket and install washer and nut, but do not tighten. Install two lower tie rods using two bolts, lock washers, and nuts.

g. Install Fan Assembly. Follow procedure outlined in paragraph 81 b to install fan.

h. Install Hood. Hold hood over the engine. Place end of hinge pin in bracket on cowl and front end of pin in front clamp on shell. Apply top clamping piece on front, insert screws, and apply nuts and lock washers. Tighten with cross-recessed screwdriver and open-end wrench.

i. Aline Radiator and Tighten Hose Clamps. Line up hood and radiator assembly and tighten lower tie rod nuts. Tighten hose clamps and fasten upper tie rod with nut, washer, bracket, lock washer, and nut in respective positions.

j. Install Lower Rubber Insulators. Place lower rubber insulators on studs with smaller side toward cross member. Place washer and nut on stud, and tighten until lower rubber insulator is $1\frac{3}{4}$ inches in diameter. Install palnut on each stud and tighten.

k. Install Radiator Shield and Strap. Install three cap screws and lock washers attaching lower radiator shield to shell. Place the steadying strap on center bottom stud on radiator, and install nut with lock washer. Apply two bolts, nuts, and lock washers to fasten strap to shell and tighten.

l. Check Cooling System. Close the drain valve in radiator outlet pipe, fill system with water, and check for leaks.

80. COOLING SYSTEM CONNECTIONS.

a. General Instructions. To replace any of the following cooling system hose, it will be necessary to drain the water from the radiator and cooling system by opening the drain cock on the lower radiator pipe. Upon completion of the replacement, fill the system with water and check for leaks.

b. Radiator to Water Pump Inlet Pipe.

(1) REMOVAL. Use screwdriver to loosen upper hose clamp, and socket wrench to loosen lower hose clamp bolt. Remove cap screw, lock washer, and clamp (with antisqueak) attached to engine. Pull

10-TON 6 x 4 TRUCK (MACK MODEL NR)

pipe out in direction of oil filter and frame, pushing between heater cables if necessary to remove. Remove drain cock from pipe, and open eye on chain with pliers to detach.

(2) INSTALLATION. Install drain cock in pipe and attach chain to lug with pliers. Place pipe in position with ends in rubber hose. Tighten upper hose clamp with screwdriver and lower hose clamp with socket wrench. Install clamp at engine with cap screw and lock washer, making sure that antisqueak is in position on bracket.

c. **Radiator Outlet Rubber Elbow.**

(1) REMOVAL. Loosen bolts on two hose clamps. Pull elbow down and slide off pipe. Loosen bolts on clamps if necessary. Pull clamps off hose.

(2) INSTALLATION. Place clamps on elbow ends and install elbow on pipes, making certain that spring which prevents its collapse is inside elbow. Tighten the bolts on the two hose clamps.

d. **Water Pump Intake Hose.**

(1) REMOVAL. Loosen two hose clamps. Slide hose back on intake elbow. Pull pipe out of way, and slide hose off pipe.

(2) INSTALLATION. With clamps on hose, slide hose on pipe from radiator. Push pipe toward engine, and slide hose onto water pump intake elbow. Tighten two hose clamps.

e. **Upper Elbow Bypass Hose.**

(1) REMOVAL. Loosen two clamps on hose. Remove two cap screws attaching bypass elbow to water pump outlet pipe, and remove bypass elbow. Pull hose off elbow and remove clamps.

(2) INSTALLATION. Place two clamps on hose and slide onto bypass elbow end. Slide other end of hose over bypass pipe, and attach elbow to water pump outlet elbow with two cap screws. Tighten clamps on hose.

f. **Radiator Upper Tank to Outlet Elbow Hose.**

(1) REMOVAL. Loosen two bolts. Slide hose to front on radiator tank pipe, and remove from water outlet elbow. Pull off upper tank pipe and remove clamps.

(2) INSTALLATION. Place clamps on hose, and slide hose onto radiator upper tank pipe. Pull hose back onto outlet elbow. Tighten bolts on clamps.

81. FAN ASSEMBLY.

a. **Removal.** Remove six cap screws and lock washers attaching fan to water pump pulley and take out fan from left-hand side.

b. **Installation.** Place fan on pulley and install the six cap screws and lock washers.

82. WATER PUMP ASSEMBLY.

a. **Removal.**

(1) DRAIN WATER. Open drain cock on lower inlet pipe. Remove radiator cap and drain water from system.

COOLING SYSTEM

(2) REMOVE FAN. Remove six bolts, cap screws and lock washers, and remove fan to left-hand side.

(3) REMOVE GENERATOR ADJUSTING LINK. Remove adjusting bolt and generator adjusting link, and remove link.

(4) REMOVE BYPASS PIPE. Remove the two cap screws and lock washers from upper elbow. Remove the two cap screws and lock washers from the lower elbow. Remove the piping and gaskets.

(5) DISCONNECT COMPRESSOR WATER LINES. Disconnect compressor water lines at water pump.

(6) REMOVE HOSE. Loosen clamps on hose connecting water pump to inlet, and push hose back to clear pump.

(7) REMOVE WATER PUMP. Remove two cap screws, with lock washers, attaching water pump to engine block. Remove fan belts from pulley, and remove pump to left side of engine.

(8) REMOVE FITTINGS. Remove the two cap screws from water pump inlet elbow with socket and open-end wrenches. Remove the 1/4-inch elbow and remove the straight pipe fitting. Remove drain cock and grease cup. Remove pipe plug.

(9) REMOVE PULLEY. Extract cotter pin and remove nut from pulley shaft. Remove pulley from shaft with puller.

b. Installation.

(1) INSTALL PULLEY. Place key in keyway on water-pump shaft and tap pulley into place on shaft with open end toward the pump. Install nut, tighten and install cotter pin.

(2) INSTALL FITTINGS. Turn grease cup into pump housing. Install straight fitting on water pump cover. Install elbow in housing at bypass connection. Install drain cock in bottom of housing. Install pipe plug.

(3) INSTALL INLET ELBOW. Place gasket between inlet elbow and pump and attach elbow with two cap screws and lock washers.

(4) INSTALL PUMP. Place pump in position on front of engine, guiding fan belts on to pulley. Place gasket between pump outlet and cylinder block. Install four cap screws, with lock washers, attaching water pump to engine block. Tighten only two screws from pump to block itself, leaving two screws on bypass connection loosely in place for alinement of pump.

(5) ATTACH WATER LINES. Connect compressor water line to straight fitting and the other compressor water line to compressor elbow. Install bypass pipe with gasket between flange at lower end and pump and gasket between upper elbow and thermostat housing, attaching with cap screws and lock washers. Install water hose between bypass pipe and elbow and tighten clamps.

(6) INSTALL GENERATOR ADJUSTING LINK. Attach generator adjusting link to pump with cap screw and insert locking bolt through link and bracket on generator strap and tighten.

(7) ADJUST FAN BELTS. Follow procedure outlined in paragraph 83 a.

(8) FILL COOLING SYSTEM. Fill cooling system with water and check for leaks.

83. FAN BELTS.

a. **Adjust.** Belts should be adjusted so that a force of 15 pounds will deflect either belt only $\frac{1}{2}$ inch on the long side of the drive. To make this adjustment loosen bolt and nut on generator adjusting link and loosen the two bolts on hinge bracket, using two wrenches. Pull generator out to tighten belts, checking for required tension and tighten bolts to hold in place.

b. **Removal.** Loosen bolt and nut on the generator adjusting link. Loosen two bolts on generator hinge bracket with two wrenches and move generator inward. With crank handle, turn engine over, guide fan belt off pulley and remove.

c. **Installation.** Remove generator adjusting link by removing cap screw from water pump bracket. Remove two rear generator mounting cap screws at engine base. Also remove toothed lock washer at both rear and center. Loosen forward cap screw and pull rear end of generator away from engine. Place belts over fan pulley and crankshaft pulley and pull over generator pulley. Pull generator out against belts. Install and aline generator (par. 87 b) and adjust belts as noted in step a (1) above.

84. WATER MANIFOLD.

a. **Removal.**

(1) DRAIN SYSTEM. Open drain cock on lower radiator pipe and drain water from system.

(2) REMOVE INTAKE MANIFOLD. Follow procedure outlined in paragraph 52 a.

(3) REMOVE THERMOSTAT HOSE. Follow procedure outlined in paragraph 85 a (2) to remove thermostat hose.

(4) REMOVE TEMPERATURE GAGE ELEMENT. Remove temperature gage unit from engine and remove adapter from engine.

(5) REMOVE PLUGS. Remove two pipe plugs from forward manifold and one from rear manifold.

(6) REMOVE MANIFOLD. Remove twelve cap screws and lock washers, and remove manifold and gaskets.

(7) SEPARATE MANIFOLD SECTIONS. Remove two bolts from the center flanges. Separate sections and remove gasket.

b. **Installation.**

(1) JOIN MANIFOLD SECTIONS. With manifold flange surfaces clean, install gasket between the two center flanges, and join with two bolts, nuts and lock washers. Do not tighten until the manifold is placed on engine.

TM 9-818
84-85

COOLING SYSTEM

(2) INSTALL MANIFOLD. Place gaskets at manifold ports and install manifold by tightening 12 cap screws and lock washers, and then tightening two bolts connecting center flanges.

(3) INSTALL PLUGS. Install two pipe plugs in front manifold and one in the rear manifold.

(4) INSTALL WATER TEMPERATURE GAGE ELEMENT. Install gage element adapter in rear of water manifold and install element in adapter.

(5) INSTALL THERMOSTAT HOSE. Follow the procedure outlined in paragraph 85 b (3), to install thermostat hose.

(6) INSTALL INTAKE MANIFOLD. Follow procedure outlined in paragraph 52 b.

(7) FILL SYSTEM. Fill system with water and check for leaks.

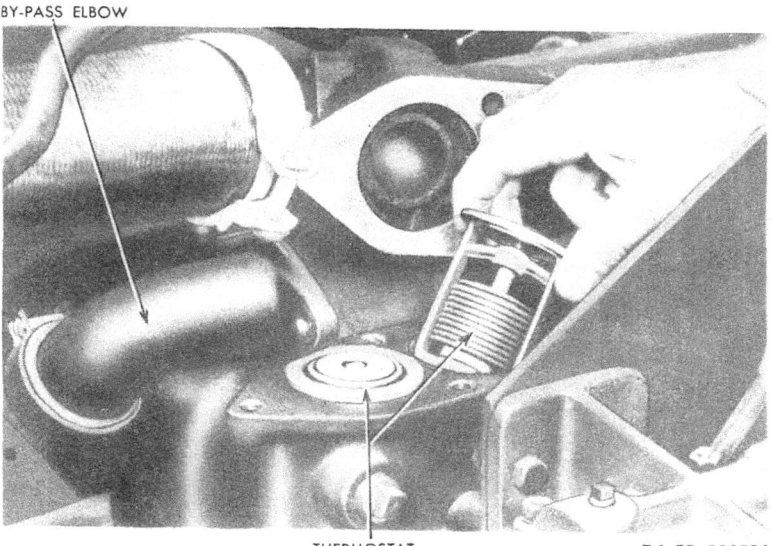

Figure 98 — Installing Thermostat

85. THERMOSTATS.

a. Removal.

(1) DRAIN WATER. Open drain plug in bottom of radiator outlet pipe and drain water from cooling system.

(2) DETACH HOSE. Loosen clamp on radiator hose and detach hose.

(3) REMOVE CAP SCREWS. Remove the two cap screws and lock washers from bypass elbow on water outlet elbow. Remove the four cap screws with lock washers from water outlet elbow.

10-TON 6 x 4 TRUCK (MACK MODEL NR)

(4) REMOVE ELBOW. Remove water outlet elbow and gasket, and lift out thermostats.

b. **Installation.**

(1) INSTALL THERMOSTATS. Install thermostats with bellows downward (fig. 78).

(2) INSTALL CAP SCREWS. Place water outlet elbow on manifold with gasket in place and install two long cap screws with lock washers on sides of elbow, the two shorter cap screws with lock washers to front and rear of elbow and tighten. Install bypass elbow on outlet elbow with gasket in place, using two cap screws with lock washers, and tighten.

(3) INSTALL HOSE. Place radiator hose on water outlet elbow and tighten clamp.

(4) FILL COOLING SYSTEM. Fill cooling system with water and check for leaks.

TM 9-818

Section XIX
STARTING AND GENERATING SYSTEM

	Paragraph
Description and tabulated data	86
Generator assembly	87
Regulator	88
Magnetic starter switch	89
Series parallel switch	90
Cranking motor	91

86. DESCRIPTION AND TABULATED DATA.

a. Description (figs. 99, 100, and 101). One part of this system, comprising the generator, its control unit and wiring, functions to keep the battery charged. The other consists of the cranking motor, control switches and cables that draw current from the battery when cranking the engine to start it. The electrical air heater system (for cold-weather starting) consists of six grid-type heaters (at the branches of the intake manifold), wiring and control switches connected to the battery. Though not interconnected with it, this system is functionally associated with the starting system, being used with it in cold weather. The generator and the cranking motor are on the left side of the engine. The generator is at the front, carried on a swinging bracket for tension adjustment of the fan belt by which it is driven. The cranking motor is fastened to the front face of the flywheel housing. Control of the generator charging is by a unit containing three elements, a cut-out relay, a voltage regulator and a current regulator. The cut-out relay "cuts out" the generator, at slow speed or when stopped, to avoid flow of current from the battery; and "cuts in" the generator when the voltage rise is sufficient to charge the battery. Either the voltage regulator or the current regulator controls charging in accordance with the battery condition, the first by limiting the voltage to the prescribed value and the second by properly limiting the current. On the instrument panel there are two ammeters to indicate separately the current of each of the two series pairs of the four 6-volt batteries. The generating system is 12-volts, as is also the lighting system. Through a series-parallel switch, the cranking motor is operated at 24 volts from the four batteries in series. Control of the cranking motor is through this switch and a magnetic switch by a push-button on the instrument panel. A push-button also controls the air heaters through magnetic relay switches. Each group of three heaters is operated at 12 volts from two batteries. The generator and cranking motor control circuits are fuse-protected. Radio interference suppression is through condensers, bonds and cable shields.

b. Data.
    Generator:
        Make .... Delco-Remy
        Model 1116658
        Voltage 12

TM 9-818
86

10-TON 6 x 4 TRUCK (MACK MODEL NR)

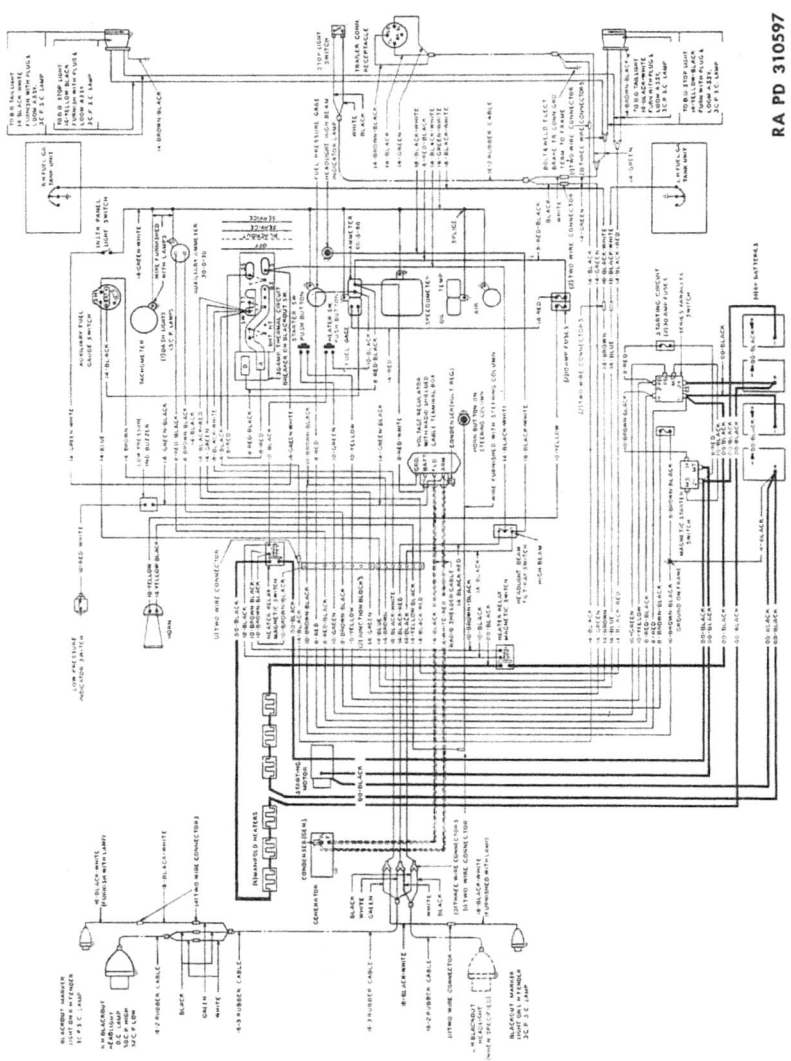

Figure 99—Complete Wiring Diagram

STARTING AND GENERATING SYSTEM

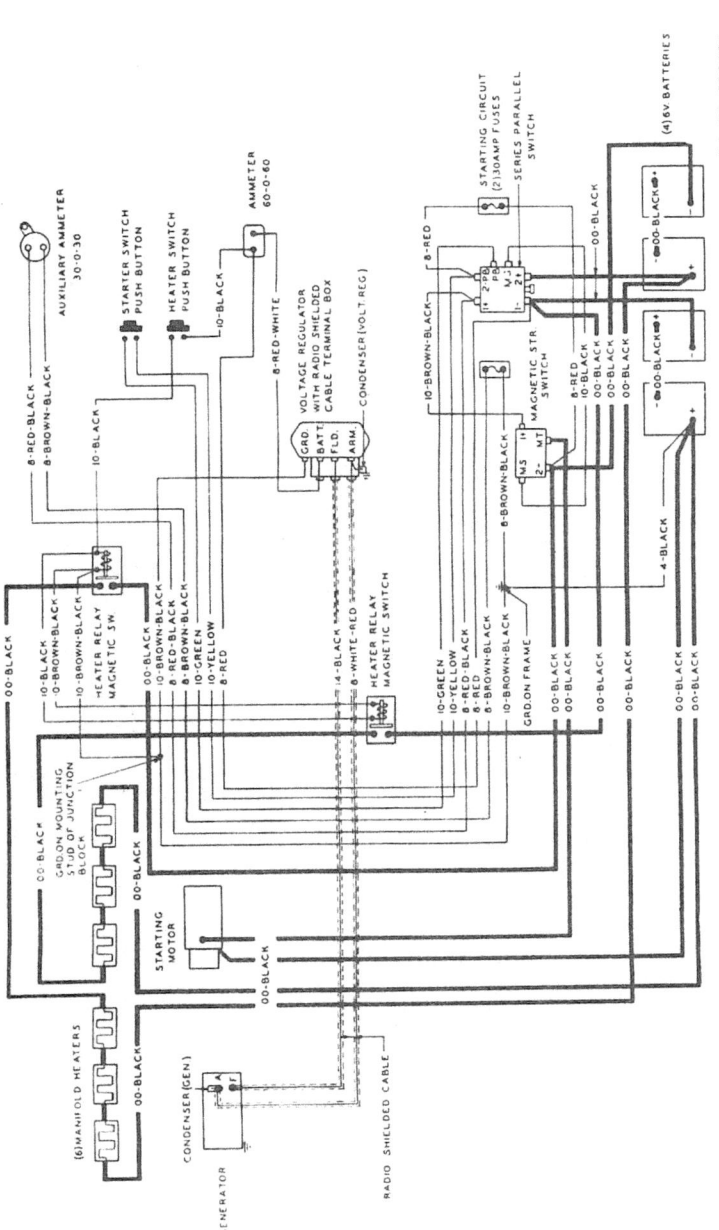

Figure 100 — Generating Circuit Wiring Diagram

TM 9-818
86

10-TON 6 x 4 TRUCK (MACK MODEL NR)

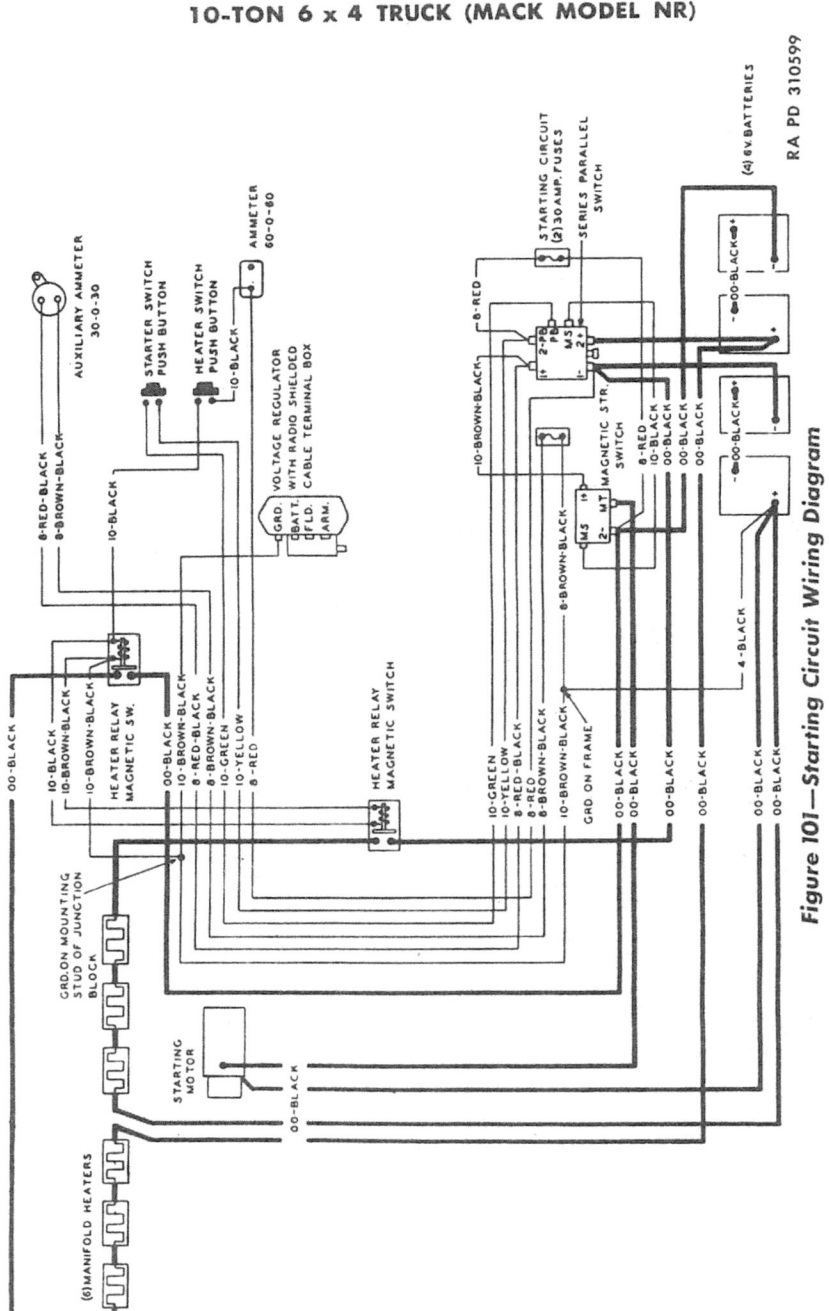

Figure 101—Starting Circuit Wiring Diagram

STARTING AND GENERATING SYSTEM

Amperes 25
Rotation, viewed at drive end .... Clockwise
Ratio .... 1.625 times crankshaft
Drive .... Fan belt
Location .... Left side of engine at front
Regulator:
  Make .... Delco-Remy
  Type .... Three-element
  Model 1118475
  Location .... Front side of dash
Cranking motor:
  Make .... Leece-Neville
  Type .... Three-bearing
  Model 1446-M
  Voltage 24
  Rotation, viewed from front .... Counterclockwise
  Ratio .... 9.83 to 1.00
  Drive .... Self-engaging, inertia
                   type with shock-absorbing,
                   multiple-disk clutch
  Location .... Left side of engine at rear
Series-parallel starting switch:
  Make .... Leece-Neville
  Type .... Relay
  Model 31-Ms-12
  Location .... On front of dash
Magnetic starting switch:
  Make .... Leece-Neville
  Type .... Relay
  Model 31-SPS
  Location .... Back of batteries
Air heaters:
  Make .... Mack
  Type .... Electrical, grid
  Number .... Six
  Location .... Between cylinder heads and
                   branches of intake manifold
Air-heater switches:
  Make .... Delco-Remy
  Type .... Relay
  Model 1422
  Number .... Two
  Location .... On front of dash

10-TON 6 x 4 TRUCK (MACK MODEL NR)

87. GENERATOR ASSEMBLY.

a. Removal.

(1) REMOVE RADIO BOND STRAP. Remove cross-recessed head screw from generator to remove radio bond strap.

(2) DETACH SHIELDED CABLES. Remove generator condenser and shielded cable terminal caps from generator by turning with pliers. Remove nut and lock washers from inside of terminals. Remove nuts attaching cables to terminal housing and detach cables.

(3) DETACH GENERATOR. Remove two hinge mounting bracket bolts and toothed lock washers, using two wrenches. Remove link-adjusting bolt and frame washer. Slip belts off pulley and remove generator.

(4) REMOVE GENERATOR ACCESSORIES. Remove generator pulley by removing nut and lock washer on end of shaft, and withdraw pulley with puller. Remove two clamping bolts and nuts, and remove generator hinge bracket and spacer.

b. Installation.

(1) INSTALL GENERATOR ACCESSORIES. Install spacer and hinge bracket, leaving clamping bolts loose until generator is alined. Install pulley on shaft with key in place and install nut and lock washer on shaft end to hold pulley.

(2) INSTALL GENERATOR. Slipping belts into place on the generator pulley, place generator against bracket and install two hinge bolts, toothed lock washers and nuts. Install adjusting link with bolt, lock washer and nut and turn only partially tight.

(3) ALINE GENERATOR. Use a straightedge at least 18 inches long to place against generator and other pulley outer surfaces. Move generator to aline face of pulley with faces of other pulleys. Tighten the two clamp bolts and install lock nuts.

(4) ADJUST FAN BELTS. Adjust the belt as outlined in paragraph 83 a.

(5) INSTALL SHIELDED CABLES. Place white and red coded No. 8 wire in heavier shielded cable assembly into the forward terminal housing on the generator and the black coded No. 14 wire from the other shielded cable in the rear terminal housing. Place terminals over studs and attach, using lock washer and nut; tighten and attach shielding to terminal housing. Install generator condenser on the forward or armature terminal by screwing into place.

(6) ATTACH RADIO BOND STRAP. Attach radio bond strap to generator, using cross-recessed head screw.

88. REGULATOR.

a. Removal.

(1) DISCONNECT BATTERY. Follow procedure outlined in paragraph 167 to disconnect batteries.

(2) REMOVE LOWER COVER. Remove four lower cover screws and lock washers, and remove cover.

TM 9-818
88-89

STARTING AND GENERATING SYSTEM

(3) DISCONNECT WIRES. Remove nuts from three terminal screws, and remove wires and cables from terminals.

(4) DISCONNECT CABLE SHIELDS. Remove cable shields from regulator lower case by removing attaching nuts.

(5) DISCONNECT GROUND WIRE. Detach ground wire from case.

(6) REMOVE REGULATOR (fig. 50). Remove four nuts and toothed lock washers attaching regulator to dash sheet and remove regulator.

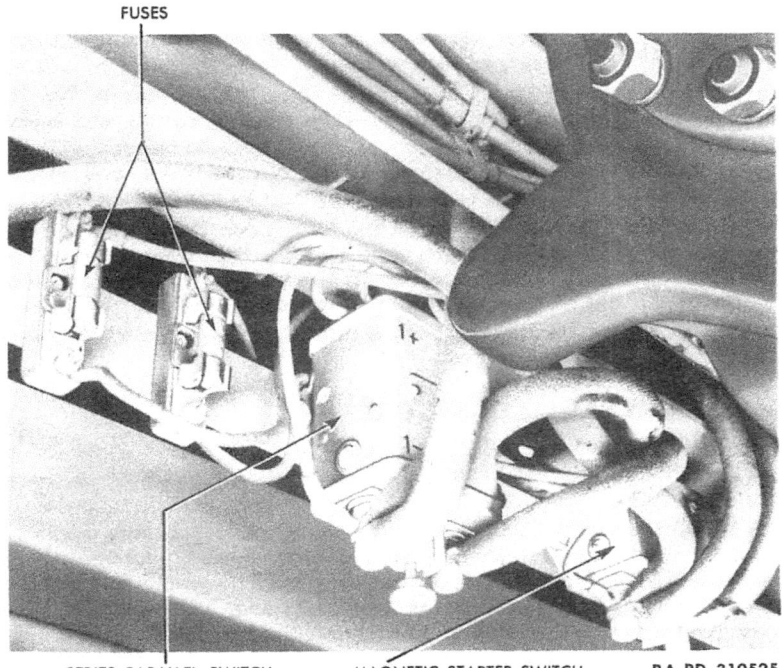

Figure 102—Magnetic Starter Switch and Series Parallel Switch

b. Installation. Installation procedure is the reverse of removal procedure. Connect wiring as follows: The No. 14 green and black wire (through grommeted hole in lower case) and the No. 8 white wire with red tracer (shielded cable) to the armature terminal; the No. 14 black wire (shielded cable) to the field terminal and the No. 8 red wire with white tracer (through grommeted hole in case) to the battery terminal. Attach the No. 10 brown wire with black tracer to the case with screw for ground connection.

89. MAGNETIC STARTER SWITCH.

a. Removal.

(1) DISCONNECT BATTERY. Follow procedure outlined in paragraph 167.

10-TON 6 x 4 TRUCK (MACK MODEL NR)

(2) DISCONNECT CABLES AND WIRES. Remove the two nuts and lock washers from bottom connections and remove the cables, and wire, tie the two No. 00 black cables when removing them from the "2—" terminal so that they may be reconnected without tracing the cables. Remove the two nuts and lock washers, one on front and one on rear of switch, and remove the wires from those terminals.

(3) REMOVE SWITCH. Remove the two bolts, nuts, and lock washers attaching switch to angle iron support, and remove switch.

b. Installation.

(1) INSTALL SWITCH. Place switch on angle iron support and attach with bolts, nuts, and lock washers.

(2) MAKE FORWARD CONNECTION (fig. 101). Connect No. 10 brown-black wire to the "1+" terminal on front of switch, and install nut and lock washer.

(3) MAKE REAR CONNECTION. Attach No. 10 black wire to MS terminal on rear of switch, and install nut and lock washer.

(4) MAKE BOTTOM CONNECTIONS. Attach No. 8 red wire and the two No. 00 black cables, which were tied together when removed to the "2—" terminal. Attach the No. 00 black cable from the cranking motor to the "MT" terminal. Install the two nuts and lock washers.

(5) CONNECT BATTERIES. Follow procedure outlined in paragraph 167.

90. SERIES PARALLEL SWITCH.

a. Removal.

(1) DISCONNECT BATTERY. Follow procedure outlined in paragraph 167.

(2) DISCONNECT CABLES AND WIRES. Remove two nuts and lock washers from bottom connections and remove wire and cables; tie the two No. 00 black cables, so that they may be reinstalled without tracing the cables. Remove the two nuts and lock washers from connections on front of switch, and remove the two cables.

(3) REMOVE SWITCH. Remove the two bolts, nuts, and lock washers from angle front support and remove switch.

(4) REMOVE TOP CONNECTION. Remove the two nuts, lock washers, and terminals on top of switch and remove the four wires.

b. Installation.

(1) MAKE TOP CONNECTIONS. Attach the No. 8 red and black wire and the No. 10 grounded brown-black wire to the "1+" terminal, and install nut and lock washer. Attach the No. 10 yellow wire and the No. 8 red wire to the "2—" PB terminal, and install nut and lock washer.

(2) INSTALL SWITCH. Place switch on angle iron support, and install the two 3/8-inch bolts, nuts, and lock washers.

(3) MAKE FRONT CONNECTIONS. Attach No. 10 green wire to PB terminal and No. 10 black wire to the MS terminal on front of switch; install the two nuts and lock washers and tighten.

STARTING AND GENERATING SYSTEM

(4) MAKE BOTTOM CONNECTIONS. Place the No. 8 red wire, the two No. 00 black cables, one from the heater relay magnetic switch and the other to the negative terminal of the second battery (which were tied together when removed) on the "1—" terminal. Place the other No. 00 black cable from the positive terminal of the third battery on the "2+" terminal. Install the two nuts and lock washers and tighten.

(5) CONNECT BATTERIES. Follow procedure outlined in paragraph 167 to connect batteries.

91. CRANKING MOTOR.

a. **Removal.** Disconnect and tie the two No. 00 black cables. Remove four cap screws and lock washers attaching cranking motor to flywheel housing. Withdraw motor from flywheel housing.

b. **Installation.**

(1) INSTALL MOTOR. Place drive end of motor in opening in flywheel housing, and attach motor to housing with four cap screws and lock washers. Install both No. 00 black cables.

TM 9-818
92

10-TON 6 x 4 TRUCK (MACK MODEL NR)

Section XX

TRANSMISSION

	Paragraph
Description and tabulated data	92
Transmission assembly	93

92. DESCRIPTION AND TABULATED DATA.

   a. Description (fig. 8). As a member of the unit power plant, the transmission is bolted to the engine rear support of the engine, through the medium of the bell housing which completes enclosure of the flywheel and clutch. The transmission affords selectivity of 10 forward and 2 reverse speeds. The gear group which gives the basic ratios for five forward speeds and one reverse is in the main case. The secondary gear group provides a reduction which results in five more forward speeds and another reverse; a slower set of speeds designed to be intermediate steps. The case which encloses the secondary group of gears is bolted to the rear end wall of the main case, unifying two sections as a compact assembly. With the exception of the first-speed and reverse basic gears, which are of the straight-spur type and have sliding engagement, all gears are helical and are engaged for driving through sliding-jaw clutches. All shafts have antifriction bearings. The gears are shifted manually by two levers which are mounted directly on the transmission and are side by side. The main shift lever, the left one, has five positions for forward speeds, and one for reverse. The selector mechanism makes it impossible for more than one speed to become engaged at a time; and a lock-out latch prevents the unintentional engagement of reverse. The shorter lever, to the right, has two positions. Moving this lever to its forward position shifts to "slow" range. "High" range is obtained with this lever in its rearward position. Lubrication of the entire transmission is assured when its case is filled with lubricant to the proper level (section VI, par. 19 c (6)).

   b. Data.

Make	Mack
Model	TRD-37, overgear
Type	Duplex
Mounting	Unified with clutch and engine
Number of forward speeds	Ten
Number of reverse speeds	Two
Shift	Manual, two levers

TRANSMISSION

Ratios	Slow Range	Fast Range
First speed	9.30	6.74
Second speed	5.27	3.82
Third speed	2.65	1.92
Fourth speed	1.38	1.00 (direct)
Fifth speed	1.08	.78 (overdrive)
Reverse	9.39	6.80

93. TRANSMISSION ASSEMBLY.

　a. **Removal.**

　(1) REMOVE FRONT PROPELLER SHAFT. Follow procedure outlined in paragraph 95 a (1).

Figure 103 — Sling and Jack Placed for Transmission Removal

　(2) REMOVE FLOORBOARDS. Follow procedure outlined in paragraph 159 a (1) through (5).

　(3) DISCONNECT SPEEDOMETER CABLE. Disconnect speedometer cable at rear of transmission.

　(4) REMOVE RUNNING BOARD CROSS BRACES. Remove the six bolts, nuts, and lock washers on rear running board cross brace and

10-TON 6 x 4 TRUCK (MACK MODEL NR)

remove cross brace. Place blocks under battery tray to support weight of battery and remove the front running board cross brace by removing seven nuts, bolts and lock washers on each side.

(5) REMOVE CLUTCH RELEASE LEVER. Remove the bolt, nut and lock washer on clutch release lever. Remove the cotter pin and clevis pin from auxiliary shifter shaft, and remove auxiliary shifter.

(6) REMOVE SHIFTER SHAFT COVER. Remove eight cap screws and lock washers and remove the shifter shaft cover.

(7) REMOVE BELL HOUSING CAP SCREWS. Remove the 12 cap screws and lock washers attaching bell housing.

(8) PLACE SLING. Place cable or rope sling around transmission and attach to overhead crane. Use additional rope or cable passing it through battery compartment and over left-hand side rail as shown in figure 103, to assist in supporting rear end of transmission by looping rope around auxiliary transmission.

(9) REMOVE PILOT STUD NUTS. Remove two nuts and lock washers from transmission pilot studs.

(10) REMOVE TRANSMISSION. Pry transmission flange away from the bell housing, and, using care not to damage the spline shaft, lower the transmission to the ground with overhead crane, steadying the assembly with rope over side rail, passing through battery compartment. NOTE: *If floor jack is available use it under the center of the transmission to lower assembly without use of hoist and cables. If floor jack is used, it is not necessary to remove the rear running board cross brace.*

b. **Installation.**

(1) PLACE SLINGS. Place sling around transmission and an additional sling around auxiliary transmission, with end of the second sling over left-hand side rail.

(2) INSTALL TRANSMISSION. With hoist raise transmission, using care not to damage spline shaft, until spline shaft can be inserted in clutch, pulling up on rope on auxiliary transmission to keep assembly righted and steady. If necessary, use screwdriver to turn transmission gears sufficiently to allow spline shaft to engage in clutch. Push transmission into place and install two nuts and lock washers on transmission pilot studs. If proper alinement cannot be obtained by using only the cable and rope, use jack and blocks under rear of transmission to aline.

(3) INSTALL BELL HOUSING CAP SCREWS. Install 12 cap screws and lock washers attaching bell housing and remove slings.

(4) INSTALL SHIFTER SHAFT COVER. Install eight cap screws and lock washers attaching shifter shaft cover to transmission.

(5) ATTACH AUXILIARY SHIFTER LEVER. Connect auxiliary shifter lever to shaft and install clevis pin and cotter pin.

TRANSMISSION

(6) INSTALL CLUTCH RELEASE LEVER. Tap clutch release lever onto shaft and apply bolt, nut, and lock washer.

(7) INSTALL CROSS BRACE. Install front running board cross brace by attaching with seven bolts, nuts, and lock washers on each side. Remove blocking from under battery tray. Install rear running board cross brace with six $\frac{3}{8}$-inch bolts, nuts, and lock washers.

(8) INSTALL SPEEDOMETER CABLE. Install speedometer cable in adapter at rear of transmission.

(9) INSTALL FLOOR BOARDS. Follow procedure outlined in paragraph 159 **b** (5) through (9).

(10) INSTALL FRONT PROPELLER SHAFT. Install front shaft as outlined in paragraph 95 a (2).

TM 9-818
94

10-TON 6 x 4 TRUCK (MACK MODEL NR)

Section XXI

PROPELLER SHAFTS

	Paragraph
Description and tabulated data	94
Propeller shaft assemblies	95
Center bearing assembly	96

94. DESCRIPTION AND TABULATED DATA.

a. Description (figs. 104 and 105). There are three propeller shafts composing the drive shaft line in this vehicle; the front shaft, rear shaft and inter-axle shaft. The front shaft consists of a tubular member, a nonslip type universal joint at the front end, and a stub shaft rearward end which projects through the self-alining, double-row ball bearing of the supporting center bearing. The rear and inter-axle shafts are also of the tubular type and each has one slip-type and one nonslip-type universal joint, the former being at the forward end of the shaft in each case. The slip-type of joint is for permitting variation of length between the connected units as caused by spring action.

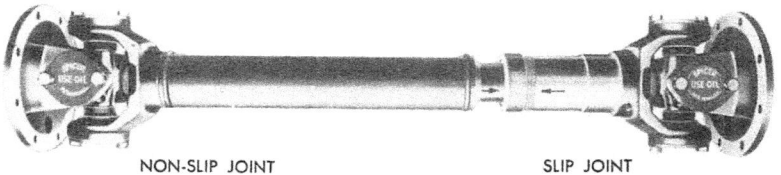

NON-SLIP JOINT   SLIP JOINT

RA PD 310614

Figure 104—Propeller Shaft

b. Data.
Front shaft:
    Make .... Spicer
    Series KR-1700
    Universal joint 1708
    Over-all length .... $36\tfrac{13}{16}$ in.
    Diameter (of tube) .... $3\tfrac{1}{2}$ in.
Rear shaft:
    Make .... Spicer
    Series KR-1700
    Universal joint (front, slip joint) 1701
    Universal joint (rear, fixed joint) 1708
    Over-all length .... $49\tfrac{3}{8}$ in.
    Diameter (of tube) .... $3\tfrac{1}{2}$ in.

TM 9-818
94-95

PROPELLER SHAFTS

Inter-axle shaft:
Make .... Spicer
Series KR-1600
Universal joint (front, slip joint) KRL-1601
Universal joint (rear, fixed joint) KRL-1608
Over-all length .... $32\frac{7}{8}$ in.
Diameter (of tube) .... 3 in.

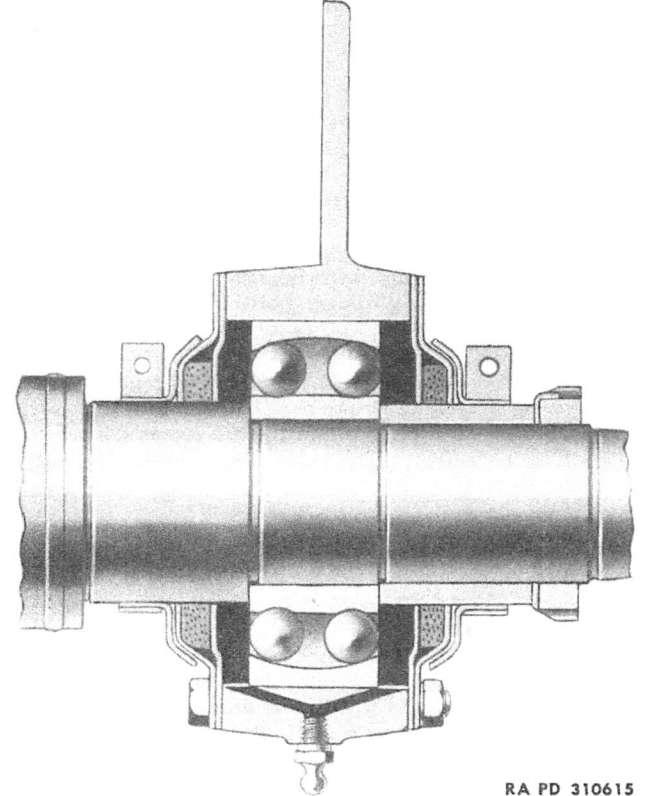

Figure 105—Cross Section of Center Bearing

95. PROPELLER SHAFT ASSEMBLIES.

a. **Front Propeller Shaft.**

(1) REMOVAL.

(a) Remove Bolts. After releasing hand brake lever, remove eight bolts, nuts, and lock washers attaching two propeller shaft flanges to brake disk, using two wrenches.

205

10-TON 6 x 4 TRUCK (MACK MODEL NR)

(b) *Loosen Brake Shoe Assemblies*. Loosen three cap screws and washers on each brake shoe anchor bracket assembly, and remove tension springs from lower corners of shoes.

(c) *Remove Hand Brake Disk*. Support brake disk and remove four bolts, nuts, and lock washers attaching bearing hanger to cross member socket and open end wrenches, and drop disk when free.

(d) *Remove Shaft* (fig. 104). Remove eight bolts, nuts, and lock washers from forward flange, supporting shaft while removing, and drop shaft to ground.

(e) *Remove Bearing* (fig. 105). If necessary to remove propeller shaft center bearing, follow procedure outlined in paragraph 96 a (2) through (4).

(2) INSTALLATION.

(a) *Install Bearing*. If center bearing has been removed above, install bearing assembly as outlined in paragraph 96 b (1) through (3).

(b) *Install Shaft*. Support shaft and install two bolts on opposite sides of forward flange, inserting from the front, to hold while installing hand brake disk.

(c) *Install Parking Brake Disk*. Support disk in position and place forward shaft in position, with bearing hanger at cross member. Insert propeller shaft bearing hanger bolts from front of cross member, with bearing housing behind cross member flange, and apply lock washers and nuts, tightening with socket and open-end wrenches.

(d) *Tighten Brake Shoe Assembly Bolts*. Tighten three cap screws with lock washers, attaching brake shoe anchor bracket assembly to cross member, and install tension springs on lower corners of shoes.

(e) *Install Bolts*. Install eight bolts, which attach propeller shaft flanges to brake disk, from the front of disk. Apply lock washers and nuts and tighten.

b. Rear Propeller Shaft.

(1) REMOVAL.

(a) *Remove Propeller Shaft Guards*. With socket and open-end wrenches, remove two bolts, lock washers, and nuts, which attach rear drive shaft guard to cross member, and remove guard. With socket and open-end wrenches remove four bolts, lock washers, and nuts which attach front guard to support and remove guard.

(b) *Remove Shaft*. Remove eight bolts from front and rear shaft flanges and remove shaft.

(2) INSTALLATION.

(a) *Check Shaft*. When installing shaft, be sure that the arrows on each side of the slip joint are alined.

(b) *Install Shaft*. Place shaft in position with slip joint toward front and install eight bolts in each flange, inserting from front. Apply lock washers and nuts and tighten.

(c) *Install Guards*. Attach forward guard to support with four bolts, lock washers, and nuts. Attach rear guard to cross member with two $\frac{1}{2}$-inch bolts, lock washers, and nuts.

PROPELLER SHAFTS

c. **Inter-axle Propeller Shaft.**

(1) REMOVAL. Remove eight bolts, nuts, and lock washers from each flange, using two wrenches, and remove shaft.

(2) INSTALLATION. When installing shaft, be sure that arrows on each side of the slip joint are alined. Place shaft in position with splined end forward and insert four bolts in each flange, from shaft side, in holes adjacent to sleeve yoke and four in each flange, from carrier side, in holes adjacent to flange yoke. Apply lock washers and nuts and tighten with two wrenches.

96. CENTER BEARING ASSEMBLY.

a. **Removal.**

(1) REMOVE FRONT SHAFT. Follow procedure outlined in paragraph 95 a (1) (a) through (d).

(2) REMOVE FLANGE. Place shaft in vise and remove cotter pin from flange end. Using socket wrench (41-W-3058-470), remove castle nut from end of shaft, and pull flange from shaft with puller. Remove key from shaft.

(3) REMOVE BEARING ASSEMBLY. Bend back tangs on locking washer and remove locking nut with spanner wrench (41-W-3250-252). Remove the locking washers and drop shaft end onto wooden block, using impact of shaft to jar bearing assembly loose. Remove bearing assembly. The forward slinger and clamp will remain on shaft.

(4) REMOVE BEARING. Remove four bolts, lock washers and nuts from housing and remove bearing felts, seal plates, and covers, as well as slinger, clamp and sleeve. Drive bearing out of housing with soft drift.

b. **Installation.**

(1) INSTALL BEARING. Place bearing in housing and apply end covers, felts, and felt seal plates, respectively, with four dowel pins in place, attaching to housing with four bolts, lock washers, and nuts.

(2) INSTALL BEARING ASSEMBLY (fig. 105). Place bearing assembly on drive shaft with machined face of housing toward front end of shaft. Drive into place with soft drift so that bearing is on high part of shaft and is seated against shaft body. Insert spacer sleeve and apply slinger and locking clamp. Tap slinger on each end, until snug against seal plate, and tighten clamp. Apply locking washers and adjusting nut. Tighten adjusting nut and bend tangs of washer over nut.

(3) INSTALL FLANGE. Place key in keyway and place flange on shaft, tapping into place. Apply castle nut and tighten. Install cotter pin.

(4) INSTALL SHAFT. Follow procedure outlined in paragraph 95 a (2) (b) through (e).

TM 9-818

10-TON 6 x 4 TRUCK (MACK MODEL NR)

Section XXII

FRONT AXLE

	Paragraph
Description and tabulated data	97
Wheel toe-in adjustment	98
Front axle assembly	99
Steering knuckle assemblies	100
Steering lever	101

97. DESCRIPTION AND TABULATED DATA.

   a. Description (fig. 106). There is no drive through the front axle. The axle is of the knuckle-steering design, with an I-section axle beam. Of reversed-Elliott design, the knuckles are bushed to turn on the knuckle pins which are taper-fitted and locked against turning in the ends of the axle beam. Wheel camber and caster are built-in characteristics, and toe-in is obtained by the adjustment of the tie-rod. The tie-rod has spring-loaded, ball-and-socket ends for connection to the cross steering levers and these connections do not require adjustment to compensate for wear. The right- and left-hand ends of the rod have, respectively, right- and left-hand threads; and it is not necessary to detach the ends to adjust the toe-in. Limiting stops prevent turning the wheels so far that the tires would rub on the frame.

   b. Data.

   Make .... Mack
   Model FA-25
   Type .... Non-driving, knuckle steer
   Design .... I-section axle beam,
                                reversed-Elliott type of knuckles
   Toe-in .... $\frac{1}{8}$ to $\frac{3}{16}$ in.
   Left-hand turning angle:
      Left wheel .... 31°
      Right wheel .... 26°
   Right-hand turning angle:
      Left wheel .... 27°
      Right wheel .... 32°

98. WHEEL TOE-IN ADJUSTMENT.

   a. Loosen Clamping Bolts. Loosen the two clamping bolts on each end of tie rod.

   b. Adjust Tie Rod.

   (1) Turn tie rod to adjust toe-in. The right end has a right-hand thread and the left end has a left-hand thread. Turning the top of the

TM 9-818
98-99

FRONT AXLE

rod toward the rear of the vehicle shortens the length. To measure toe-in, make a mark with chalk on the inside of each front tire, at the point of greatest section, at the hub level ahead of the axle. Measure the distance between the marks, then move the vehicle forward until marks are at the same height behind axle, and measure distance between marks. If second measurement exceeds the first by $\frac{1}{8}$ to $\frac{3}{16}$ inch, toe-in is correct. Measure toe-in with the wheels in straight-ahead position.

(2) If wheel-alining gage is available, use it for greater convenience and accuracy, to measure the toe-in by placing it between the tires behind the axle, with the chains just touching the floor and mov-

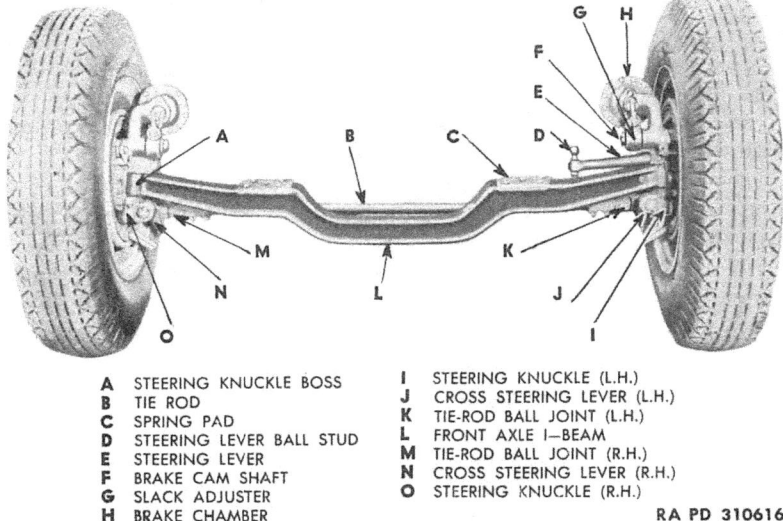

A STEERING KNUCKLE BOSS	I STEERING KNUCKLE (L.H.)
B TIE ROD	J CROSS STEERING LEVER (L.H.)
C SPRING PAD	K TIE-ROD BALL JOINT (L.H.)
D STEERING LEVER BALL STUD	L FRONT AXLE I—BEAM
E STEERING LEVER	M TIE-ROD BALL JOINT (R.H.)
F BRAKE CAM SHAFT	N CROSS STEERING LEVER (R.H.)
G SLACK ADJUSTER	O STEERING KNUCKLE (R.H.)
H BRAKE CHAMBER	RA PD 310616

Figure 106—Front Axle

ing vehicle back until gage is forward of axle and at same height as indicated by chains. The decreased reading of gage will indicate toe-in. Refer to step b (1) above for tolerance and method of adjustment.

c. **Tighten the Clamping Bolts.** When above procedure has been completed, tighten the two clamping bolts on each end of tie rod.

99. FRONT AXLE ASSEMBLY.

a. **Removal.**

(1) DETACH DRAG LINK. Remove cotter pin from drag link ball stud and remove nut. Turn wheel slightly to left. Place jack under stud and elevate slightly, striking drag link arm with hammer to free stud.

209

10-TON 6 x 4 TRUCK (MACK MODEL NR)

(2) DISCONNECT BRAKE HOSES. Detach brake chamber hoses at frame, using open-end wrench to hold fitting, and open-end wrench to turn.

(3) REMOVE SPRING CLIPS. Remove spring clip nuts by using open-end wrench with extension lever, long handle, or pipe slipped over wrench handle for leverage, on front nuts, and socket wrench with handle extension on rear nuts. Remove spacer washers from front clips. Tap spring clips upward with hammer to dislodge from axle, and move shock absorber linkage and bracket forward and out of the way. Return wheels to straight-ahead position.

(4) RAISE FRONT END. Place jack under each of two front spring rear hanger shock insulator caps and, using blocking to support chassis, elevate front end high enough for fenders to clear tires. If hoist is available, use the following procedure: Place sling or chain from hoist on two front tow hooks and take up slack. Raise front end with hoist, following elevation with jacks or blocking for safety, until fenders clear tires.

(5) REMOVE AXLE. Roll axle out from under truck, with man on each wheel to keep wheels running straight.

(6) REMOVE CASTER PLATES. Lift caster plates from axle.

(7) DETACH BRAKE HOSE. Remove brake hose and fitting from each brake chamber.

(8) REMOVE TIRES. If replacement axle does not already have tires mounted, remove rims with tires, following instructions in paragraph 148 a (2).

b. **Installation.**

(1) INSTALL TIRES. Mount tires and rims as instructed in paragraph 148 b (1).

(2) ATTACH BRAKE HOSES TO CHAMBERS. Attach air brake chamber hoses to chambers.

(3) INSTALL AXLE. Roll axle into position under truck, guiding wheels to keep them running straight. Place caster plates on axle spring pad, with thicker edge toward front. Lower truck onto axle, using pry-bar in spring pad hole to right axle while lowering. Line up holes in spring pad and replace spring clips in holes.

(4) INSTALL CLIPS. Place shock absorber brackets under front axle pads, with spacer washers between front bracket and axle, and with clips through holes. Replace nuts and tighten.

(5) REMOVE HOIST OR JACKS. Lower front end and remove hoist chain and jacks.

(6) ATTACH BRAKE HOSES AT FRAME. Attach brake chamber hoses at frame and tighten with one wrench holding fitting with another wrench, making sure the hose is not twisted or strained.

(7) CONNECT DRAG LINK. Place drag link stud in steering lever hole. Turn wheels to left to enable replacing nut on drag link stud. Tighten and replace cotter pin.

TM 9-818

FRONT AXLE

100. STEERING KNUCKLE ASSEMBLIES.

a. Removal.

(1) REMOVE FRONT WHEELS. Remove front wheels following procedure outlined in paragraph 130 a (1) through (5).

(2) DETACH BRAKE CHAMBER HOSE. Using an open-end wrench to hold the fitting at the frame, detach brake chamber hose, freeing fitting by use of another open-end wrench.

(3) REMOVE BRAKE SPIDER. Remove four nuts from brake spider bolts; holding bolts with one wrench, and turning nuts with another. Tap out bolts and remove spider.

(4) REMOVE SPACER. Grease retainer spacer will probably be loose on spindle. In case it is tight, dislodge with soft drift and remove spacer.

(5) DISCONNECT TIE ROD. Remove cotter pin from tie rod ball stud and remove nut from stud. Place jack under stud and elevate slightly, tapping tie rod arm to dislodge stud. Remove stud from arm.

(6) DISCONNECT DRAG LINK (left-hand only). Turn wheel slightly to left, remove cotter pin from drag link forward stud and remove nut. Place jack under stud and elevate slightly, tapping drag link arm to dislodge. Remove drag link stud from arm.

(7) REMOVE STEERING LEVER (left-hand only). Remove the steering lever by following procedure in paragraph 101 a (1).

(8) REMOVE CROSS STEERING LEVER. Follow procedure outlined in paragraph 101 b (1).

(9) REMOVE KNUCKLE PIN DUST CAP. Remove nut from knuckle pin dust cap clamping bolt. Remove bolt and pull off cap.

(10) REMOVE KNUCKLE PIN. Remove cotter pin at bottom of knuckle and remove adjusting screw. The thrust button will generally adhere to this screw. If it does not, remove it also. Use a heavy hammer on knuckle pin boss to dislodge knuckle pin. After several blows with hammer, tap top of pin to ease removal. Do not use sledge or other hammer on top of pin to drive it out. Remove knuckle pin.

(11) REMOVE KNUCKLE. Remove knuckle by pulling from axle boss.

(12) REMOVE GREASE FITTINGS. Remove two pressure lubrication fittings from knuckle.

(13) REMOVE STOP SCREW. Loosen nut and knuckle stop screw and remove screw and nut.

b. Installation.

(1) INSTALL NEW KNUCKLE. Place knuckle in position on axle knuckle boss, after making sure that hole and boss faces are clean.

(2) INSTALL KNUCKLE PIN. Insert knuckle pin from below, with smaller end toward top, and tap into place. Insert adjusting nut with thrust button in place on top of nut, putting a small quantity of grease on it. Turn nut into knuckle with heavy screwdriver and

10-TON 6 x 4 TRUCK (MACK MODEL NR)

tighten. Place jack under knuckle, elevate slightly and strike top of axle beam several heavy blows with hammer to lodge pin solidly in position. Leave jack in place with weight on the knuckle, check clearance between top face of axle and knuckle to insure 0.008 to 0.012 inch clearance, turning adjusting nut counterclockwise to increase, or clockwise to decrease clearance. Install cotter pin.

(3) INSTALL DUST CAP. Tap dust cap into place on top of knuckle pin with light hammer blows and insert clamping bolt and lock washers, making sure that bolt fits in groove on pin, and tighten.

(4) INSTALL GREASE FITTINGS. Place two grease fittings in holes in knuckle.

(5) INSTALL CROSS STEERING LEVER. To install tie rod arm to knuckle, follow procedure outlined in paragraph 101 b (2).

(6) INSTALL STEERING LEVER. Follow procedure outlined in paragraph 101 a (2).

(7) CONNECT DRAG LINK. Place drag link stud in drag link arm and apply 1-inch nut Tighten nut and install cotter pin.

(8) CONNECT TIE ROD. Place tie rod ball stud in tie rod arm, and install nut. Tighten and install cotter pin.

(9) INSTALL BRAKE SPIDER ASSEMBLY. Place brake spider assembly in position on knuckle, and insert bolts from outer side; install and tighten lock washers and nuts on inside.

(10) CONNECT BRAKE CHAMBER HOSE. Attach brake chamber hose to fitting on frame and tighten connection, making sure there is no strain or twist on the hose.

(11) INSTALL GREASE RETAINER SPACER. Place grease retainer spacer on spindle, flat side out with pin in hole, and tap home.

(12) INSTALL WHEEL. Install wheel and adjust brakes, following procedure outlined in paragraph 130 b (5) through (10).

(13) INSTALL KNUCKLE STOP SCREW. Place knuckle stop screw in knuckle, locking nut turned back to head, and turn into knuckle until angle of wheel with frame, stop against axle, is 27 degrees for left wheel and 26 degrees for right wheel.

(14) ALINE WHEELS. Follow procedure outlined in paragraph 98.

101. STEERING LEVER.

a. Replace Steering Lever.

(1) REMOVAL. Turn front wheels to left. Remove cotter pin from forward drag link stud and remove nut. Place jack under stud and exert slight pressure only to ease removal of stud from steering lever. Use hammer on steering lever to shock stud free. Remove cotter pin from steering lever end, and remove nut with socket wrench. Place jack under center of axle and turn wheels to right. Place pry-bar between the axle and lever, lower jack and use hammer on knuckle to shock stud free. Tap out lever. Remove key by gripping key in vise and pulling arm endwise away from key.

TM 9-818
101

FRONT AXLE

(2) INSTALLATION. Tap key solidly into lever. Replace lever in knuckle, with key in keyway, and tap into place. Replace nut, tighten, and install cotter pin. With jack under axle center, elevate slightly, line up wheels and place drag link stud in lever. Place nut on drag link stud and tighten. Replace cotter pin. Lower jack and remove.

b. Replace Cross Steering Levers (right and left sides).

(1) REMOVAL. Remove cotter pin from tie rod stud and loosen nut on stud. Place jack under tie rod close to ball stud and use hammer on arm to shock stud free. Remove nut and remove tie rod stud from lever. Remove cotter pin from knuckle end of lever. Remove nut. Use hammer on knuckle to shock lever free. Place key in vise and pull lever to remove key.

(2) INSTALLATION. Tap key into place in lever. Place lever in knuckle with key in keyway and tap with hammer. Install nut, tighten and install cotter pin. Place tie rod stud in lever, making sure felt and metal washers are in proper place on stud. Place nut on stud, tighten and install cotter pin.

10-TON 6 x 4 TRUCK (MACK MODEL NR)

Section XXIII
BOGIE (REAR AXLE ASSEMBLY)

	Paragraph
Description and tabulated data	102
Bogie assembly	103
Torque rod assemblies	104
Rear spring seat assembly	105

102. DESCRIPTION AND TABULATED DATA.

a. **Description** (fig. 107). The two rear axles are attached by rubber shock insulators to the ends of the two longitudinal springs, and these are mounted at their centers on a transverse trunnion tube on which they are free to oscillate. A wide bracket on each frame rail secures axle assemblies to the chassis frame.

b. **Data.**

Attachment of axles	At spring ends, rubber shock insulators with positive locks
Type of springs	Inverted semi-elliptic, leaf, center-mounted on trunnion
Assembly arrangement	Axles held in assembly through springs
Inter-axle drive	Shaft with two universal joints

103. BOGIE ASSEMBLY.

a. **Removal.**

(1) BLOCK FRONT WHEELS. Place blocks under front wheels at both front and rear to prevent movement of truck.

(2) DISCONNECT FRONT DRIVE SHAFT. Using two wrenches, one to turn and the other to hold, remove eight bolts from front propeller shaft flange, and drop shaft.

(3) DISCONNECT AIR LINES. Remove flexible air lines from junction fittings on each axle housing, using an open-end wrench to hold fitting and another to turn.

(4) DISCONNECT TORQUE RODS. Remove torque rod studs from axle carrier bosses as outlined in paragraph 104 a (1) (c).

(5) REMOVE REAR SPLASH GUARDS. Remove nuts from the four carriage bolts securing rear splash guards, and from two mounting bracket bolts on each side. Remove splash guards.

(6) REMOVE TRUNNION MOUNTING BOLTS. Remove four nuts from trunnion bearing bracket bolts on each side, using a wrench to hold bolt heads. Remove bolts, nuts and lock washers.

(7) RAISE TRUCK. Place chain or sling around rear tow hooks and attach to hoist or overhead crane. Lift the chassis to clear bogie mounting and carriers when rolling bogie out. Place blocking under chassis flange, as close to the rear axle installation as possible, as

BOGIE (REAR AXLE ASSEMBLY)

a safety precaution. If no crane or hoist is available, lift rear of chassis by using blocks under flange ends, using same blocking in front of axle to facilitate rolling out bogie.

(8) REMOVE BOGIE. Roll bogie out from under the truck (fig. 108).

(9) REMOVE AIR LINES AND FITTINGS. Remove cap screws on axle and remove fitting and bracket. Remove air line and 45-degree elbow at brake chambers. Remove two clamps attaching air line to each axle, using screwdriver and open-end wrench.

RA PD 310617

Figure 107—Bogie Assembly

(10) REMOVE INTER-AXLE PROPELLER SHAFT. Remove eight bolts, nuts and lock washers from each flange of inter-axle propeller shaft. Use two wrenches, one to turn and one to hold and remove shaft.

(11) REMOVE TIRES. Remove tires and rims, following procedure outlined in paragraph 148 a.

b. Installation.

(1) INSTALL TIRES. Mount tires and rims on wheels as outlined in paragraph 153 b.

(2) INSTALL AIR LINES, ELBOWS AND FITTINGS. Install air lines by attaching with two clamps on each axle and fasten with cross-recessed head screws, using screwdriver and open-end wrench to tighten. Install the 45-degree elbows on brake air chambers. Attach brake lines at brake-chamber elbows. Place cap screw through

10-TON 6 x 4 TRUCK (MACK MODEL NR)

bracket attaching Y-fitting on each axle and install in carrier housing with lock washer in place. Tighten securely.

(3) INSTALL INTER-AXLE PROPELLER SHAFT. With slip joint forward, mount inter-axle propeller shaft, attaching to carrier flanges with eight bolts. Apply lock washers and nuts and tighten, using two wrenches, one to turn and one to hold.

(4) INSTALL BOGIE. Roll bogie into position under truck and lower chassis onto bogie, alining trunnion mounting bracket with drift pins. Install four trunnion bracket mounting bolts on each bracket. Apply lock washers and nuts and tighten, using socket wrench while holding bolts with open-end wrench.

(5) INSTALL TORQUE RODS. Connect torque rods to carrier housings as outlined in paragraph 104 b (1) and (2).

(6) CONNECT AIR LINES. Attach the air lines to the Y-fittings on each carrier, using one wrench to turn fitting and another to hold.

(7) ATTACH FRONT PROPELLER SHAFT. Using two wrenches, one to turn and one to hold, attach forward propeller shaft to power-divider flange with eight bolts, nuts and lock washers.

(8) INSTALL SPLASH GUARDS. Attach splash guards to second body cross beam from the rear with four carriage bolts, and attach brackets to next forward cross beam with two carriage bolts. Install and tighten lock washers and nuts.

104. TORQUE ROD ASSEMBLIES.

a. Removal.

(1) REMOVE AIR LINE. Detach air line from two clamps on torque rod.

(2) PLACE JACK. To ease strain on rod assemblies in removal of forward rod, place jack under power divider housing and elevate slightly. For the rear rod removal, place jack under projection below filler plug and elevate slightly.

(3) DISCONNECT RODS. Remove the cotter pins on end of torque rod ball stud and remove castle nuts. Hammer on ball stud bosses until studs are loose. Use jacks, if available, to force stud from boss with impact on boss, being careful not to damage air lines or wires on left-hand frame member.

(4) REMOVE RODS. Remove torque rods with spring and stud washers and remove spring from stud (fig. 109).

b. Installation.

(1) INSTALL ROD. Place spring on ball stud, with larger side toward rod, and place studs in bosses on cross member and axle with air line clamp clips upward.

(2) INSTALL NUTS. Place two castle nuts on studs, tighten and install cotter pins.

c. Service. Maintain tightness of ball sockets and spring tension by removing cotter pin on end of torque rod and turning end plug clockwise to tighten. Removal of the rod is necessary to perform

TM 9-818
104-105

BOGIE (REAR AXLE ASSEMBLY)

this operation on the cross member end of the rod. See step a (4) above.

105. REAR SPRING SEAT ASSEMBLY.

a. **Removal.**

(1) PLACE JACK. Place jack under trunnion bracket, and elevate enough to take some weight off spring.

(2) REMOVE CLIPS. Follow procedure outlined in paragraph 141 a (4) to remove spring clips.

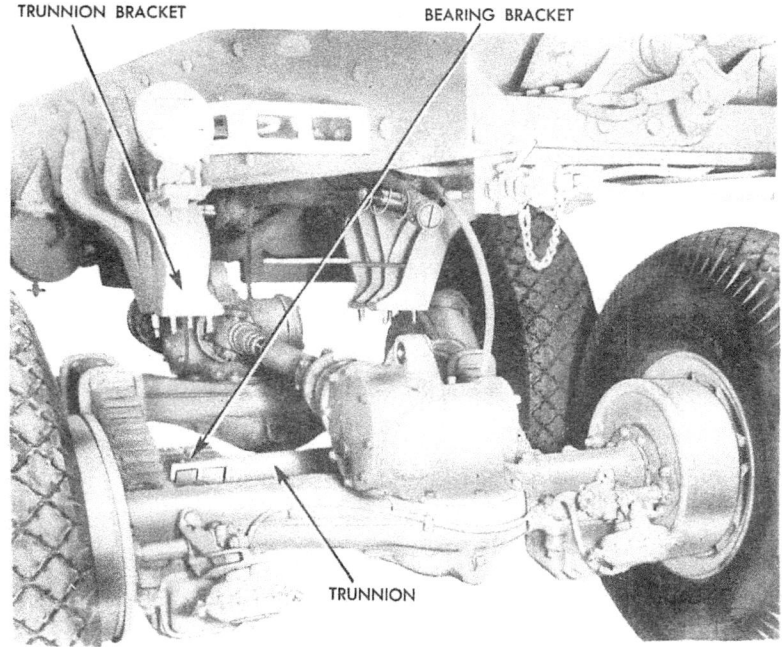

Figure 108—Bogie Cleared for Rolling Out

(3) JACK UP AXLES. Jack up both rear axles with jacks under shock insulator lower caps until spring clears seats.

(4) REMOVE TRUNNION CAP. Remove trunnion cap by removing remaining four bolts with two wrenches, using one to turn and one to hold. Place a container to catch oil draining out of the trunnion bearing.

(5) REMOVE ADJUSTING NUTS AND WASHERS. Straighten tang of locking washer and remove jam nut with socket wrench (41-W-639-390). Remove locking washer and slotted washer.

(6) REMOVE ADJUSTING NUT. Tap adjusting nut with soft drift and hammer to loosen and remove, using wrench (41-W-639-390).

TM 9-818
10-TON 6 x 4 TRUCK (MACK MODEL NR)

(7) REMOVE SPRING SEAT ASSEMBLY. Pull seat assembly off trunnion shaft.

b. **Installation.**

(1) INSTALL SEAT ASSEMBLY. Place spring seat assembly on trunnion shaft, making certain seal and trunnion washer are in place in seat. Engage dowel pin in oil-seal spacer on trunnion shaft in hole in trunnion washer. Apply adjusting washer and tighten. Mount slotted washer and mark so that adjusting washer can be backed off

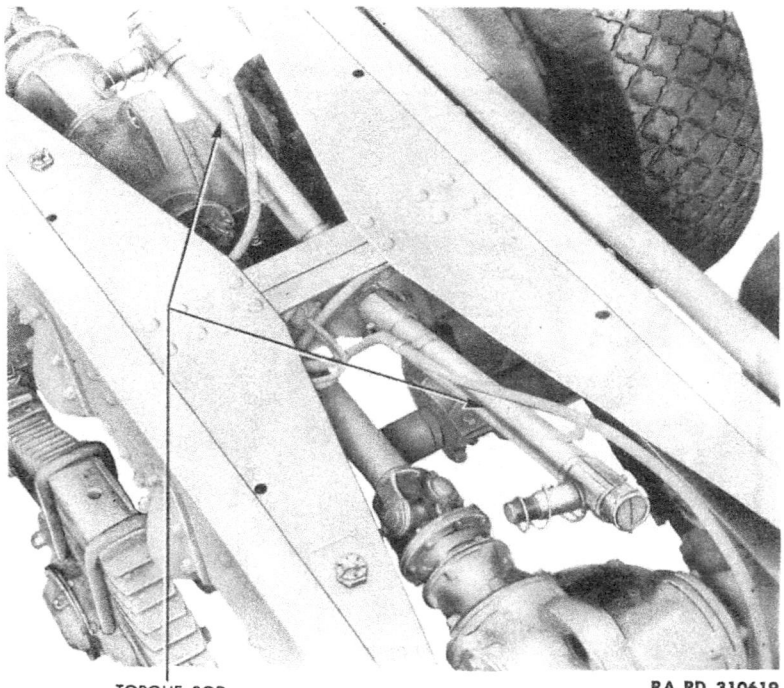

TORQUE ROD
RA PD 310619

Figure 109 — Torque Rods Disconnected

three notches, then reapply slotted washer and locking washer with tang toward outside, and turn on locking nut. Tighten with socket wrench (41-W-639-390) and bend over tang on locking washer.

(2) INSTALL TRUNNION CAP. Install trunnion bearing cap and gasket with three upper and one bottom bolts, making sure that filler hole in cap is at top, and tighten, using two wrenches, one to turn and one to hold.

(3) LOWER WHEELS. Lower each wheel to ground and remove jacks.

(4) INSTALL SPRING CLIPS. Raise jack under trunnion shaft bracket, and install spring clips as outlined in paragraph 141 b (5).

(5) REMOVE JACK. Remove jack from under trunnion bracket.

TM 9-818
106-107

Section XXIV

REAR AXLES

	Paragraph
Description and tabulated data	106
Forward rear axle assembly	107
Rearward rear axle assembly	108
Axle shafts	109

106. DESCRIPTION AND TABULATED DATA.

a. Description (fig. 10). Drive of the truck is through the two rear axles which, in assembly with the suspension parts, constitute the rear axle assembly, generally known as the bogie. The two axles are the same except that the gear carrier of the forward one has the through drive for the second axle, and carries the power divider unit incorporated with the through drive. A shaft with universal joints transmits the power from the forward axle to the rearward one. The axles are of the full-floating type, with flat-type banjo and top-mounted gear carrier, within which is the double-reduction gear set. Being an in-line arrangement, the power divider is incorporated as a quill and shaft drive, with the carrier of the forward rear axle.

b. Data.

Make	Mack
Model:	
Forward carrier	CR-32
Rearward carrier	CR-33
Type	Full-floating, double reduction (dual reduction)
Gears:	
First reduction	Spiral bevel
Second reduction	Straight spur
Ratio	9.02 to 1.00

107. FORWARD REAR AXLE ASSEMBLY.

a. Removal.

(1) PLACE BLOCK. Place 4-inch block between the rearward rear axle housing and the axle bumper on each side (fig. 110).

(2) DISCONNECT PROPELLER SHAFTS. Follow procedure outlined in paragraph 95 b (1) and c (1) to disconnect rear and inter-axle propeller shafts at flanges on power divider and carrier respectively.

(3) DISCONNECT TORQUE ROD. Follow procedure outlined in paragraph 104 a (1) (c) to disconnect torque rod at carrier only.

(4) DISCONNECT AIR LINES. Remove flexible air lines from junction fitting at axle housing, using one open-end wrench to hold the fitting and another to turn.

10-TON 6 x 4 TRUCK (MACK MODEL NR)

(5) ADJUST SLING. Place a rope over body sills directly above power divider, with a loop down around power divider at forward flange to prevent axle from turning over when disconnected from spring ends.

(6) REMOVE TIRES AND RIMS. Remove tires and rims as outlined in paragraph 148 **a**, and rest wheels on block approximately 8 inches high.

(7) REMOVE SHOCK INSULATORS. Follow procedure outlined in paragraph 141 **a** (2), (3) and (5) to remove cap and insulators from axle housing.

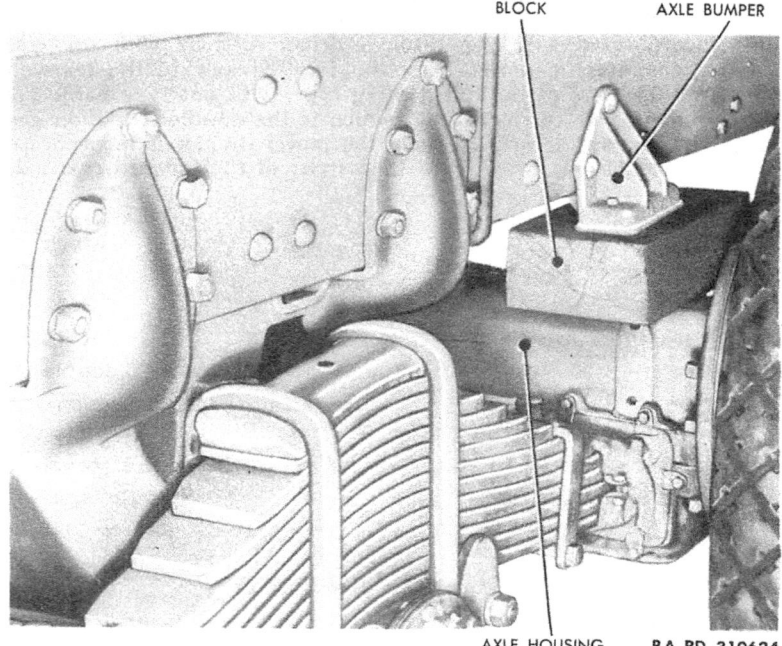

Figure 110—Block Inserted Between Housing and Bumper

(8) REMOVE AXLE. With sling around power divider at forward flange, allow axle to turn over and lower power divider. Support axle, remove blocking and lower axle to ground. Remove from either side (fig. 111).

(9) REMOVE AIR LINES AND FITTINGS. Remove cap screw from housing and remove fitting and bracket. Remove air lines and 45-degree elbows from air brake chambers. Remove two clamps attaching air line to axle housing, using screwdriver and wrench.

TM 9-818
107

REAR AXLES

b. Installation.

(1) INSTALL AIR LINES AND FITTINGS. Install air lines by attaching two clamps to axle housing with cross-recessed head screws, using cross-recessed screwdriver and wrench to tighten. Install the 45-degree elbows in brake air chambers, using wrench to tighten. Attach air lines to brake chamber elbows. Place cap screw through bracket attaching Y-junction fitting to axle housing, and install with lock washer in place. Tighten with socket wrench.

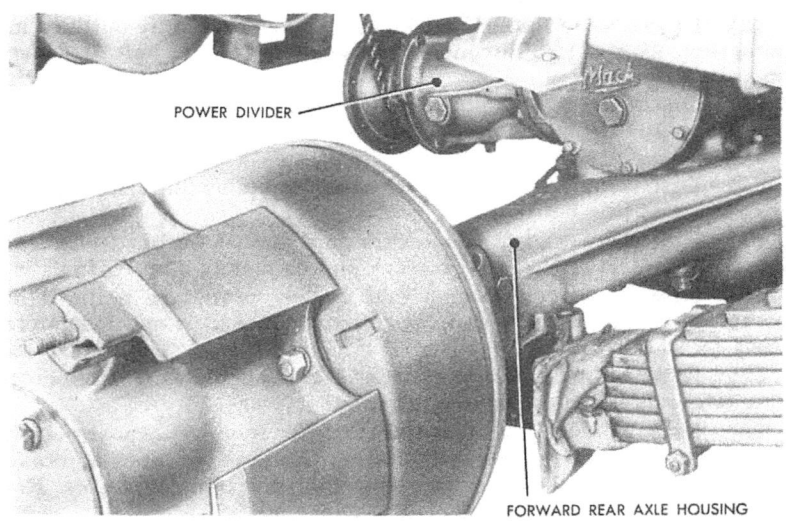

Figure 111—Removing Forward Rear Axle

(2) INSTALL AXLE. Place axle in position under chassis with power divider forward and elevate so that it may be placed on spring ends, guiding axle with rope around power divider flange to steady it.

(3) INSTALL SHOCK INSULATORS AND CAPS. Follow procedure outlined in paragraph 141 b (2), (3) and (4) to install shock insulators and caps, excepting that it is not necessary to jack up the chassis but only to drop the axle housing on the spring ends.

(4) INSTALL TORQUE ROD. Follow procedure outlined in paragraph 104 a (2).

(5) INSTALL PROPELLER SHAFTS. Follow procedure outlined in paragraph 95 b (2) and c (2) to connect propeller shafts at power divider and carrier flanges.

(6) CONNECT AIR LINES. Attach the air lines to the Y-junction fittings on axle housing, using one wrench to turn fitting and another to hold.

10-TON 6 x 4 TRUCK (MACK MODEL NR)

(7) INSTALL TIRES. Follow procedure outlined in paragraph 148 b to install tires and rims.

(8) LUBRICATION. Refer to section 19 c (6) for amount and type of lubricant. Remove block from between rear axle housing and bumper, and adjust brakes at slack adjuster as outlined in paragraph 111 a.

108. REARWARD REAR AXLE ASSEMBLY.
a. Removal.

(1) PLACE BLOCK. Place 4-inch block between front rear axle housing and axle bumper on each side.

(2) DISCONNECT PROPELLER SHAFT. Follow procedure outlined in paragraph 95 c (1) to disconnect inter-axle propeller shaft at flange on carrier.

(3) DISCONNECT TORQUE ROD. Follow procedure outlined in paragraph 104 a (1) *(c)* and *(d)* to disconnect torque rod at carrier only.

(4) DISCONNECT AIR LINES. Remove flexible air lines from junction fitting at axle housing.

(5) REMOVE TIRES. Remove tires and rims as outlined in paragraph 153 a, and rest wheels on block approximately 8 inches high.

(6) REMOVE SHOCK INSULATORS. Follow procedure outlined in paragraph 141 a (2), (3) and (5) to remove the caps and insulators from the axle housing.

(7) REMOVE AXLE. Support axle, removing blocking, and drop axle to ground. Remove from either side.

(8) REMOVE AIR LINES AND FITTINGS. Remove cap screw on housing and remove fitting and bracket. Remove air lines at brake chamber and remove 45-degree elbow from air brake chamber. Remove two clamps attaching air line to axle housing.

b. Installation.

(1) INSTALL AIR LINES AND FITTINGS. Install air lines by attaching two clamps to axle housing with cross-recessed head screws. Install the 45-degree elbows in brake air chambers. Attach brake lines at brake chamber elbows. Place cap screw through bracket attaching Y-junction fitting to axle housing and install with lock washer in place.

(2) INSTALL AXLE. Place axle in position under chassis and elevate so as to place axle on spring ends, with upper shock insulators in sockets.

(3) INSTALL SHOCK INSULATORS AND CAPS. Follow procedure outlined in paragraph 141 b (2), (3) and (4) to install shock insulators and caps, except that it is not necessary to jack up chassis but only to drop axle housing on spring ends.

(4) INSTALL TORQUE ROD. Follow procedure outlined in paragraph 106 a (2) *(a)*.

(5) INSTALL PROPELLER SHAFT. Follow procedure outlined in paragraph 95 c (2) to connect propeller shaft at carrier flange.

REAR AXLES

(6) CONNECT AIR LINES. Attach air lines to Y-junction fittings on axle housing, using two wrenches.

(7) INSTALL TIRES. Follow procedure outlined in paragraph 149 b (2) to install tires on rims.

(8) LUBRICATION. See paragraph 19 c (6) for amount and type of lubricant to furnish. Remove block from between front rear axle housing and bumper, and adjust brakes at slack adjusters as outlined in paragraph 111 a.

109. AXLE SHAFTS.

a. **Removal.** Remove six cap screws and lock washers from each hub cap, and remove hub caps. Pull out axle shafts.

b. **Installation.** Place axle shaft in housing with end marked "OUT" to the outside. The shorter shaft belongs in the right-hand, and the longer in left-hand side. Place gasket and hub cap on wheel, install six cap screws and lock washers and tighten.

10-TON 6 x 4 TRUCK (MACK MODEL NR)
Section XXV

BRAKE SYSTEM

	Paragraph
Description and tabulated data	110
Front brake shoe assemblies	111
Rear brake shoe assemblies	112
Air compressor assembly	113
Air pressure governor	114
Front brake chambers	115
Rear brake chambers	116
Slack adjuster assembly	117
Brake application valve	118
Relay valve	119
Quick release valve	120
Safety valve	121
Pressure indicator	122
Air supply valve	123
Flexible air hose, lines and connections	124
Air reservoirs	125
Parking brake shoe assemblies	126
Parking brake disk	127
Parking brake controls and linkage	128

110. DESCRIPTION AND TABULATED DATA.

a. **Description.**

(1) SERVICE BRAKES (figs. 14, 15, and 112). All six wheels have air-actuated brakes. These are internal, two-shoe brakes, each operated by an axle-mounted air chamber through a slack adjuster on the brake camshaft. The air system includes reservoirs and automatic control and governing units. There are couplings for making connections to a towed or towing vehicle. In addition to pressure gage, on instrument panels there is a low-pressure buzzer alarm. The two-cylinder air compressor is flange-mounted on the engine with built-in drive, and is water-cooled and pressure-lubricated from the engine systems. The compressor is governed by the air-governor control of its "unloading" type of head. Brake adjustment is by means of the several slack adjusters.

(2) PARKING BRAKE (fig. 16). The brake for parking is of the disk type, with four brake shoes. It is on the propeller shaft adjacent to the center bearing, and the cross member carries the shoe assembly. Application of the brake is by a hand lever in the cab through a rod and lever linkage. Adjustment is at the brake shoes.

BRAKE SYSTEM

b. Data.

(1) SERVICE BRAKES.

Make	Mack
Type	Internal-expanding, two-shoe
Number	Six
Actuation	Westinghouse, air
Size:	
Front	17¼ x 4 in.
Rear	17¼ x 5 in.

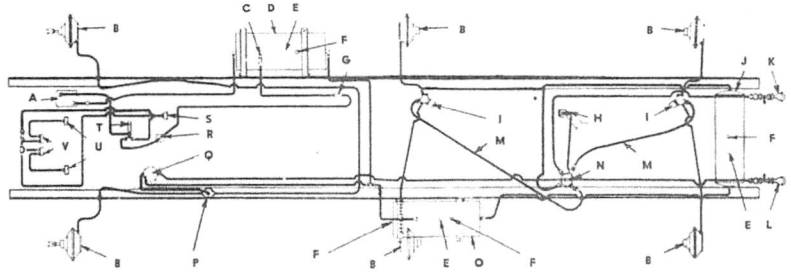

A	AIR COMPRESSOR	L	SERVICE COUPLING
B	BRAKE CHAMBER	M	FLEXIBLE HOSE
C	SAFETY VALVE	N	RELAY VALVE
D	AIR RESERVOIR NO. 1	O	AIR RESERVOIR NO. 2
E	RESERVOIR DRAIN COCK	P	QUICK RELEASE VALVE
F	PIPE PLUG	Q	BRAKE APPLICATION VALVE
G	AIR SUPPLY VALVE	R	LOW PRESSURE INDICATOR
H	STOP LIGHT SWITCH	S	AIR PRESSURE GAGE
I	TEE-FITTING	T	AIR PRESSURE GOVERNOR
J	AIR RESERVOIR NO. 3	U	WINDSHIELD WIPER
K	EMERGENCY COUPLING	V	WINDSHIELD WIPER VALVE

RA PD 332262

Figure 112—Service Brake System

Lining:	
Front	Moulded, ¾ in. thick
Rear	Moulded, ¾ in. thick
Adjustment	Slack adjuster
Brake drums.	
Make	Mack
Type	Cast
Brake surface nominal diameter, front	17¼ in.
rear	17¼ in.

10-TON 6 x 4 TRUCK (MACK MODEL NR)

Air compressor:
  Make .... Westinghouse
  Model .... 2UE 7¼ FW
  Type .... Two-cylinder, unloading-head
  Mounting .... Flange
  Cooling .... Water (from engine)
  Lubrication .... Pressure (from engine)

Air reservoirs:
  Shape .... Cylindrical
  Number .... Three
  Size, small (one) .... 8 x 26 in.
        large (two) .... 9½ x 27 in.

Brake chambers:
  Make .... Westinghouse
  Size, front .... 7 in.
        rear .... 9 in.
  Minimum stroke, front .... ⅝-¾ in.
                  rear .... ¾-⅞ in.
  Maximum stroke, front .... 1¾ in.
                  rear .... 2¼ in.
  Maximum recommended stroke, front ..... 1⅜ in.
                              rear ...... 1¾ in.
  Slack adjusters, front .... 5 in.
                   rear .... 5 in.

Application valve:
  Make .... Westinghouse
  Type .... B-48 (pedal on top of valve)

Trailer air connections:
  Make .... Westinghouse
  Type .... Self-locking and quickly detachable
  Number .... Two (service and emergency)
  Location .... Rear end of chassis frame

(2) PARKING BRAKE.
  Make .... American Chain & Cable Co.
  Model 65D
  Type .... Four-shoe disk
  Location .... At propeller shaft center
  Size .... 16-in. diameter
  Lining type .... Moulded block
  Lining width .... 3 in.
  Lining thickness .... ¼ in.

TM 9-818
111

BRAKE SYSTEM

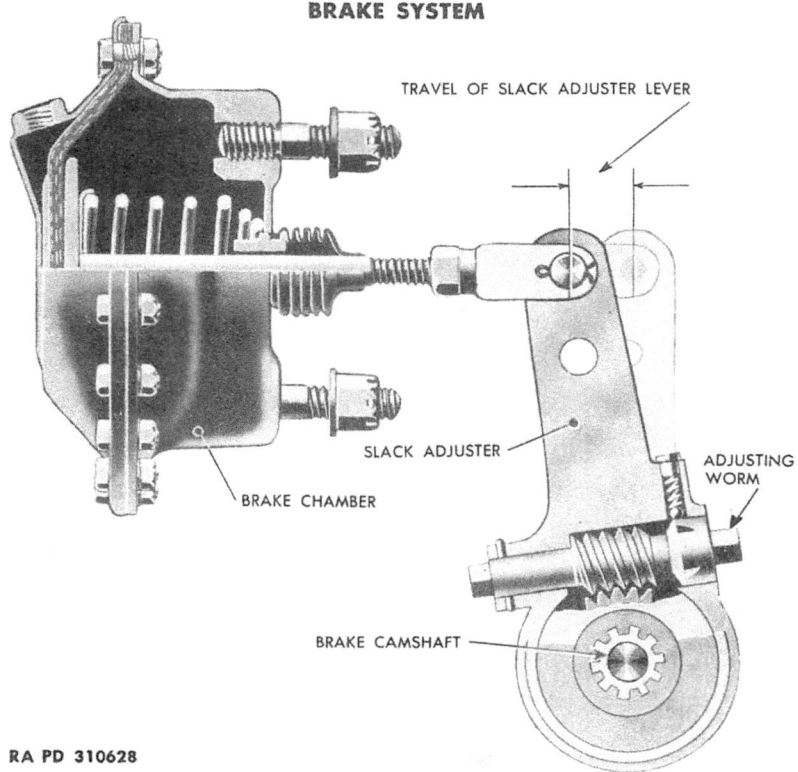

Figure 113—Brake Chamber and Slack Adjuster

111. **FRONT BRAKE SHOE ASSEMBLIES.**

a. **Adjustment.** Turn slack adjusting worm (fig. 113) until lining is tight against drum, then back off three notches. This will provide sufficient clearance to prevent dragging brakes. Make similar adjustment at all four wheels.

b. **Removal.**

(1) REMOVE WHEEL. Follow procedure as outlined in paragraph 130 a (1) through (5).

(2) REMOVE BRAKE ANCHOR PIN LOCK SCREW. Using socket wrench, remove cap screw from brake anchor pin lock.

(3) REMOVE UPPER OIL CAP. Using adjustable open-end wrench, remove oil cup from upper anchor pin by turning counterclockwise.

(4) REMOVE ANCHOR PIN LOCK. Drive out anchor pin lock with brass drift.

(5) REMOVE LOWER OIL CUP. Turn lower oil cup counterclockwise to loosen. Because it is impossible to remove the lower oil cup with pin in place, drive out lower pin far enough to permit removal of oil cup and remove cup.

TM 9-818

10-TON 6 x 4 TRUCK (MACK MODEL NR)

(6) REMOVE ANCHOR PINS. Using drift and hammer, drive out the two brake-anchor pins far enough to free the brake shoes, retaining pin in outer spider mounting for greater ease in alining pin lock when reinstalling.

(7) REMOVE SHOE ASSEMBLIES. Raising up brake shoes from each side so that eye clears spider boss, remove shoes.

(8) REMOVE SPRINGS. Place shoe assemblies on floor and detach retracting springs.

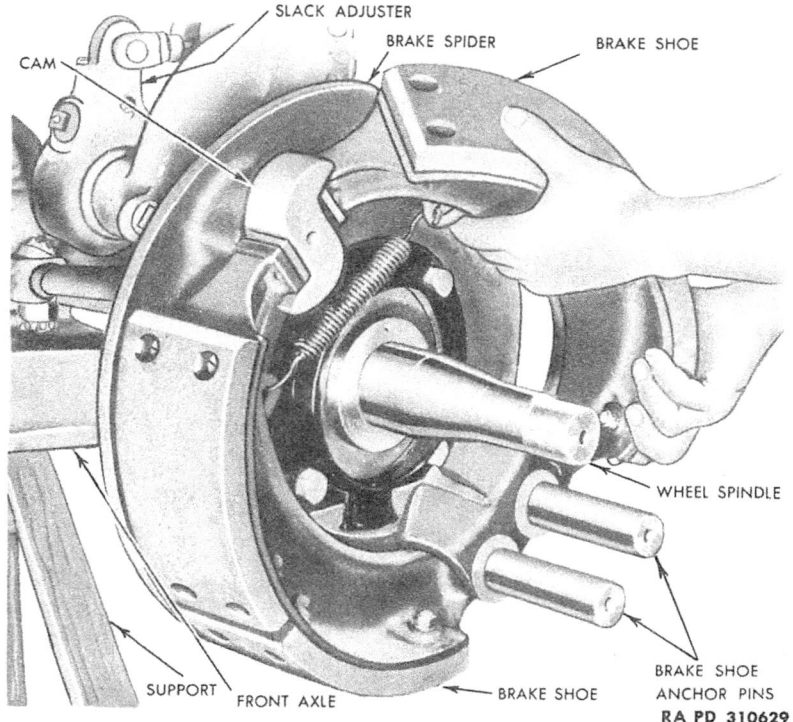

Figure 114—Placing Front Brake Shoe on Spider

c. Installation.

(1) INSERT SPRINGS. Place new shoe assemblies on floor with flat wear plates together at same end, and insert spring hooks in eyes at wear plate end.

(2) PLACE SHOES ON SPIDER. Lift each shoe assembly with spring in position, and place lower shoe with wear plate on cam and eye in spider-boss opening. Tilt upper shoe and lift into place on upper side of spider. The spring should hold wear plates snugly against the flat of brake cam (fig. 114).

TM 9-818
111-112

BRAKE SYSTEM

(3) INSTALL ANCHOR PINS. Alining holes in spider bosses with eyes on shoe assemblies from inside of spider, drive anchor pins into shoe eyes enough to engage shoes. Drive each upper pin into position.

(4) INSTALL LOWER OIL CUP. With lower pin partially driven into position, place lower oil cup on inner end of pin, and turn partially into place. Drive pin into position.

(5) ALINE ANCHOR PINS. Line up anchor pins for locking by placing bolts in outer holes and turning clockwise to turn pin and aline grooves on inner end for locking plate.

(6) INSTALL LOCK. Place locking plate with ends in anchor pin grooves, install cap screw and lock washer and tighten.

Figure 115—Rear Brakes

RA PD 310630

(7) INSTALL UPPER OIL CUP. Place upper oil cup in upper anchor pin, and turn into place. Tighten both oil cups so that they are up in final position.

(8) INSTALL WHEELS. Install wheels and adjust brakes as outlined in paragraph 130 b (5) through (10).

112. REAR BRAKE SHOE ASSEMBLIES (fig. 115).

a. **Adjustment.** Follow procedure outlined in paragraph 111 a.

b. **Removal.**

(1) REMOVE WHEEL. Follow procedure outlined in paragraph 134 a (1) through (7), to remove wheel. It is unnecessary to remove the tire in this case.

(2) REMOVE ANCHOR PIN LOCKING PLATE. Remove two cap screws and lock washers from brake-anchor pin locking plate and remove plate.

10-TON 6 x 4 TRUCK (MACK MODEL NR)

(3) REMOVE UPPER BRAKE SHOE. With pry-bar, pry brake shoe eye off anchor pin. Lift shoe and disengage spring. Remove shoe.

(4) REMOVE LOWER BRAKE SHOE. Remove lower shoe by prying off anchor pin with pry-bar.

c. Installation.

(1) INSTALL NEW SHOES. Place lower shoe with eye on anchor pin and tap into place. Connect spring to hook at wear plate end and connect other end of spring to upper shoe hook. Drop upper shoe into position and place on anchor pin.

(2) ANCHOR PIN ADJUSTMENT. Loosen each anchor pin nut with one wrench, holding the pin with another wrench. Turn both anchor pins, which are eccentric, to effect the greatest clearance between the shoes and the drum.

(3) INSTALL WHEEL. Follow procedure outlined in paragraph 134 b (4) through (7) only as far as replacing the bearing adjusting nut.

(4) ANCHOR PIN ADJUSTMENT. Turn the upper anchor pin so as to bring brake shoe against drum and back off until the drum is freed. Repeat for lower pin. Tighten anchor-pin nuts with one wrench, holding eccentric pins with another wrench to maintain adjustment.

(5) COMPLETE WHEEL INSTALLATION. Complete installation of wheel and adjust brakes as outlined in paragraph 134 b (8) through (10).

113. AIR COMPRESSOR ASSEMBLY.

a. Service.

(1) AIR CLEANER. Check the compressor air cleaner every inspection period or oftener if dusty atmosphere prevails. Wash filter in solvent and blow dry with air hose. Fill to the bead with engine oil upon reinstallation.

(2) VALVE CLEARANCE. Discharge valve lift should be 0.042 to 0.075-inch. To make valve adjustment remove dust cap and felt as outlined below, and use small screwdriver and wrench to make adjustment of the tappet screws.

b. Removal.

(1) DRAIN WATER. Open drain valve in radiator outlet pipe and drain water from cooling system.

(2) DISCONNECT LINES. Use two wrenches to disconnect flexible air line leading from the compressor to the reservoir line. Remove two nuts and lock washers from compressor discharge elbow flange with socket wrench, and remove discharge elbow and line. Disconnect line to governor at elbow.

(3) REMOVE OIL FILLER SPOUT. Remove the three cap screws and lock washers attaching oil filler-spout elbow to engine and remove spout.

(4) REMOVE COMPRESSOR AIR CLEANER ASSEMBLY. Remove wing

BRAKE SYSTEM

nut from top of air cleaner assembly, and remove cleaner. Remove the two cap screws and lock washers from the air-cleaner mounting bracket with socket wrench. Loosen hose clamp at cleaner fitting, and remove hose and bracket.

(5) DISCONNECT WATER LINES. Disconnect the two water lines at the elbows.

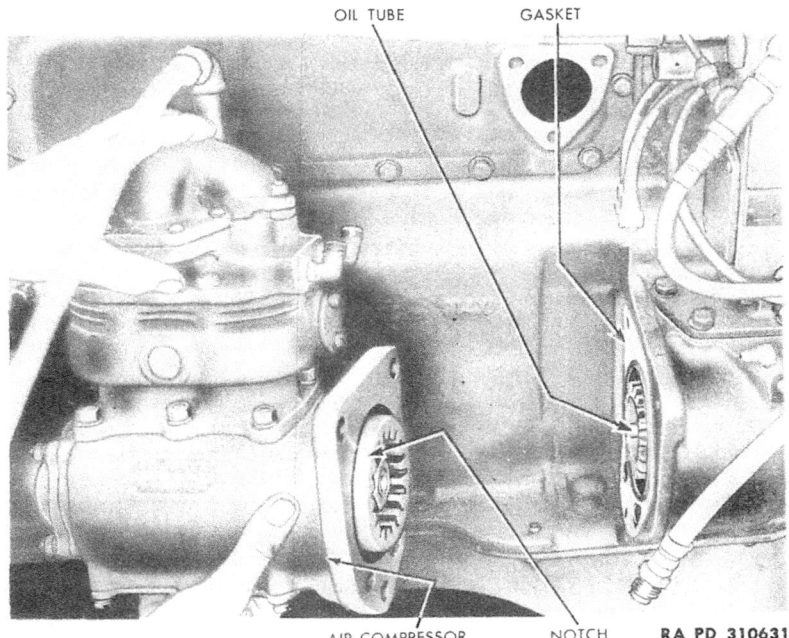

Figure 116—Installing Air Compressor

(6) REMOVE AIR INTAKE FITTING. Remove two cap screws and lock washers attaching intake fitting to compressor. Remove fitting and gasket.

(7) REMOVE MUD PAN. Remove three nuts attaching right-hand mud pan, and remove bolt, nut and lock washer at front end of pan. Remove pan.

(8) REMOVE COMPRESSOR. Cut locking wire from three cap screws, and remove the cap screws with socket wrench having short socket. If not available, remove the two water-line elbows and use open-end and box wrenches to remove cap screws. Lower compressor and remove.

(9) REMOVE FITTINGS. Remove the two water-line elbows with adjustable open-end wrench. Remove drain cock with same wrench. Remove reducer elbow.

10-TON 6 x 4 TRUCK (MACK MODEL NR)

(10) REMOVE DRIVE COUPLING. With hammer and chisel, open lock washer tangs on compressor drive coupling and remove nut with socket wrench. Remove lock washer, and pull coupling from shaft with a puller.

(11) REMOVE COVER. Remove two nuts and lock washers from the cover studs, and remove the cover and felt.

(12) REMOVE STUDS. Remove the two special studs by which the cover was fastened.

c. Installation.

(1) INSTALL COVER. Install the two special cover studs. Install cover with felt beneath it, and apply two nuts and lock washers to cover studs.

(2) INSTALL FITTINGS. Install drain cock and two water line elbows on compressor, unless it is necessary to leave elbows off until after mounting flange cap screws are installed. Install reducer and elbow on top of compressor head.

(3) INSTALL COUPLING. Place key in shaft and tap coupling onto shaft with key in keyway. Install locking washer with inner tang in puller screw hole, and remaining three tangs pointing outward. Install nut. Tighten down nut and lock it by bending over three outer tangs of the washer.

(4) INSTALL COMPRESSOR. Be sure that the oil tube between compressor and accessory shafts is inserted in latter before mounting compressor. Bring No. 1 engine piston to upper dead center. The notch on the compressor coupling should then be in line with upper outer mounting flange hole. Place gasket on flange and mount compressor with three cap screws and flat washers, and tighten with socket wrench having short socket (see a (8) above). Place locking wire through cap screws and twist to secure it (fig. 116).

(5) CONNECT WATER LINES. Connect and tighten water lines to compressor elbows.

(6) INSTALL AIR CLEANER FITTINGS. Attach air cleaner fitting bracket flange to compressor with gasket in place, using two cap screws and lock washers. Attach air-cleaner mounting bracket to engine with two cap screws and lock washers. Tighten hose clamps.

(7) INSTALL OIL FILLER SPOUT. Place gasket on oil filler spout elbow and attach to engine with three cap screws and lock washers. Tighten with socket wrench.

(8) INSTALL DISCHARGE PIPE FLANGE. Place gasket on compressor discharge-line flange, and attach to compressor with two nuts and lock washers. Install pipe in flange with pipe wrench.

(9) ATTACH FLEXIBLE LINE. Connect flexible line to compressor with elbow on pipe and tighten with two wrenches, being careful not to twist the line.

(10) INSTALL GOVERNOR LINE. Connect governor line to compressor elbow.

(11) INSTALL AIR CLEANER. Place air cleaner on mounting. Apply wing nut and tighten.

BRAKE SYSTEM

(12) REFILL SYSTEM. Close drain valve in radiator outlet pipe, and check water and drain plugs for tightness. Fill radiator with water and test for leaks.

(13) INSTALL MUD PAN. Attach mud pan with three nuts and lock washers and install bolt, nut and lock washer.

d. **Replace Compressor Head.**

(1) REMOVAL. Following compressor replacement procedure, drain water from radiator and compressor head, disconnect flexible air line from line to reservoir, also governor line and water lines, and remove cover and felt as outlined in b (1), (2), (5) and (11) above.

(2) REMOVE STUDS. Remove the four nuts and special cover studs and remove head.

(3) REMOVE FITTINGS. Follow procedure outlined in b (9) above.

e. **Installation.**

(1) INSTALL FITTINGS. Follow procedure as outlined in c (2) above to install fittings.

(2) INSTALL HEAD. Make sure that faces of compressor and head are clean before installing. Place head on compressor with gasket in place and install four nuts and two special cover studs.

(3) FOLLOW COMPRESSOR INSTALLATION PROCEDURE. To install dust cover, flexible line, line to reservoir, water lines, governor line and drain valve, follow the procedure outlined in c (1), (9), (10) and (12) above.

(4) CHECK FOR LEAKS. Check water drain plugs for tightness. Fill radiator and test for leaks.

114. AIR PRESSURE GOVERNOR.

a. **Removal.**

(1) REMOVE AIR LINES. Loosen and detach air lines at tee fitting on governor. Also air line to fitting on side of governor.

(2) REMOVE MAT. Pull dash mat away from dash on the right-hand inside of cab, and remove two nuts from governor-mounting bolts, with socket wrench to turn and open-end wrench inside the cab to hold. Remove nuts and toothed lock washers, and remove governor.

(3) REMOVE FITTINGS. Remove reducer and tee fitting from governor. Remove adapter fitting on side of governor.

b. **Installation.**

(1) INSTALL FITTINGS. Place reducer in lower hole on side of governor and install tee fitting in reducer with open-end wrench. Install adapter fitting in upper hole on side of governor and tighten.

(2) INSTALL GOVERNOR. Place governor on dash. Insert two bolts from cab side and apply toothed lock washers and nuts and tighten with one wrench, holding bolt heads with another.

(3) CONNECT AIR LINES. Attach air line from compressor to adapter fitting on governor. Attach line from low-pressure indicator

to outer connection on tee fitting. Attach line from pressure gage on instrument board to inner connection on tee fitting.

115. FRONT BRAKE CHAMBERS.
a. Removal.

(1) DISCONNECT CHAMBER ROD. Back off slack adjuster by turning adjusting worm with wrench. Remove cotter pin from chamber rod clevis pin and remove pin.

(2) DETACH AIR LINE. Disconnect air line at frame, using one wrench to turn and another to hold fitting.

(3) REMOVE CHAMBER. Remove two nuts and lock washers and remove brake chamber.

(4) REMOVE AIR LINE. Remove air hose from chamber.

b. Installation.

(1) INSTALL AIR HOSE. Attach air hose to chamber.

(2) INSTALL CHAMBER. Place brake chamber in position, with the air hose fitting at top, and attach by installing the two nuts with lock washers on studs.

(3) CONNECT CHAMBER ROD. Turn slack adjuster worm to aline adjuster with chamber rod. Replace clevis pin and cotter pin.

(4) ATTACH AIR HOSE. Connect air hose at frame fitting, using one wrench to turn and another to hold fitting.

116. REAR BRAKE CHAMBERS.
a. Removal.

(1) DISCONNECT CHAMBER ROD. Back off slack adjuster by turning adjusting worm with wrench. Remove cotter pin from clevis pin and remove pin, detaching rod. Turn slack adjuster worm further to clear rod.

(2) DETACH AIR LINE. Disconnect air line at elbow in chamber cover.

(3) REMOVE CHAMBER. Remove cotter pins from studs on chamber and remove the castle nuts. Remove the washers from the outer studs and remove the chamber.

(4) REMOVE ELBOW. Detach 45-degree elbow from chamber with wrench.

b. Installation.

(1) INSTALL CHAMBER. Place chamber on wheel with air-line fitting at lower or inner side. Place washer on outer stud and attach chamber by means of castle nuts. Install cotters.

(2) INSTALL ELBOW. Place 45-degree elbow in chamber cover and tighten with wrench, lining up so that air hose may be easily attached.

(3) CONNECT BRAKE HOSE. Attach brake hose to chamber.

(4) CONNECT CHAMBER ROD. Turn slack adjuster worm to aline push rod with adjuster, and replace clevis pin and cotter pin.

(5) ADJUST BRAKES. Adjust brakes as noted in paragraph 111 a

BRAKE SYSTEM

117. SLACK ADJUSTER ASSEMBLY.

a. Removal.

(1) DISCONNECT CHAMBER ROD. Back off brake adjustment by turning square nut. Remove cotter pin from clevis pin on chamber rod and remove pin.

(2) REMOVE NUT. Remove cotter pin from brake camshaft and remove castle nut from shaft with adjustable open-end wrench.

(3) REMOVE SLACK ADJUSTER. Back off adjuster by again turning square nut to clear chamber rod, and tap adjuster off camshaft.

b. Installation.

(1) INSTALL SLACK ADJUSTER. Fit slack adjuster on brake camshaft so that adjuster slides easily on splines. The adjuster should not be tight on the shaft.

(2) INSTALL NUT. Apply castle nut to brake camshaft and tighten with adjustable open-end wrench. Back off one-quarter inch, and tap brake cam outward. Make sure cam is free moving. Install cotter pin in shaft end.

(3) ATTACH CHAMBER ROD. Turn slack adjuster worm to line up adjuster with chamber rod clevis. Install clevis pin and cotter pin.

(4) ADJUST BRAKES. Make brake adjustments as outlined in paragraph 115 a.

118. BRAKE APPLICATION VALVE.

a. Removal.

(1) REMOVE PEDAL. Remove three cap screws and lock washers. Remove three countersunk machine screws and remove pedal.

(2) DISCONNECT ACCELERATOR PEDAL. Disconnect accelerator pedal at ball joint by removing nut and lock washer.

(3) REMOVE FLOORBOARD. Remove the six cap screws and washers from floorboard, and remove floorboard.

(4) DETACH AIR LINES. Disconnect the air line to relay valve from elbow on brake valve. Detach line to quick release valve from the straight fitting on brake valve. Disconnect supply line from bottom fitting on brake valve.

(5) REMOVE VALVE. After the air lines are disconnected, remove the valve.

(6) REMOVE FITTINGS. Remove the two straight fittings with socket wrench, and remove the elbow with small wrench.

b. Installation.

(1) INSTALL FITTINGS. Install straight fitting in lower hole of brake valve. Install elbow in hole at rear of brake valve with small pipe wrench, and the straight fitting in the forward hole with socket wrench.

(2) CONNECT AIR LINES. Connect supply line to lower fitting. Connect air line from quick release valve to the straight fitting, and attach the line from relay valve.

10-TON 6 x 4 TRUCK (MACK MODEL NR)

(3) INSTALL FLOORBOARDS. Place floorboard over shifting levers, and attach with six cap screws and washers.

(4) ATTACH VALVE. Place pedal plate on valve and attach, using three flathead, countersunk machine screws. If necessary have another man support the valve from underneath to facilitate the insertion of screws. Install the three cap screws and lock washers.

(5) CONNECT ACCELERATOR PEDAL. Attach accelerator ball joint by placing pedal on stud and installing lock washer and nut.

119. RELAY VALVE.

a. Removal.

(1) DISCONNECT AIR LINES. Detach air lines from relay valve at junction on each axle housing, using one wrench to turn and another to hold. Remove hose from two clamps on rear torque rod, two clamps from front torque rod and one from frame cross member, using cross-recessed screwdriver, with wrench to hold nut. Remove air line to stop light switch at tee, using open-end wrench. Remove air line from rear chambers with tee fitting, using pipe wrench. Remove line from front chambers, guiding hose so that it is not strained or damaged. Remove air line from No. 3 reservoir. Remove the air line to right-hand trailer valve with open-end wrench, holding fitting with another wrench. Remove the line to left-hand trailer valve, and foot valve from tee on top of relay valve.

(2) REMOVE VALVE. Using two wrenches, one to hold and the other to turn, remove two nuts, lock washers and bolts attaching valve to cross member.

(3) REMOVE FITTINGS. Remove the tee from the top of the relay valve with a monkey wrench, the 45-degree elbow with an adjustable open-end wrench, and the straight fitting with an open-end wrench.

b. Installation.

(1) INSTALL FITTINGS. Install tee fitting on top of relay valve with monkey wrench. Install the 45-degree elbow on lower right-hand side of relay valve, and tighten with an adjustable open-end wrench. Install straight fitting on lower left-hand side of relay valve with an open-end wrench.

(2) INSTALL VALVE. Place relay valve on cross member between axles, and insert two bolts. Apply lock washers and nuts, but do not tighten until all air line connections are secured.

(3) CONNECT AIR LINES. Attach the air line from No. 3 reservoir to 45-degree elbow on lower right-hand side of relay valve. Attach the air line from the right-hand or emergency coupling to fitting on lower, left-hand side of relay valve. Attach line for rear chambers, with tee fitting on it, to relay valve on the upper right-hand side with pipe wrench. Attach the air line from the forward brake chambers to the upper, left-hand side of relay valve. Attach stop light switch air line to the tee on rear brake chamber line fitting. Connect the air line from the brake application valve to the forward tee connection on top of the relay valve, and also connect the air line from the left-hand or service coupling to the rear tee connection on

BRAKE SYSTEM

the relay valve. Attach chamber hoses to torque arms and cross member with clamps and cross-recessed head screws, nuts and lock washers, using screwdriver, with wrench to hold nuts. Attach air lines from relay valve to junction on each axle housing, using one wrench to turn and another to hold fitting.

120. QUICK RELEASE VALVE.

a. Removal.

(1) DISCONNECT AIR LINE. Detach air lines from forward and rear fittings on quick release valve, which lead to left and right front brake chambers respectively. Also detach air line from upper elbow connection on valve which leads to brake application valve, using open-end wrench on all three fittings.

(2) REMOVE BATTERY COVER. Remove the two cap screws from battery cover on left-hand side, and remove cover.

(3) REMOVE VALVE. Use socket wrench from inside of battery compartment to remove two bolts which attach valve to frame, and remove valve.

(4) REMOVE FITTINGS. Remove elbow on top of valve with small pipe wrench. Remove the two end fittings with socket wrench.

b. Installation.

(1) INSTALL FITTINGS. Install the two end fittings in valve with socket, and the elbow on top of valve with small pipe wrench, turning elbow so that open end is forward.

(2) INSTALL VALVE. Install valve inside of frame side rail with the elbow up and forward. Insert the two bolts from the battery compartment. Apply lock washers and nuts, and tighten bolts with socket wrench, holding nuts with open-end wrench.

(3) CONNECT BRAKE LINES. Attach the air line from left-hand brake chamber to forward fitting, that from right-hand brake chamber to rear fitting, and the line from the brake application valve to the elbow on top of quick release valve, using open-end wrench for all three fittings.

(4) INSTALL BATTERY COVER. Place cover over batteries and attach to apron with the two cap screws, using socket wrench to tighten.

121. SAFETY VALVE.

a. Removal. Hold tee fitting with monkey wrench, and turn safety valve with open-end wrench on the end closest to fitting. Remove valve.

b. Installation. Place valve in tee fitting, and turn into place with open-end wrench, holding the tee fitting with a monkey wrench.

122. PRESSURE INDICATOR.

a. Removal.

(1) REMOVE MAT. Pull dash mat away from dash at right-hand end.

10-TON 6 x 4 TRUCK (MACK MODEL NR)

(2) DISCONNECT AIR LINES. Detach air lines at tee connection on low-pressure indicator.

(3) DISCONNECT WIRES. Remove top and bottom wires from low pressure indicator.

(4) REMOVE INDICATOR. Using one wrench to hold and another to turn, remove the two mounting bolts attaching indicator to the dash. Remove toothed lock washers and ground wire, and remove indicator.

(5) REMOVE FITTINGS. Remove reducer and tee fitting.

b. Installation.

(1) INSTALL TEE FITTING. Install tee fitting and reducer in indicator.

(2) INSTALL INDICATOR. Install indicator on dash with mounting bracket down. Insert two mounting bolts with toothed lock washers at head, between indicator and dash, and under nuts. Place ground wire on bolt between indicator and dash. Apply nuts and tighten.

(3) CONNECT AIR LINES. Attach air line from governor to one side of tee fitting, and air line from air supply valve to other side of tee fitting.

(4) CONNECT WIRES. Attach red and white coded wire from harness to top connection.

123. AIR SUPPLY VALVE.

a. Removal.

(1) DETACH AIR LINES. Disconnect two air lines from air supply valve fittings.

(2) REMOVE FITTINGS. Remove the straight fitting and loosen reducer.

(3) REMOVE VALVE. Remove the two mounting bolts attaching valve to tire carrier, using two wrenches, one to hold and one to turn. Remove valve, and remove reducer and elbow from valve.

b. Installation.

(1) INSTALL FITTINGS. Turn elbow into reducer and turn into valve, with elbow in down position. Install straight fitting in right-hand end of valve.

(2) INSTALL VALVE. Place valve on tire carrier and insert mounting bolts. Apply lock washers and tighten, using two wrenches.

(3) ATTACH AIR LINES. Connect air line from tank to straight fitting on valve and attach the other line to the elbow on valve.

124. FLEXIBLE AIR HOSE, LINES AND CONNECTIONS.

a. When it is necessary to replace air lines be sure to install them just as they were removed, with the same size tubing and the same connections and layout. When cutting tubing for new lines, use pipe cutter, and clean burs from pipe ends, blowing chips out of lines.

TM 9-818
124-125

BRAKE SYSTEM

Do not make sharp bends in tubing. Be sure to install loom in the places originally furnished. Make certain that no vibration occurs in air lines during operation.

125. AIR RESERVOIRS.

a. Air Reservoir No. 1.

(1) REMOVAL.

(a) Detach Air Lines. Disconnect line from compressor to reservoir at forward elbow and supply line to reservoir No. 2 at rear elbow. Detach line from reservoir to air supply valve at tee fitting on top of reservoir.

(b) Remove Clamps. Remove nuts and lock washers from clamp bolts on outer ends of clamps. Remove two lower side-rail clamp bolts, nuts and lock washers, using two wrenches, one to hold and one to turn. Remove the lower clamps and drop the reservoir.

(c) Remove Fittings. Remove 90-degree elbow at each end of reservoir. Remove drain cock. Remove safety valve and tee fitting. Remove pipe plug from reservoir.

(2) INSTALLATION.

(a) Install Fittings. Install the elbows at each end of No. 1 reservoir, with the longer elbow at the rear. Install safety valve and tee fitting on top of reservoir, with safety valve facing forward and tighten. Install pipe plug in reservoir. Install drain cock in bottom of reservoir using adjustable open-end wrench to tighten.

(b) Install Reservoir. Place reservoir in brackets with safety valve at forward end, and drain cock on bottom. Put clamps in place and install bolts, lock washers and nuts in lower frame bracket assembly and tighten. Attach outer clamp bolts, nuts and lock washers and tighten.

(c) Connect Lines. Attach compressor air line at forward elbow and tighten. Attach supply line to reservoir No. 2. Attach line from air supply valve to tee fitting on top of reservoir.

b. Air Reservoir No. 2.

(1) REMOVAL.

(a) Disconnect Air Lines. Detach air line from elbow on top of reservoir. Detach air line from fitting at rear of reservoir.

(b) Remove Reservoir. Remove clamp stud nuts and lock washers using two wrenches, one to turn and one to hold. Remove nuts and lock washers from bolts on lower clamp brackets. Remove brackets and lower reservoir.

(c) Remove Fittings. Remove drain cock. Remove elbow on top of reservoir. Remove fitting at the rear of reservoir. Remove two pipe plugs.

(2) INSTALLATION.

(a) Install Fittings. Install drain cock in hole at bottom of reservoir in center. Install fitting in rear end of reservoir. Install pipe plugs in front end of reservoir and in rearmost hole on top of reservoir

10-TON 6 x 4 TRUCK (MACK MODEL NR)

with adjustable open-end wrench. Install elbow in forward hole on top of reservoir with open-end wrench. The final elbow position must be with the open end forward.

(b) Install Reservoir. Place reservoir against upper bracket clamps, with end having pipe plug toward front. Place lower brackets, installing bolts, lock washers and nuts on the frame brackets and tighten. Install two studs in outer bracket holes, and apply lock washers and nuts.

(c) Connect Air Lines. Attach air line from reservoir No. 1 through tee fitting to the elbow and tighten. Connect supply line to reservoir No. 3 to fitting on rear of reservoir and tighten.

c. **Air Reservoir No. 3.**

(1) REMOVAL.

(a) Disconnect Air Lines. Loosen clamping bolts on top of reservoir so that the reservoir may be slid to each side to disconnect air lines. Detach air supply line from No. 2 reservoir, on left end of reservoir, and line from relay valve to right end of reservoir.

(b) Remove Reservoir. Remove bolts from top of clamps and bolts from front clamps only where attached to frame cross member brace. Remove clamps and drop reservoir.

(c) Remove Fittings. Remove reducer and elbow at each end of reservoir. Remove drain cock and pipe plug.

(2) INSTALLATION.

(a) Install Fittings. Install drain cock in pipe plug in reservoir. Install reducers and elbows in ends of reservoir so that elbows will face forward when reservoir is installed with drain cock at bottom.

(b) Install Reservoir. Place reservoir against clamping brackets remaining on frame member support with drain cock at bottom and elbows on ends facing forward. Place forward clamps and install, using bolts in frame cross member support. Apply lock washers and nuts and tighten. Install clamping bolts through holes in top of clamps. Apply lock washers and nuts, but do not tighten.

(c) Connect Air Lines. Attach air supply line from reservoir No. 2 to left-hand elbow, and air line from relay valve to right-hand elbow, and tighten. Reservoir may be moved to each side to assist in this operation, after which tighten upper clamping bolts and nuts.

126. PARKING BRAKE SHOE ASSEMBLIES.

a. **Removal.**

(1) DISCONNECT PUSH RODS. After releasing parking brake lever, remove cotter pins from clevises on push rods.

(2) REMOVE SHOE AND BRACKET ASSEMBLY. Remove three cap screws and lock washers attaching each brake anchor bracket to frame cross member. Drop shoe and bracket assembly, guiding with brake operating rod.

(3) REMOVE SPRING. Remove tension spring at bottom of shoe.

TM 9-818
126

BRAKE SYSTEM

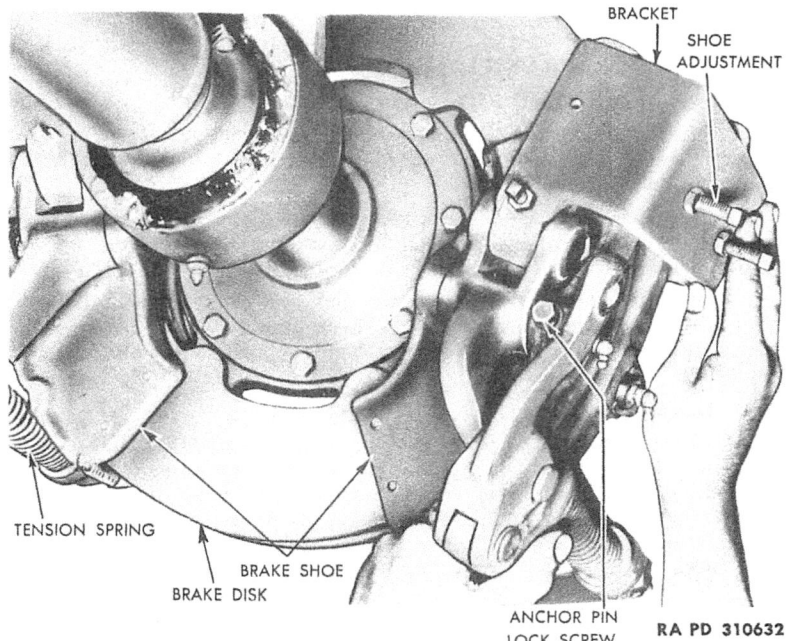

Figure 117—Removing Parking Brake Shoe Assembly

(4) REMOVE SHOE ASSEMBLIES (fig. 117). Remove two anchor pin locking cap screws from each assembly. Remove lock washers and locks. Drive out pins with hammer and drift, and remove shoe assemblies.

b. **Installation.**

(1) LOOSEN ADJUSTMENTS. Loosen square-head adjusting screws and jam nuts on side of bracket, using one wrench to turn screw and another to turn and hold nut.

(2) INSTALL SHOES. Install shoe assemblies with pins through bosses. Line up holes with drift and drive in pins with locking groove to outer side. Slip locks in grooves and attach to bracket with cap screws and lock washers.

(3) INSTALL SHOE AND BRACKET ASSEMBLIES. Lining up the three holes in each bracket with those in frame cross member, attach brackets with cap screws and lock washers.

(4) INSTALL SPRINGS. Hook tension springs in holes at lower end of brake shoes with pliers.

(5) INSTALL PUSH RODS. Place push rods with adjustable clevises to rear. Insert clevis pins and install cotter pins.

241

10-TON 6 x 4 TRUCK (MACK MODEL NR)

(6) ADJUSTMENT. Follow procedure outlined in paragraph 128 a to adjust brake.

127. PARKING BRAKE DISK.

a. Removal.

(1) REMOVE BOLTS. After releasing parking brake lever, remove the eight bolts, nuts and lock washers attaching two propeller shaft flanges to brake disk.

(2) LOOSEN SHOE ASSEMBLIES. Loosen three cap screws and washers on each brake shoe anchor bracket assembly, and remove tension springs from lower corners of shoes.

(3) REMOVE DISK. Support brake disk and remove the four bolts, nuts and lock washers attaching bearing hanger to cross member with socket and open-end wrenches. Remove disk when free of shoe assemblies.

b. Installation.

(1) INSTALL DISK. Support disk in position and place the forward shaft in position with bearing hanger at cross member. Insert propeller shaft bearing hanger bolts from front of cross member, with housing behind cross member flange, apply lock washers and nuts, and tighten with socket and open-end wrenches.

(2) TIGHTEN SHOE ASSEMBLY BOLTS. Tighten the three cap screws, with lock washers, attaching brake shoe anchor bracket assembly to cross member, and install tension springs on lower corners of shoes.

(3) INSTALL BOLTS. Install the eight bolts to attach propeller shaft flanges to brake disk, from front of disk. Apply lock washers and nuts.

128. PARKING BRAKE CONTROLS AND LINKAGE.

a. Adjustment. With parking brake lever in free position place 0.016-inch long feeler gage under each shoe on same side in vertical position. Loosen two square head screws and jam nuts in each upper bracket. Turn screws clockwise until feeler gages have slight drag. Take up on lever arm tie rod nut until there is slight drag. Hold the square head screws and tighten check nuts without disturbing assembly. Check shoes for 0.016-inch clearance at each shoe. Tighten locking nut on tie rod, maintaining the 0.016-inch clearance. Repeat on opposite side.

b. Replacement.

(1) SHOE ASSEMBLIES. Remove and install shoe assemblies as outlined in paragraph 127.

(2) BRAKE CROSS SHAFT ASSEMBLY.

(a) Removal. Remove cotter pins from three clevis pins on cross shaft levers. Remove clevis pins. Remove two bolts from each cross shaft bracket. Remove cross shaft assembly.

BRAKE SYSTEM

(b) Installation. Install assembly by attaching with two bolts on each side and install clevis pins to make lever connections, locking pins with cotter pins. Make sure that the cross shaft is free before connecting the rods.

(3) CLUTCH SHAFT BRACKET LEVER. Remove and install bell crank on clutch shaft bracket as outlined in paragraph 62 a (3) and b (5).

(4) PARKING BRAKE LEVER ASSEMBLY.

(a) Complete Assembly.

1. Removal. With the driver's seat assembly as far forward as possible for convenience, remove the cotters, rod and clevis pins at each end of brake rod. Remove the clevis pins at each end of brake shoe. Remove the two bolts and two cap screws from underneath the chassis, and remove brake lever.

2. Installation. Install brake lever by placing bolts and two cap screws from underneath chassis, attach cross shaft rod linkage and attach brake lever rod with clevises and clevis pins, inserting cotter pins to lock pins.

(b) Parking Brake Latch.

1. Removal. Loosen locking nut on parking brake lever latch rod. Remove two cotter pins and two castle nuts. Remove bolts and washers. Remove latch. Pull out rod and unscrew latch.

2. Installation. Install latch by compressing spring. Place rod in lower notch of ratchet. Install two nuts and flat washers. Replace castle nuts and check latch, making sure that it is free in movement. Replace cotter pins. Latch position may be adjusted by removing latch rod only from lever. Remove top nut and replace after making a few turns of rod.

(c) Parking Brake Ratchet.

1. Removal. Remove ratchet by removing two bolts and lock washers using two wrenches, one to turn and one to hold.

2. Installation. To install ratchet hold in place with stop at bottom, install two bolts and lock washers, and tighten.

10-TON 6 x 4 TRUCK (MACK MODEL NR)

Section XXVI

WHEELS

	Paragraph
Description and tabulated data	129
Front wheel and hub assemblies	130
Front wheel grease retainers	131
Front wheel bearings	132
Front brake drums	133
Rear wheel and hub assemblies	134
Rear wheel grease retainers	135
Rear wheel bearings	136
Rear brake drums	137

129. DESCRIPTION AND TABULATED DATA.

a. **Description** (figs. 118 and 119). The steel wheels are of the integral hub type, designed to accommodate standard 28-degree taper rims. Both front and rear wheels are carried on two opposed, fully adjustable roller bearings.

b. **Data.**

Make	Dayton
Type of rim	R
Wheel size	20x11
Bearing make	Timken
Front inner cup No.	563
Front inner cone No.	565
Front outer cup No.	432
Front outer cone No.	438
Rear inner cup No.	652
Rear inner cone No.	663
Rear outer cup No.	742
Rear outer cone No.	748S

130. FRONT WHEEL AND HUB ASSEMBLIES.

a. **Removal** (fig. 137).

(1) SUPPORT FRONT OF TRUCK. Place jack under front axle end, and elevate so tire is off ground. Place block or other support under axle.

(2) REMOVE HUB CAP. Remove eight cap screws from hub cap, and remove cap and gasket.

(3) REMOVE BEARING ADJUSTING NUT. Remove locking wire from cap screw, and bolt on bearing adjusting nut. Remove clamping bolt and locking screw from bearing adjusting nut. Remove bearing adjusting nut.

WHEELS

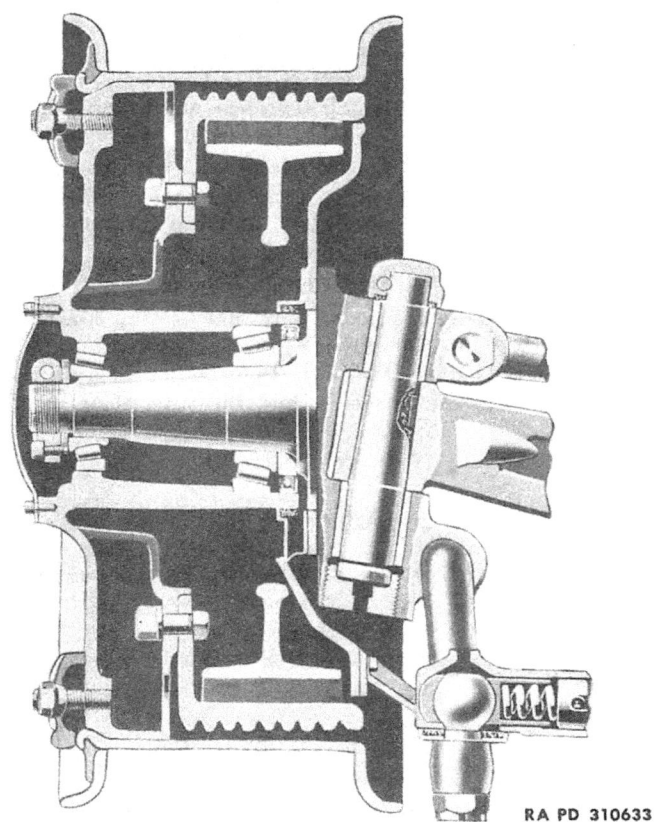

Figure 118 — Cross Section of Front Wheel

(4) REMOVE SLOTTED WASHER. Remove slotted locking washer by lifting out.

(5) REMOVE WHEEL. Loosen brake adjustment by turning slack adjuster worm enough to increase the clearance between drum and shoes, and ease wheel removal. Pull on wheel, vibrating so as to remove outer bearing cone. Remove wheel with inner bearing cone, which is retained in wheel by the grease seal.

(6) REMOVE INNER BEARING AND SEAL. Drive out grease retainer and complete bearing by use of soft drift on bearing cup, using care not to damage it.

(7) REMOVE OUTER BEARING CUP. Remove outer bearing cup by use of puller. If puller is not available, drive out with soft drift.

10-TON 6 x 4 TRUCK (MACK MODEL NR)

b. Installation.

(1) INSTALL BEARING CUPS. Examine bearings and cups to determine if they can be reused. Refer to paragraph 136 b. Install cup in hub, driving into place with soft drift and hammer. Make sure that the larger diameters of the tapered cups are toward the respective hub ends. Remove all dirt or grit from hub before installing races.

(2) LUBRICATE BEARING. See paragraph 19 c (8).

(3) INSTALL INNER BEARING CONES. Replace inner bearing cone to match taper of cup, install seal, and tap into place with brass drift, being sure to guide parts into wheel properly.

(4) LUBRICATE HUB. See paragraph 19 c (8).

(5) INSTALL WHEEL. Making sure that bearing spacer is in place on spindle with pin in hole, place wheel on spindle.

(6) INSTALL OUTER BEARING. Place outer bearing on spindle, with smaller diameter to inside, and tap into place in wheel with soft drift.

(7) REPLACE SLOTTED WASHER. Replace slotted washer on spindle.

(8) INSTALL BEARING ADJUSTING NUT. Place bearing adjusting nut on spindle and turn into place. Turn down until tight and then turn back three notches of slotted washer. Install locking screw and tighten clamping bolt, placing locking wire through hole in each.

(9) INSTALL HUB CAP. With gasket in place, put hub cap back on wheel and tighten down cap screws.

(10) ADJUST BRAKES. Adjust brakes in accordance with instructions in paragraph 115 a.

131. FRONT WHEEL GREASE RETAINERS.

a. Remove Wheel. Follow procedure outlined in paragraph 130 a to remove front wheels.

b. Remove Retainer. Using soft drift and hammer, drive grease retainer out of wheel hub.

c. Install Retainer. Place grease seal on wheel in hub and tap into place with soft drift, being sure that seal is properly seated.

d. Install Wheel. Follow procedure in paragraph 130 b to install wheel and adjust brakes.

132. FRONT WHEEL BEARINGS.

a. Remove Wheel and Bearings. Follow procedure outlined in paragraph 130 a to remove wheel and bearings.

b. Clean and Inspect Bearings. Follow procedure outlined in paragraph 136 b.

c. Install Bearings and Wheel. Follow procedure outlined in paragraph 130 b to install bearings and wheel and adjust brakes.

TM 9-818
133

WHEELS

133. FRONT BRAKE DRUMS.

a. **Remove Wheel.** Remove wheel, following procedure outlined in paragraph 130 a.

b. **Remove Brake Drum.** Use open-end wrench inside the drum to hold bolts and socket wrench to turn nuts. Remove six nuts and lock washers, and remove bolts in drum, using soft drift and hammer.

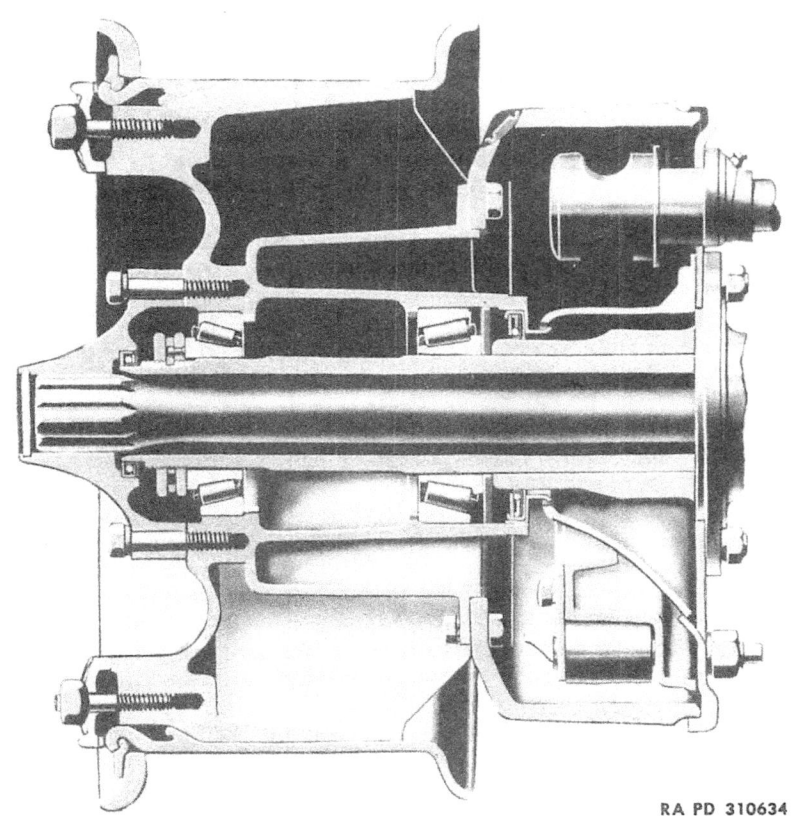

RA PD 310634

Figure 119 — Cross Section of Rear Wheel

c. **Install Brake Drum.** Making sure that drum and wheel mounting faces are clean, place drum on wheel, lining up holes with pry bar. Insert six bolts from inside of drum, place lock washers and nuts, and tighten opposite nuts progressively.

d. **Install Wheel.** Install wheel in accordance with procedure outlined in paragraph 130 b.

247

10-TON 6 x 4 TRUCK (MACK MODEL NR)

134. REAR WHEEL AND HUB ASSEMBLIES (fig. 119).

a. **Removal.**

(1) RAISE AXLE. After placing blocks under front and rear of tires to keep truck from moving, place jack under rubber shock insulator cap and elevate so that tire clears ground.

(2) REMOVE TIRE. Follow procedure outlined in paragraph 130 a to remove tire from wheel.

(3) LOOSEN BRAKES. Turn slack adjuster worm sufficiently for brake shoes to have enough clearance to ease wheel removal.

(4) REMOVE HUB CAP AND GASKET. Remove six cap screws and lock washers from hub cap. Remove hub cap and gasket by tapping lightly and pulling off.

(5) REMOVE AXLE SHAFT. Pull out axle shaft. If shaft will not pull free easily, insert bar from other side and tap lightly to free. If both shafts stick, tap outer end of one with block of wood to drive other free.

(6) REMOVE LOCKING AND ADJUSTING NUTS. Loosen outer nut, using wrench (41-W-639-390), and remove nut. Remove locking ring and remove bearing adjusting nut.

(7) REMOVE WHEEL AND OUTER BEARING. Remove wheel by withdrawing from axle. Outer bearing will come off with wheel, and should be lifted out of hub.

(8) REMOVE BRAKE DRUM. Using an open-end wrench to hold the six bolts, remove nuts with a socket wrench. Remove lock washers and bolts and remove drum by tapping lightly with hammer on soft drift. Lift drum and dust shield off of hub.

(9) REMOVE INNER BEARING AND SEAL. Remove inner bearing race and oil seal by use of universal puller. If puller is not available, tap out carefully with soft drift. This will also remove the inner bearing oil seal retainer and the spacer. Use care in performing this operation in order not to damage the bearing or other parts.

(10) REMOVE OUTER BEARING RACE. Remove outer bearing race by use of puller. If puller is not available, drive out with soft drift.

b. **Installation.**

(1) INSTALL BEARINGS. Examine bearings and cups to determine whether they can be reused. (Refer to par. 136 b). Install bearing races in hub, driving into place with soft drift and hammer. Make sure that race tapers toward outside for the outer and inside for the inner races. Be careful to remove all dirt or grit from hub before installing races. Pack inner bearing with grease, and place in wheel hub. Install washer with flat side against bearing.

(2) INSTALL OIL SEAL AND RETAINER. Place oil seal in retainer, and drive into place on inner side of hub with a soft drift.

(3) INSTALL DRUM. Clean wheel and brake drum outer surfaces. Using pry bar to line up holes, place drum on wheel with inspection slot in one of the two positions where the space between spokes is

WHEELS

smaller than the others. Place dust shield on inside of drum with holes alined and insert six bolts from inside of drum. Apply lock washers and nuts and tighten opposite sides progressively.

(4) INSTALL WHEEL. Lift wheel into place on axle and tap lightly until inner bearing is seated.

(5) LUBRICATION. See paragraph 19 c (8).

(6) INSTALL OUTER BEARING. Pack outer bearing with grease and install on axle by tapping lightly into place with hammer handle.

(7) INSTALL ADJUSTING NUT AND LOCKING WASHER. Place adjusting nut on axle with locking lug to outside and tighten down with wrench (41-W-639-390). Place locking washer on axle and indicate position to turn adjusting nut back three holes. Remove washer and turn back adjusting nut the indicated distance. Place locking washer on axle with lug in hole. Apply outer locking nut and tighten.

(8) INSTALL AXLE SHAFT. Place axle shaft in axle with end marked "OUT" to outside and tap into place.

(9) INSTALL HUB CAP. Inspect oil seal in hub cap and if necessary remove and replace. Place gasket on wheel and place hub cap on shaft, tapping into place so holes line up with holes in wheels. Apply six cap screws and lock washers and tighten with socket wrench.

(10) ADJUST BRAKES. Follow procedure outlined in paragraph 111 a to adjust brakes.

(11) INSTALL TIRE. Follow procedure outlined in paragraph 148 b.

135. REAR WHEEL GREASE RETAINERS.

a. **Remove Wheel.** Follow procedure outlined in paragraph 134 a (1), (3), (4), (5), (6), (7) to remove wheel. It is not necessary to remove the tire in this case.

b. **Remove Retainer.** Remove inner grease retainer by driving out of hub by use of puller. Retainer shield will also come off wheel.

c. **Install New Retainer.** Place new retainer in outer retainer shield, place assembly in wheel and drive into place, using wooden block or soft drift.

d. **Install Wheel.** Follow procedure outlined in paragraph 134 b (4) through (10) to install wheel and adjust brake.

e. **Replace Outer Retainer.** If outer retainer alone needs attention, it is not necessary to remove the wheel. Remove only the hub cap, drive out seal from cap, install new seal by tapping into cap and install cap as outlined in paragraph 134 b (9).

136. REAR WHEEL BEARINGS.

a. **Remove Wheel and Bearings.** Follow procedure outlined in paragraph 139 a to remove wheels and bearings.

10-TON 6 x 4 TRUCK (MACK MODEL NR)

b. **Clean and Inspect Bearings.**

(1) Place bearing cone and roller assemblies in dry-cleaning solvent and permit parts to soak for a few minutes. Agitate the parts up and down in fluid several times. Use a brush to clean parts thoroughly.

(2) Remove bearing cone and roller assembly, and strike larger side of bearing flat against a wooden block to knock out heavier portions of old lubricant and dirt. Wash parts in solvent again.

(3) Wipe bearings dry with clean rags. If compressed air is available, blow rollers dry. Direct air jet across bearings. Never spin dry bearings with air jet or with hands.

(4) Inspect rollers and races carefully for worn spots and cracks, and discard if defective, using new bearings for installations.

(5) Wash out hubs and cups thoroughly with dry-cleaning solvent. Wipe dry with clean rags. Inspect cups carefully for cracks and worn spots.

(6) Thoroughly lubricate bearing cone and roller assemblies with engine oil. With hand or bearing lubricator, force lubricant (par. 19 c (8)) into bearings. Apply at large end of rollers until grease comes out between cage and race at small end. Smear grease around outside of cage and rollers.

c. **Install Bearings and Wheels.** Follow procedure outlined in paragraph 134 b to install bearings and wheels, and adjust brakes.

137. REAR BRAKE DRUMS.

a. **Remove Wheel and Drum.** Follow procedure outlined in paragraph 134 a (1) (3) (4) (5) (6) (7) (8), to remove wheel and brake drum. It is not necessary to remove the tire in this case.

b. **Install Drum.** Follow procedure outlined in paragraph 134 b (3) to install drum.

c. **Install Wheel.** Follow procedure as outlined in paragraph 134 b (4) through (10) to install wheel and adjust brakes.

Section XXVII

SPRINGS AND SHOCK ABSORBERS

	Paragraph
Description and tabulated data	138
Front spring assemblies	139
Shock absorber assemblies	140
Rear spring assemblies	141

138. DESCRIPTION AND TABULATED DATA.

a. Description.

(1) FRONT SPRINGS (fig. 11). The chassis front springs are all of the semielliptical type, and rest on spring pads on the axle I-beam. The ends of the springs are attached to the side frame members by rubber shock insulators enclosed in housings formed at the lower ends of rigid brackets which are riveted to the frame. The moving metal parts are insulated by the rubber blocks, and they must not be lubricated.

(2) REAR SPRINGS (figs. 12 and 13). Two inverted semielliptic springs carry the load on the rear axles, equalize the load between the front and rear axles of the bogie, and act as radius rods maintaining the position of the rear axles. Each spring end is retained by two large rubber shock insulator blocks confined in a housing under the axle at the location of the usual spring pad.

b. Data.

(1) FRONT SPRINGS.

Load center to load center	50 in.
Load center to center bolt	25 in.
Width	3½ in.
Number of rebound clips	4
Number of leaves	11
Grading thickness	11 at ⅜ in.
Total pack thickness	4⅛ in.
Spring clip diameter	¾ in.
Spring clip tightening torque	350 ft-lb

(2) AXLE TO FRAME CLEARANCE.

Top of spring clips to frame stop (empty)	4 in.
Top of spring clips to frame stop (loaded)	2¾ in. (min)

(3) REAR SPRINGS.

Load center to load center	55 in.
Load center to center bolt	27½ in.
Width	4 in.
Number of rebound clips	2
Number of leaves	13
Grading thickness	13 at ⅝ in.
Total pack thickness	8⅛ in.

TM 9-818
138-139

10-TON 6 x 4 TRUCK (MACK MODEL NR)

Spring clip diameter..................................1 in.
Spring clips ....Tighten to 800 ft-lb
Spring seat clamp studs..............Tighten to 250 ft-lb
(4) SHOCK ABSORBERS.
Make .... Houde Mfg. Corp.
Model ....BBCLL

139. FRONT SPRING ASSEMBLIES.

a. Removal.

(1) REMOVE ATTACHED PARTS. Follow procedure outlined in step c (1) below to remove shock absorber linkage and insulator caps and insulator.

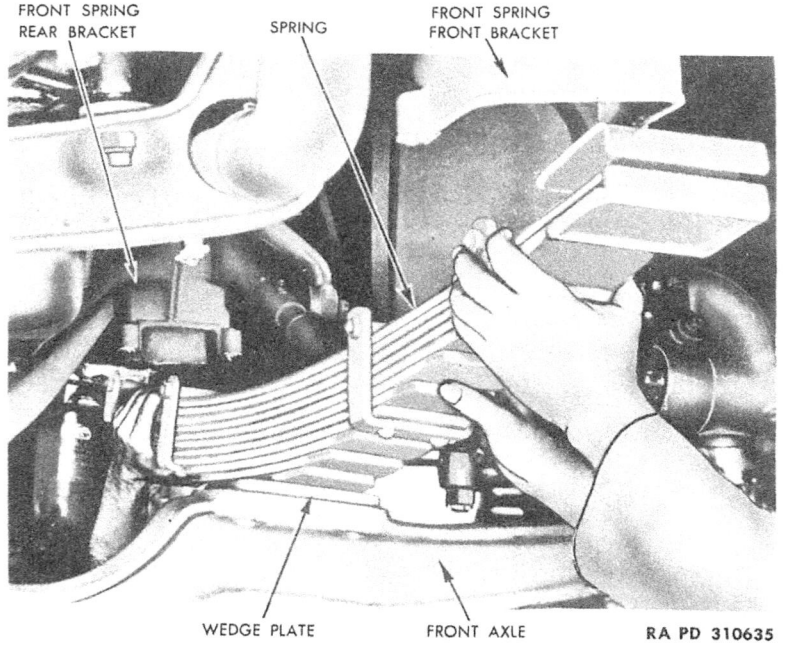

Figure 120—Removing Front Spring

(2) REMOVE SPRING CLIPS. Remove four nuts and remove spring clips. Shock absorber bracket and spacer washers will now be free.

(3) REMOVE SPRING. Lift spring and remove. The wedge plate will now be free (fig. 120).

b. Installation.

(1) INSTALL SPRING. Place wedge plate on axle spring pad with thicker edge at front. Place spring on plate and install spring clips

252

TM 9-818
139

SPRINGS AND SHOCK ABSORBERS

with spring clamping plate on top of spring, the longer clip at the front, with shock absorber bracket on forward clip beneath axle pad, with eye to left and spacer washers between bracket and axle. Attach four nuts.

(2) INSTALL ATTACHED PARTS. Install rubber shock insulator and complete attachment of parts as outlined in step c (2) below.

c. **Rubber Shock Insulators.**

(1) REMOVAL.

(a) *Disconnect Shock Absorber Linkage.* Remove nut and lock washer from shock absorber ball joint with socket wrench, using bar between spring and shock absorber rod. Pry linkage loose, tapping ball joint socket to free linkage.

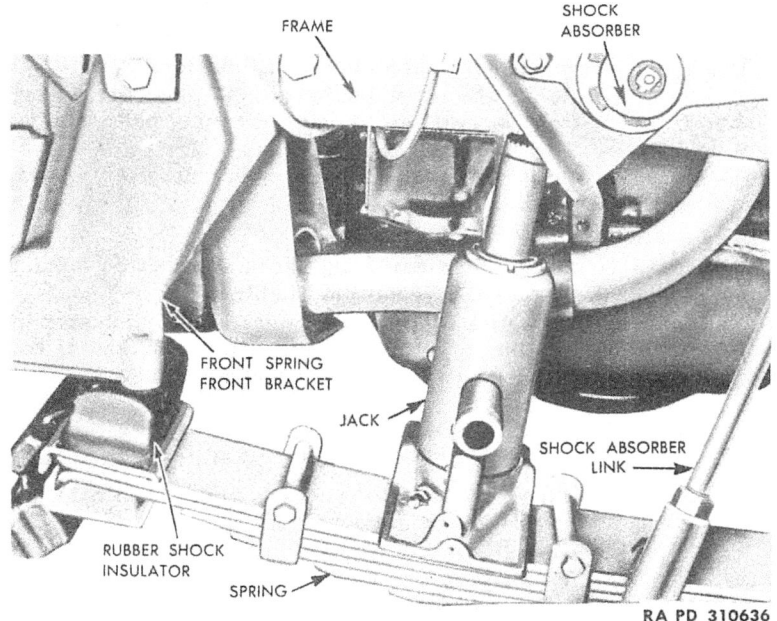

Figure 121 — Jack Placed Between Frame and Spring

(b) *Remove Lower Insulator Cap.* Remove three cap screws and lock washers from each insulator lower cap and remove caps.

(c) *Place Jacks.* Place jack under frame at rear of front spring and elevate sufficiently to place another jack between spring and frame (fig. 121).

(d) *Remove Rubber Shock Insulator.* Raise second jack sufficiently for rubber block to clear seat at front, and remove front insulator. Repeat for rear insulator, placing jack between frame and spring to rear of axle.

253

10-TON 6 x 4 TRUCK (MACK MODEL NR)

(2) INSTALLATION.

(a) Install Front Shock Insulators. Lubricate insulators with liquid soap. Place heavier block rubber, marked "TOP," on top of spring and, with the smaller block in its housing, lower jack between frame and spring, guiding upper block into proper position in upper seat. Repeat for installation of forward insulator.

(b) Remove Jacks. Remove jack from between frame and spring, and then remove jack at rear of spring.

(c) Install Insulator Caps. Place each cap under lower rubber block and attach by means of three cap screws and lock washers.

140. SHOCK ABSORBER ASSEMBLIES.

a. Removal.

(1) DETACH LINKAGE. Loosen nut on lower shock absorber linkage bracket. Insert bar between link rod and spring, and use hammer on bracket to loosen. Remove nut and remove link from bracket.

(2) DETACH SHOCK ABSORBER MOUNTING. With socket wrench remove two nuts from shock absorber body fastening bolts, another man holding bolt if necessary.

(3) REMOVE SHOCK ABSORBER. Remove shock absorber, leaving attaching bolts in place.

b. Installation.

(1) INSTALL SHOCK ABSORBER. Place shock absorber on bolts in frame. Replace lock washers and nuts and tighten.

(2) CONNECT LINKAGE. Place lower linkage ball joint stud in bracket on axle. Replace lock washer and nut and tighten. If ball joint stud should turn, hold with screwdriver in slotted end.

141. REAR SPRING ASSEMBLIES.

a. Removal.

(1) REMOVE TIRES AND RIMS. Place jack under each axle and elevate enough to remove tires. Remove tires and rims as outlined in paragraph 148 **a** (2). Block up each axle in this position.

(2) REMOVE SPRING ANCHOR CAPS. Remove four cap screws from spring-end anchor cap at each end of spring. Remove caps.

(3) DETACH BOOTS. Remove six cap screws, lock washers and clamps attaching dustproof boot to each shock insulator housing. Remove two bolts from each spring boot clamp on spring, using two wrenches, one to hold and one to turn.

(4) REMOVE SPRING CLIPS. Remove two bolts from trunnion bearing cover to provide clearance for socket wrench on outer spring clip nuts. Place jack under rear of frame, and support weight of chassis on jack. Do not place blocking on this jack too high, as frame will be dropped to rest spring bumpers on axles in order to remove springs. Remove spring clip nuts, using special long opening socket. Remove washers and remove spring clips, tapping upward with hammer and remove clip spacer.

TM 9-818
141

SPRINGS AND SHOCK ABSORBERS

(5) REMOVE RUBBER SHOCK INSULATOR LOWER CAPS. Place jack under cup and elevate only enough to support cap. Remove four cap screws and lock washers from each cap. Remove jack and drop cap. If lower insulating rubber blocks do not come out with cap, remove them.

(6) PLACE BLOCKS. Place blocking about 16 inches square and 40 inches long at end of trunnion (fig. 140).

(7) LOWER JACK. Lower jack under rear of frame until spring bumpers rest on axle housing. Remove upper rubber insulator blocks and slide boots off ends of spring.

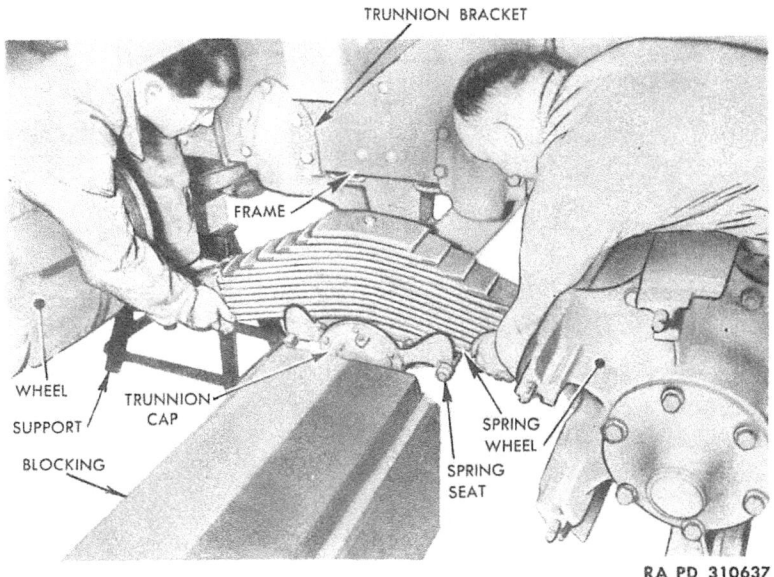

Figure 122—Removing Rear Spring

(8) REMOVE SPRING. Lift spring out of seat, place on blocking (fig. 122) and slide out.

b. Installation.

(1) INSTALL SPRING. Place spring on blocking and push into trunnion end (fig. 124). Lift spring onto spring seat. Slide boots onto spring ends with frames toward ends of spring.

(2) INSTALL UPPER RUBBER INSULATOR BLOCKS. Lubricate rubber insulator blocks with liquid soap. Do not use engine oil or grease on rubber. Place smaller blocks on top of spring ends with metal projection in socket on top of spring. Jack up chassis with jack at rear of frame to raise spring and rubber block into place in axle. Use

255

TM 9-818
141

10-TON 6 x 4 TRUCK (MACK MODEL NR)

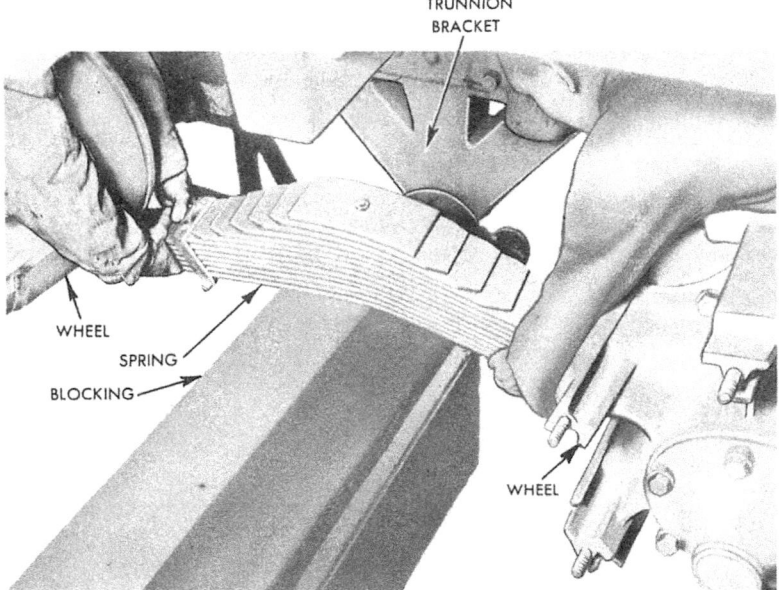

Figure 123—Rear Spring Removed

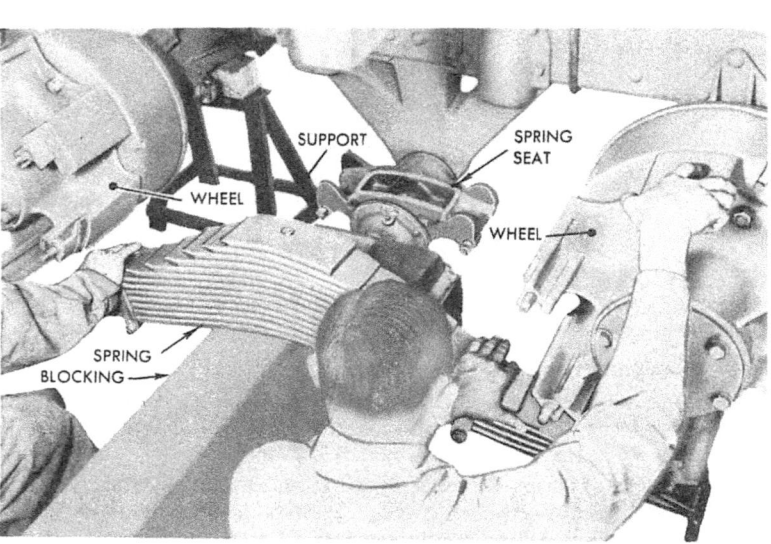

Figure 124—Installing Rear Spring

SPRINGS AND SHOCK ABSORBERS

screwdriver or other tool to ease rubber into socket. Do not use fingers. Remove blocking at end of trunnion.

(3) INSTALL LOWER INSULATOR BLOCKS AND CAPS. With lower rubber insulator block in cap, place cap on top of jack and elevate into position beneath spring end. Aline holes with drift. Install four cap screws with lock washers in each cap. Tighten alternately and remove jacks.

(4) INSTALL ANCHOR CAPS. Place anchor caps in position over spring-end anchor and install four cap screws and lock washers on each.

(5) INSTALL SPRING CLIPS. Place clip spacer on top of spring and install spring clips, hammering down into position. Apply lock washers and nuts, and tighten. Install two bolts, removed from trunnion cover, by use of two wrenches, one to hold and one to turn.

(6) INSTALL TIRES. Follow procedure outlined in paragraph 148 b (1) to install rear tires. Remove blocking supporting axles and lower to ground.

10-TON 6 x 4 TRUCK (MACK MODEL NR)

Section XXVIII

STEERING

	Paragraph
Description and tabulated data	142
Steering wheel	143
Steering gear assembly	144
Steering (Pitman) arm	145
Drag link assembly	146

142. DESCRIPTION AND TABULATED DATA.

a. Description (figs. 125 and 127). The steering gear is of the type in which the shaft turns a worm, imparting angular movement to a worm gear sector plate, with a shaft to which the steering (Pitman) arm is attached. The latter imparts steering movement to the front wheels through a drag link, a steering lever on the left-hand steering knuckle, and cross-steering levers on both knuckles, which are connected by a tie rod. The drag link has ball sockets on both ends which automatically compensate for wear by means of sliding wedges actuated by springs, thereby eliminating backlash between the sockets and the ball studs with which they engage. No adjustment of the sockets is required.

b. Data.

Type	Archimoid (worm and sector plate)
Make	Mack
Model	SG-16
Gear ratio	21 to 1
Steering wheel diameter	22 in.
Drag link, type	Automatic sliding wedge take-up
Wear take-up	Automatic
Adjustment	Threaded plug
Tube	Seamless steel

Clearances:

Jacket tube bushing to worm shaft	0.001-0.003
Worm shaft deflection, button to worm	0.0055-0.0155
Sector deflection, button to sector	0.015-0.021

143. STEERING WHEEL.

a. Removal.

(1) DISCONNECT HORN WIRE. Remove strap clamp at lower end of steering post to free horn wire, remove tape from connection and detach wires by removing screw and lock washer.

(2) REMOVE HORN BUTTON. Lift rubber horn button cover, turn button about one-third turn, and remove button and cover. Also remove contact cup, horn button spring and contact washer.

TM 9-818
143

STEERING

(3) REMOVE BASE PLATE AND WIRE. Remove the three wood screws from the base plate and remove base plate, horn wire ferrule, contact spring and washer.

(4) REMOVE NUT. Remove the nut from the top of the worm shaft.

(5) REMOVE STEERING WHEEL. Use puller to remove steering wheel.

b. Installation.

(1) INSTALL WHEEL. Place wheel on steering post. Install nut and tighten.

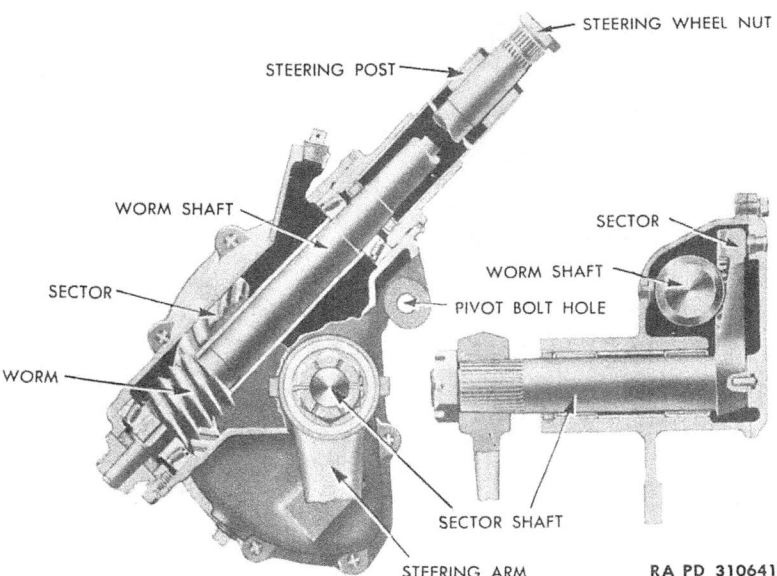

Figure 125—Sectional View of Steering Gear Assembly

(2) INSTALL HORN WIRE. Place washer in wheel, with contact spring and base plate at end, and pull wire with ferrule on it through the steering post. Attach base plate by means of three wood screws.

(3) INSTALL HORN BUTTON. Place contact washer on top of horn wire ferrule. Add horn button spring and contact cup and place horn button with rubber cover in wheel. Press down on button and turn one-third turn to engage button.

(4) ATTACH HORN BUTTON WIRE. Attach the horn button wire at the lower end of the steering post to the No. 14 yellow and black coded wire with screw, lock washer and nut, and tighten. Tape to insulate and attach to steering post by means of clamp.

259

TM 9-818
144

10-TON 6 x 4 TRUCK (MACK MODEL NR)

144. STEERING GEAR ASSEMBLY (fig. 125).

a. Removal.

(1) REMOVE STEERING WHEEL. Remove steering wheel as outlined in paragraph 148 a.

(2) REMOVE STEERING (PITMAN) ARM. Follow procedure outlined in paragraph 145 a (2), (3), without disconnecting the drag link from the arm.

(3) REMOVE CRANKING MOTOR. Follow procedure outlined in paragraph 91 a.

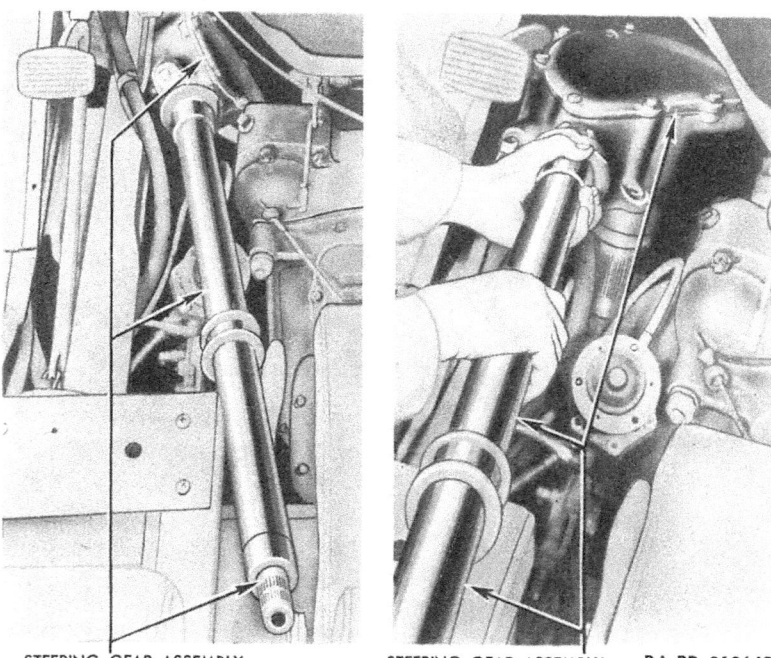

STEERING GEAR ASSEMBLY   STEERING GEAR ASSEMBLY   RA PD 310642

Figure 126 — Removing Steering Gear Assembly

(4) REMOVE FLOORBOARD. Disconnect accelerator pedal at ball joint by removing nut and pull pedal away from joint. Remove six cap screws from floorboards. Detach brake application valve pedal by removing three flathead, countersunk screws and remove pedal. Remove three cap screws from pedal mounting bracket and remove floorboard.

(5) REMOVE TOEBOARD. Remove eight screws from steering post draft plate and remove plate and felt. Remove nine cap screws from toeboard and remove toeboard.

(6) DETACH STEERING GEAR (fig. 126). Remove nut and lock washer from pivot bolt, using socket and open-end wrenches. Remove

STEERING

bolt with tapered washer at head. Remove two cap screws and lock washers from frame bracket. Remove two nuts, lock washers and bolts from dash clamp bracket. Remove speedometer cable from adapter, and remove cap screw and lock washer attaching cable clamp to transmission. Place hand brake lever in full "OFF" position. Remove cap screw from lower corner of steering gear cover for clearance. Remove steering post by placing it in position shown by left-hand view in figure 146. The right-hand view shows assembly being withdrawn.

145. STEERING (PITMAN) ARM.

a. Removal.

(1) FREE DRAG LINK. Loosen the drag link rear ball socket stud as outlined in paragraph 146 a (1) and (3). Turn front wheels completely to left and remove drag-link stud from the steering arm.

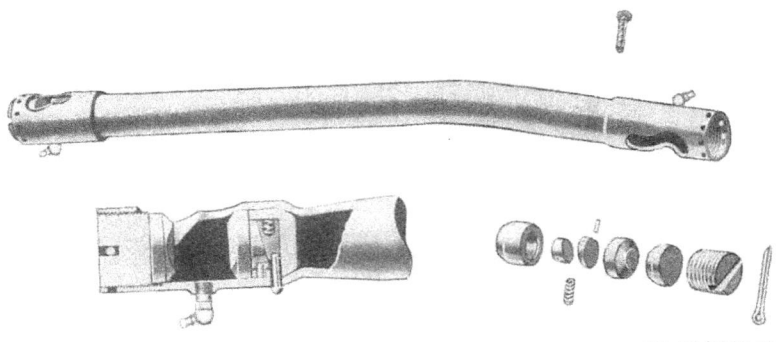

Figure 127 — Drag Link and Details of Socket End

(2) REMOVE NUT. Remove cotter pin and castle nut from steering sector shaft.

(3) REMOVE STEERING (PITMAN) ARM. Using universal puller, remove steering arm from steering sector shaft.

b. Installation.

(1) ALINE STEERING ASSEMBLY. Place front wheels in a straight-ahead position. Turning steering wheel from stop-to-stop positions, count the number of turns of wheel and return sector shaft to position attained by turning steering wheel one-half the number of turns from either stop position.

(2) INSTALL STEERING (PITMAN) ARM. Place rear drag link stud in steering arm socket and apply nut loosely. Install steering arm on sector shaft in a vertical position, and tap securely into place.

(3) INSTALL NUTS. Place castle nut on sector shaft and tighten down. Install cotter. Tighten nut on the drag link stud and install cotter pin.

10-TON 6 x 4 TRUCK (MACK MODEL NR)

146. DRAG LINK ASSEMBLY (fig. 127).

a. **Removal.**

(1) REMOVE NUTS. Turn wheels slightly to left to give access to drag link and remove cotter pins and castle nuts from the ball socket studs on each end of drag link.

(2) DISLODGE FORWARD STUD. Place jack underneath forward stud and elevate slightly. Use hammer on steering arm to free stud from socket. Remove stud from arm.

(3) REMOVE DRAG LINK. Use hammer on steering arm to dislodge rear stud, and remove drag link.

b. **Installation.**

(1) INSTALL DRAG LINK. After making sure that studs and sockets on drag link assembly are clean, place drag link in position with studs in sockets in steering lever (Pitman) arm. If lubrication fitting on forward end of shaft is to outside when stud is installed in steering lever, link is in proper position, but if fitting is to inside, drag link should be turned end-for-end.

(2) INSTALL NUTS. Tap ball studs securely into place with hammer and apply castle nuts. Tighten and install cotter pins.

TM 9-818
147

Section XXIX

TIRES

	Paragraph
Description and tabulated data	147
Tire and rim assembly	148
Tire casings and tubes	149

147. DESCRIPTION AND TABULATED DATA.

a. **Description** (fig. 147). Tires on both the front and rear wheels are single pneumatic, balloon type. There are two sizes, the rears being

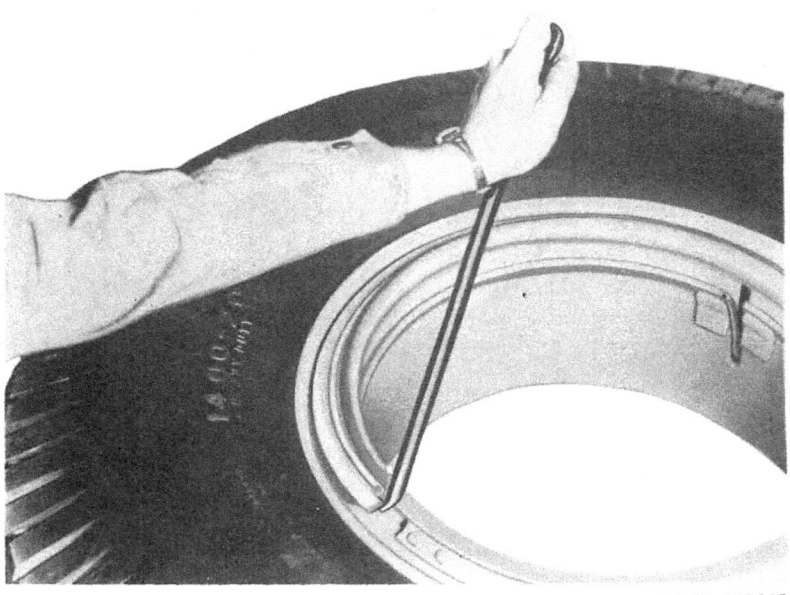

RA PD 310645

Figure 128—Removing Locking Ring

of larger section than the front tires, but of smaller base diameter. The tires are mounted on continuous base, demountable rims, having a locking ring to secure the tire. The rims seat on the wide surfaces of the spoke ends, which are machined for firm bearing of the rim and for clamping of its beveled head. Six lugs, one at each spoke, secure the rim to the wheel. A stud and nut hold each lug which bears against the spoke face and the edge of the rim. Two spare tires, one for the front and one for the rear wheels, are mounted on rims and carried across the back of the cab in a carrier which is supported by the chassis frame.

TM 9-818
147

10-TON 6 x 4 TRUCK (MACK MODEL NR)

b. Data.

Tire type	Pneumatic, balloon
Tire size:	
Front	11.00/24 (12-ply), single
Rear	14.00/20 (16-ply), single
Tire air pressure:	
Front	70 lb
Rear	90 lb

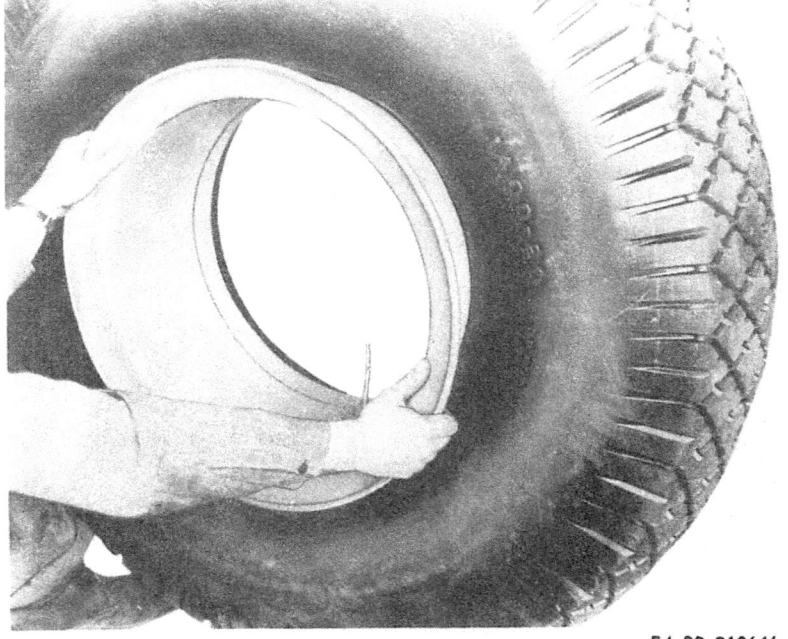

Figure 129—Removing Rim

Tube valve stem type:	
Front	Right-angle offset
Rear	Right-angle offset
Tube valve stem number:	
Front	T.R. 78 F 12
Rear	T.R. 79 A 38
Rim type, front and rear	R (continuous-base with locking ring)
Rim size:	
Front	24 x 9-10
Rear	20 x 11

TM 9-818
148-149

TIRES

148. TIRE AND RIM ASSEMBLY.

a. Removal.

(1) RAISE AXLE. After placing blocks under tires to prevent movement of vehicle, place jack under axle and elevate until tire clears ground.

(2) REMOVE TIRE AND RIM. Using wheel rim nut wrench in tool kit, remove the six lug nuts and remove lugs. Pull tire and rim off wheel, using tire iron and hammer to assist.

b. Installation.

(1) INSTALL TIRE AND RIM. Place tire and rim on wheel, driving into place with hammer and tire iron. Install lugs and lug nuts and tighten with wheel rim nut wrench.

(2) LOWER AXLE. Lower tire to ground with jack, and remove jack.

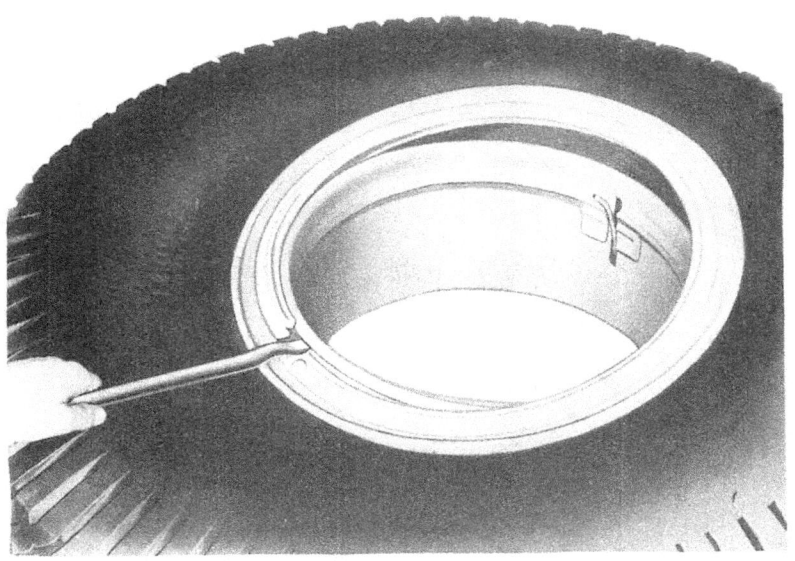

RA PD 310647

Figure 130—Installing Locking Ring

149. TIRE CASINGS AND TUBES.

a. Removal.

(1) REMOVE TIRE AND RIM ASSEMBLY. Follow procedure outlined in paragraph 148 a (figs. 128 and 129).

(2) REMOVE TIRE FROM RIM. Deflate tire, using valve cap inserted in valve stem to remove valve core. Place tire on flat surface with locking ring on upper side. With a tire iron, furnished with

TM 9-818
149

10-TON 6 x 4 TRUCK (MACK MODEL NR)

truck, loosen and remove locking ring and side ring from rim. Hold tire upright with valve at bottom and push rim free.

(3) REMOVE TUBE. Pull tube out of casing.

b. Installation.

(1) INSTALL TUBE. Insert tube (slightly inflated to have form) and flap in tire and inflate until tube is barely rounded out. Too much air will make mounting difficult.

(2) INSTALL TIRE ON RIM. Place rim on flat surface with gutter side up, and remove side ring and locking ring. Place tire over rim

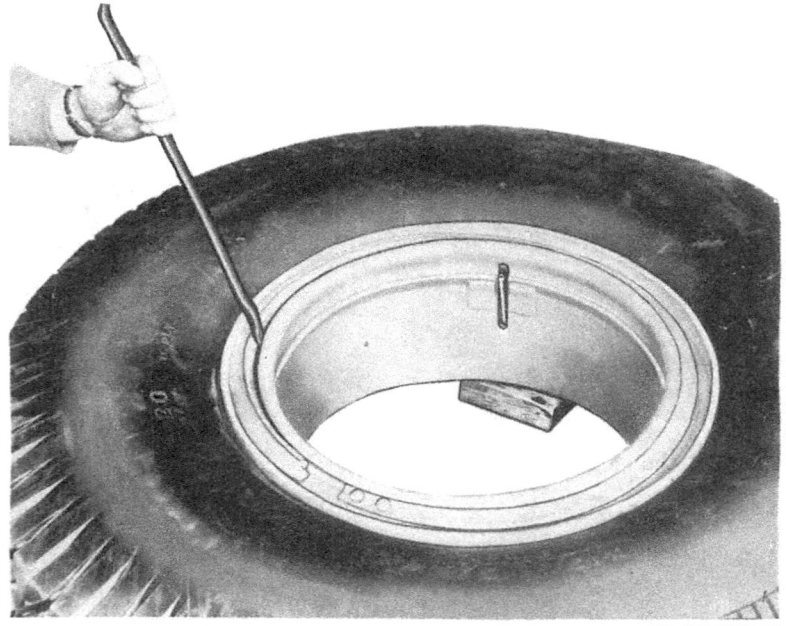

Figure 131—Seating Locking Ring

RA PD 310648

with valve stem in line with opening in rim and valve in line with balance mark on tire, if any. Insert valve and slide tire over rim. Install side ring and locking ring. Seat ring in gutter, for safety reasons, being absolutely sure locking ring seats properly. Inflate tire (again for safety reasons, locking ring side should be away from operator) and deflate to allow tube to become adjusted, then inflate to recommended pressure, 70 pounds front, 90 pounds rear (figs. 130 and 131).

(3) Install Tire and Rim. Follow procedure in paragraph 153 b (1).

TM 9-818
150

Section XXX

BODY AND CAB

	Paragraph
Description and tabulated data	150
Troop seats	151
Bow assemblies	152
Tarpaulin	153
Cab top and curtains	154
Cab seat cushions	155
Windshield assembly	156
Windshield frame assembly	157
Windshield wiper assembly	158
Floor and toeboards	159

150. DESCRIPTION AND TABULATED DATA.

a. Description.

(1) BODY (figs 1, 2, 3 and 4). The body is a cargo type, of wood construction reinforced with steel built upon heavy subframing. It is securely fastened to the chassis on oak sills, and is clamped to chassis frame at front and rear with long bolts which run through thick cross bars over the body sill and chassis rail bottom flange. The side rack may be used for troop seats by folding down approximately three-fourths of rack length at either side of body.

(2) CAB. The cab is of open-type, all-steel construction with a three-point mounting, the front two corners being solid, while the rear is flexible. The folding-type top with side curtains, together with a folding one-piece windshield serve to fully enclose upper half of cab. The windshield is equipped with slotted arms and thumbscrews, permitting it to be swung open to within 15 degrees of the horizontal position. The windshield is mounted in a movable framework which can be laid down flat, when the cab top curtain is untied at the back, by loosening the large thumbscrews on either side at the bottom. When in folded-down position, thumbscrews should be tightened, and top curtain used as a windshield cover to prevent reflected glare from the glass being visible to aircraft. Rear vision is obtained by means of two outside, rear-vision, swiveled mirrors, or if desirable, the rear curtain may be rolled up and strapped to the rear bow. A canvas bag container for the door and side curtains is hung in back of the driver's seat. A tool box is provided under the driver's seat cushion, and a larger compartment for other items is furnished under the right-hand seat. The cab doors are fitted with a long hook-and-eye arrangement whereby the doors can be held in a partly open position to increase ventilation at the cab floor level. The driver's seat is adjustable toward front and rear by means of a slotted lever and fixed stop. Operated with a handle located on the front of the seat frame just below bottom of the cushion, it permits a close adjustment of seat position to suit individual driver. The right-hand seat and back are hinged so that they can be quickly

10-TON 6 x 4 TRUCK (MACK MODEL NR)

folded up out of the way. Dual air-operated windshield wipers are mounted at the top of the windshield frame on the inside, and are controlled by valves on the instrument panel.

b. **Data.**

Body .... Perfection Steel Body Co.
Cab .... Mack
Hardware .... C. Cowles and Co.
Windshield wiper—make .... Trico
Model 4923
Cab cushions .... Duck covered, spring construction

151. TROOP SEATS.

a. **Remove Hinge Pins.** Take cotter pins out of hinge pins with pliers. Drive out hinge pins with hammer and soft drift. It may be necessary to lower tail gate to remove hinge pin at rear and/or remove front piece to remove front hinge pins.

b. **Remove Troop Seats.** Remove troop seats.

c. **Seat Supports.** If it is necessary to replace seat supports, remove bolt at seat end and remove support, installing new support by inserting bolt and nut.

d. **Install Troop Seats.** Line up hinges with drift pin and install hinge pins, locking in place by installing cotter pins.

152. BOW ASSEMBLIES.

a. **Replace Bow Assemblies.** Raise bow assemblies out of socket on each side and install new bow assemblies by dropping into sockets.

153. TARPAULIN.

a. **Removal.**

(1) LOOSEN TIE ROPES. Untie knots at the 16 body lashing hooks on sides and front of the body and remove ropes from hooks.

(2) REMOVE TARPAULIN. Fold up sides of tarpaulin in approximately three-foot widths onto top of bows on each side. Make an additional fold to consolidate tarpaulin in approximately three-foot wide roll, of the length of body. Fold from front and rear, so that finished bundle is approximately three feet square. Remove tarpaulin.

(3) FRONT AND REAR PIECES. If necessary to remove front or rear pieces, unlash rope and detach straps from bows and remove.

b. **Installation.**

(1) INSTALL FRONT AND REAR PIECES. Hold front or rear pieces in place, and thread lashing rope around poles and through grommets successively.

(2) INSTALL TARPAULIN. Place folded tarpaulin on two center bows, unfold on bows, positioning so that there is approximately one foot overlap on each end. Unfold to sides so that strap buckles on

BODY AND CAB

each side line up with bows. Make certain end marked "FRONT" is forward and that buckles are on outside. Lash ropes to hooks, crossing front stay ropes, and tie fast.

154. CAB TOP AND CURTAINS.

a. Replace Door Curtain.

(1) REMOVE DOOR CURTAIN. Release eyelet fastener by removing leather strap from eyelet and pull curtain free. Raise curtain, prying with screwdriver under metal frame if necessary to free support iron from socket, and raise out of track on end of windshield.

(2) INSTALL CURTAIN. Thread bead on forward end of curtain into channel on end of windshield and pull curtain down, inserting support iron into socket in door. Push down until iron is seated properly and fasten back corner of curtain over eyelet on door, passing leather strap through eyelet to hold.

b. Replace Rear Side Curtains.

(1) REMOVE CURTAIN. Unfasten the eight eyelet fasteners by removing straps from eyelets. Detach curtain and remove.

(2) INSTALL CURTAIN. Place curtain with fasteners over eyelets, the middle flap under or inside of the top bow supports and the top and bottom flaps outside of supports, and fasten with straps in eyelets.

c. Curtain Storage. Curtains when not in use are stored in the duck bag in back of driver's seat.

d. Replace Cab Top Back.

(1) REMOVE BACK. Untie stay ropes at each side of cab. Remove rope from hook fastenings along top of cab back. Detach eyelet fastenings at each end. Remove fifteen roundhead cross-recessed screws from rear top bow. Remove back.

(2) INSTALL BACK. Place back on rear top bow so that 15 holes in bead line up with those in bow, with seam toward top and against bow. Install 15 roundhead cross-recessed screws. Lace rope through hooks on top of cab back, tieing rope at ends on tops of handles on sides of cab.

(3) TO ROLL UP BACK. Unfasten back from cab as noted in step d (1) above, and fold in ends and rope. Roll up back with roll forward and fasten in place with web strap around roll and over top of bow. Buckle strap.

e. Replace Top.

(1) REMOVE TOP. Untie rope on each side and unhook rope from fastening in back of cab. Bring top forward over hood. Using a screwdriver, pry open each end of channel frame on top of windshield, so that top bead may be slid out of channel. Remove top by sliding out of channel.

(2) INSTALL TOP. Slide bead on front end of new top into channel on top of windshield with seam up or forward. When in place, peen ends of channel slightly with hammer, to clinch. Pull top back over

TM 9-818
10-TON 6 x 4 TRUCK (MACK MODEL NR)

bows. Attach rope loop to hook in center of back of cab and tie rope from each end around handle on each side.

(3) TO ROLL TOP. Detach ropes on side and back of cab as outlined in step d (1) above. Fold in rope and sides and roll forward, roll to bottom, fastening roll on top of windshield assembly with web strap passed through eyes on front of windshield frame and returning to buckle.

(4) WINDSHIELD COVER. When windshield is in horizontal position, top may be laid back over windshield to form protective cover.

f. **Replace Bow Assembly.**

(1) REMOVE ASSEMBLY. Remove nuts from two flathead screws on each side of top of cab back. Remove screws and lock washers. Remove the nuts from two bolts on each side bracket and remove bolts and lock washers. Raise bow assembly to remove.

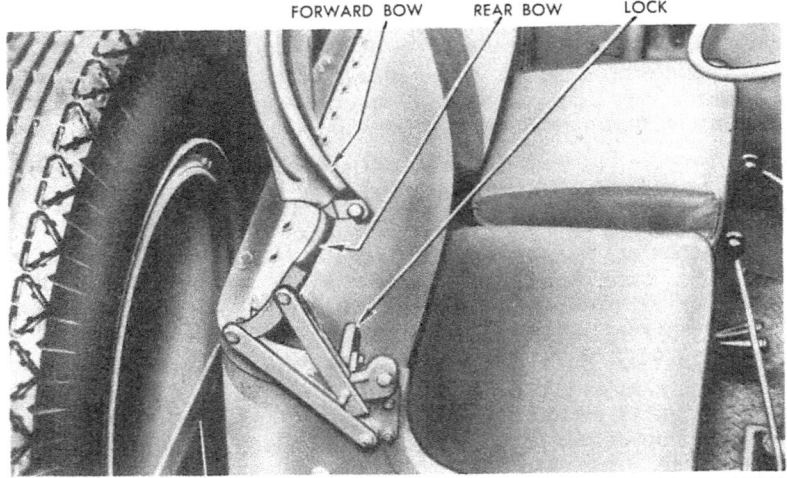

Figure 132—Cab Top Bows Folded

(2) INSTALL ASSEMBLY. Place bow assembly brackets on top of seat back with bracket holes over holes in cab side. Insert two bolts at side from the inside, attach lock washers and nuts and tighten. Insert two flathead cross-recessed screws from the top. Attach lock washers and nuts and tighten.

(3) TO FOLD BOWS. Turn lock in back of bows support toward inside of cab to fold bows support down. With back rolled into position on rear bow, fold forward bow back against rear bow and force bows toward rear, which will start to fold bows down. Drop bow supports until channel rests on leather-covered pin on each side of cab (fig. 132).

TM 9-818
154-156

BODY AND CAB

(4) To ELEVATE BOWS. Pull up on bows, with some force toward rear, to unfold supports and push support assembly forward to clear lock. Turn locking piece into plate in back of supports to hold.

155. CAB SEAT CUSHIONS.

a. **Replace Right-hand Cushion.**

(1) REMOVE CUSHION. Raise right-hand seat back, raise right-hand cushion and remove six flathead screws from hinges. Remove cushion.

(2) INSTALL CUSHION. Place new cushion on hinges in upward position and attach with three flathead screws in each hinge. Lower into place.

b. **Right-hand Seat Back.**

(1) REMOVE SEAT BACK. Raise right-hand seat back and remove the four roundhead screws from hinges. Remove seat back.

(2) INSTALL SEAT BACK. Place new seat back on hinges in up position and attach to hinges, using two roundhead screws in each hinge. Lower into place.

c. **Left-hand Seat Cushion.**

(1) REMOVE SEAT CUSHION. Lift out driver's seat cushion to remove.

(2) INSTALL CUSHION. Place new cushion on left-hand seat frame, making certain that dowels in the bottom of the forward part of cushion enter the holes in seat carriage.

d. **Left-hand Seat Back.**

(1) REMOVE SEAT BACK. Remove six square nuts from back of seat back frame and remove cushion.

(2) INSTALL SEAT BACK. Place cushion on seat back frame with six bolts through holes in frame and attach with six square nuts.

156. WINDSHIELD ASSEMBLY.

a. **Removal Procedure.**

(1) DISCONNECT BATTERY. Follow procedure outlined in paragraph 167 f (1) (2).

(2) REMOVE TOP. Follow procedure outlined in paragraph 154 e (1).

(3) REMOVE RIFLE HOLDER IF IT IS MOUNTED AT RIGHT SIDE OF COWL. Remove rifle holder bracket from windshield assembly by removing nut. Remove lock washer, bolt and washer. Remove nut from bottom rifle holder bracket at holder. Remove lock washer and bolt. Remove cap screw from bottom of rifle holder with socket wrench and remove holder.

(4) REMOVE INSTRUMENT BOARD SCREWS. Remove nine cross-recessed head screws from top of instrument board.

TM 9-818
156

10-TON 6 x 4 TRUCK (MACK MODEL NR)

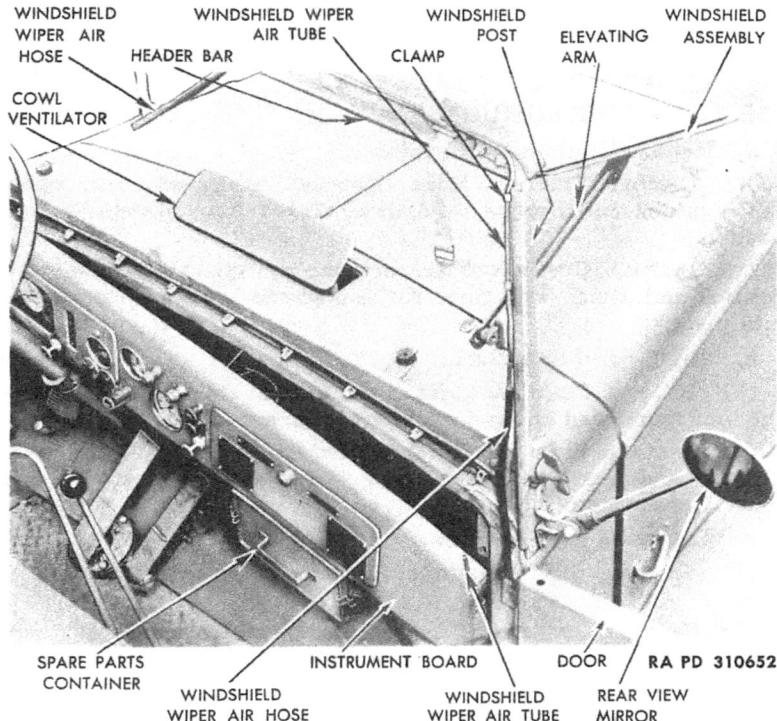

Figure 133—Instrument Board Partially Removed

(5) REMOVE WINDSHIELD WIPER AIR HOSE. Pull windshield wiper hose off tube through each end of instrument board. Remove hose clamps by removing four cross-recessed head screws from right-hand clamps, and three from left-hand clamps.

(6) REMOVE INSTRUMENT BOARD BOLTS. Remove two bolts on each side under cowl, which attach instrument board to cowl, using two wrenches, one to hold bolt and the other to turn nut.

(7) FREE INSTRUMENT BOARD. Pull instrument board free from cowl, using screwdriver as wedge, if necessary. Do not remove instrument board but leave in this free position.

(8) REMOVE WINDSHIELD WIPER ASSEMBLIES. Follow procedure outlined in paragraph 158 a to remove windshield wiper assemblies.

(9) REMOVE REAR-VIEW MIRRORS. Remove six cross-recessed head screws from each rear-view mirror bracket on the windshield posts.

(10) REMOVE COWL BAR BOLTS. Remove nuts from the three left-hand and two remaining right-hand bolts in windshield cowl bar. Remove five cross-recessed head screws, holding cowl bar to

272

BODY AND CAB

cowl, using screwdriver under cowl. It will be necessary to use a short screwdriver on the right side, and a long screwdriver on the left.

(11) REMOVE WINDSHIELD POST BOLTS. Remove the nuts, bolts, and lock washers from windshield post on top of cowl at each side, using two wrenches, one to hold and one to turn. Drive out bolts. Remove the three bolts attaching lower end of each windshield post to side of cowl, using two wrenches, one to hold and the other to turn. Remove lock washers and bolts as well as door stop plate from upper bolt.

(12) REMOVE WINDSHIELD ASSEMBLY. Elevate the windshield assembly to remove from cowl. The cowl bar will come off with the assembly.

(13) REMOVE COWL BAR. Remove the wing nut from the windshield elevating arm on one side, and on the same side remove the two screws from the upper end of windshield post, using screwdriver to turn screw and wrench to hold nut. Remove the windshield post on this side. This will free cowl bar. Remove cowl bar.

(14) REMOVE TOP STRAPS AND LOOPS. Pull cab top straps out of loops to remove. Remove the six strap loops from windshield header bar by removing the six screws and nuts.

b. Installation.

(1) INSTALL COWL BAR. Remove one windshield post by removing the two nuts from cross-recessed head screws with wrench, and removing screws. Remove thumbscrew from elevating arm assembly. Install cowl bar on other windshield post and reinstall post, attaching at windshield header bar with the cross-recessed head screws and nuts, and tighten. Attach thumbscrew in elevating arm assembly.

(2) INSTALL LOOPS AND STRAPS. Replace three strap loops on front, and three on rear of header bar with head screws and nuts, and install straps in loops.

(3) MOUNT WINDSHIELD ASSEMBLY. Place antisqueak on cowl, and place windshield post in holes in cowl with windshield lock toward rear. Drop into place.

(4) INSTALL POST AND COWL BAR FASTENINGS. Fasten cowl bar to cowl with five cross-recessed head screws in under cowl, using short screwdriver for screws at right and long screwdriver for screws at left. Install bolts in windshield post on top of cowl at each side. Place lock washers and nuts, and tighten. Install three bolts from outside in each windshield post to attach to cowl, with door stop plate attached by top bolt on each side. Place lock washers and nuts and tighten. Line up cowl bar holes with holes in post, by use of drift. Insert bolts from the cab side, three on each side, and apply lock washers and nuts to all excepting the right-hand upper bolt which holds rifle holder. Tighten nuts.

(5) INSTALL WINDSHIELD WIPER AIR SUPPLY TUBES. Hold windshield wiper air supply tubes in place on windshield frame and attach with four cross-recessed head screws on right, and three on left side.

10-TON 6 x 4 TRUCK (MACK MODEL NR)

(6) INSTALL REAR-VIEW MIRRORS. Holding mirrors in place on windshield post, insert six cross-recessed head screws in each mirror bracket, and tighten.

(7) INSTALL INSTRUMENT BOARD. Push instrument board into place and install nine cross-recessed head screws across top of board. Insert the two bolts in cowl bracket at each end of instrument board, from the rear. Install lock washers and nuts, and tighten with two wrenches, one to hold and one to turn.

(8) INSTALL RIFLE HOLDER. Place rifle holder against brackets on right-hand end of cowl, making sure antisqueak fabric is placed under holder. Insert the bolt in upper bracket from inside of cowl bar with plain washer under bolt head. Install lock washer and nut, and tighten. Install cap screw through bottom of holder, and tighten. Install bolt at bottom bracket, inserting from rear. Install lock washer and nut.

(9) CONNECT WINDSHIELD WIPER HOSE. Connect windshield wiper hose by sliding rubber tube onto tubes extending through top of instrument board at ends.

(10) INSTALL WINDSHIELD WIPER ASSEMBLY. Follow procedure outlined in paragraph 158 b to install wiper assemblies.

(11) INSTALL CAB TOP. Follow procedure outlined in paragraph 154 e (2) to install top of cab.

(12) LOWER WINDSHIELD. Release windshield elevating arm thumbscrew, and lower windshield into place.

157. WINDSHIELD FRAME ASSEMBLY.

a. Removal.

(1) REMOVE WINDSHIELD WIPERS. Follow procedure outlined in paragraph 158 a to remove wiper assemblies.

(2) REMOVE ELEVATING ARM WING NUT. Remove elevating arm wing nut.

(3) LOOSEN POST SCREWS. Loosen the two screws at one end of windshield header bar.

(4) REMOVE WINDSHIELD. Open windshield shoe and slide out to side, supporting outer end to ease sliding (fig. 134).

(5) REMOVE ELEVATING ARM. Remove elevating arm and bracket with wrench and screwdriver. Remove screws in elevating arm brackets, and remove arms.

b. Installation.

(1) INSTALL ELEVATING ARMS. Place elevating arms in position on windshield, and insert two cross-recessed head screws from rear. Install lock washers and nuts, and tighten.

(2) INSTALL WINDSHIELD. Holding windshield assembly in horizontal plane, engage windshield frame at top in sliding track in header bar, and slide windshield into place.

(3) INSTALL THUMBSCREWS. Install elevating arm thumbscrew through arm, and turn into bracket on windshield post.

TM 9-818
157-158

BODY AND CAB

(4) INSTALL WINDSHIELD WIPERS. Follow procedure outlined in paragraph 158 b to install wipers.

(5) TIGHTEN HEADER SCREWS. Tighten two screws on head or bar on top of windshield post.

158. WINDSHIELD WIPER ASSEMBLY.

a. Removal.

(1) DISCONNECT HOSE. Using wrench to hold fitting, turn connection with another wrench, and disconnect hose. Remove fitting with first wrench.

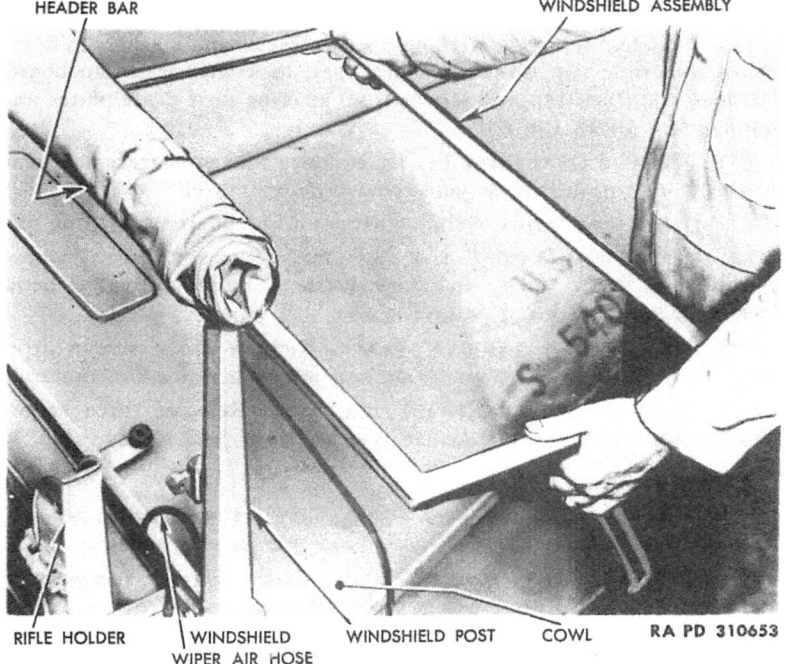

Figure 134—Removing Windshield

(2) REMOVE BLADE AND ARM. Remove nut from end of blade shaft and remove blade and arm shaft.

(3) REMOVE WIPER MOTOR. Remove two wiper bracket screws from windshield frame on forward side with screwdriver. Remove wiper motor.

b. Installation.

(1) INSTALL WIPER MOTOR. Place motor on inner side of windshield frame so that two screws can be inserted through holes in front of windshield. Insert screws in wiper motor and tighten.

275

10-TON 6 x 4 TRUCK (MACK MODEL NR)

(2) INSTALL WIPER ARM AND BLADE. Mount wiper arm and blade on motor shaft. Apply nut to shaft and tighten.

(3) INSTALL AIR HOSE. Install air fitting on wiper motor, and attach hose to fitting with one wrench, using another to hold fitting.

159. FLOOR AND TOEBOARDS.

a. Removal.

(1) REMOVE ACCELERATOR PEDAL. Follow procedure outlined in paragraph 176 d (1) *(a)*.

(2) REMOVE BRAKE APPLICATION VALVE PEDAL. Follow procedure outlined in paragraph 118 a (1).

(3) REMOVE SCREWS. Remove six cap screws and washers in floorboard and nine cap screws, washers, and lock washers in toeboard. Remove eight self-tapping screws from steering post draft plates and remove two plates and felt.

(4) REMOVE GEARSHIFT PLATE. Remove four cross-recessed head screws from gearshift plate, and remove plate and felt.

(5) REMOVE BOARDS. Remove floor- and toeboards by lifting out.

(6) REMOVE VALVE PEDAL SPACER. Remove three flathead countersunk nuts, bolts, and lock washers attaching valve pedal spacer to floorboard and remove spacer.

(7) REMOVE ACCELERATOR PLATE. Remove three screws from accelerator plate and remove plate and felt or composition material.

(8) REMOVE ACCELERATOR PEDAL STOP. Remove three screws which attach accelerator pedal stop to toeboard, and remove stop.

b. Installation.

(1) INSTALL ACCELERATOR PEDAL STOP. Attach pedal stop to floorboard with three cross-recessed head self-tapping screws.

(2) INSTALL ACCELERATOR PLATE. Place felt or composition material over accelerator hole in toeboard and attach plate, fastening with three screws.

(3) INSTALL VALVE PEDAL SPACER. Place spacer on floorboard, thin edge to the rear, and attach with three flathead countersunk bolts, the longer bolts forward. Install lock washers and nuts and tighten with screwdriver above and wrench below.

(4) ATTACH ACCELERATOR PEDAL TO FLOORBOARD. Place spacer on floorboard in proper position, and attach accelerator hinge bracket and spacer to floorboard, using two flathead, countersunk screws.

(5) INSTALL BOARDS. Place floorboard in floor supports, and attach with six cap screws and washers. Put toeboard in place and attach with nine cap screws, washers, and lock washers. Retain one cap screw to install after steering post plates are installed.

BODY AND CAB

(6) INSTALL GEARSHIFT PLATE. Place felt over shift levers and install plate, inserting four self-tapping screws.

(7) INSTALL STEERING POST DRAFT PLATE. Place split plates in position around steering post, after applying felt, and attach with eight self-tapping screws. Insert cap screw with washer and lock washer and tighten.

(8) ATTACH BRAKE APPLICATION VALVE PEDAL. Follow procedure outlined in paragraph 118 b (4).

(9) CONNECT ACCELERATOR PEDAL. Follow procedure outlined in paragraph 175.

TM 9-818
160

10-TON 6 x 4 TRUCK (MACK MODEL NR)

Section XXXI

FRAME, FENDERS AND RUNNING BOARDS

	Paragraph
Description and tabulated data	160
Front bumper	161
Towing hooks	162
Pintle hook	163
Front fenders	164
Running boards	165

Figure 135—Frame, Towing Hooks and Pintle Hook

160. DESCRIPTION AND TABULATED DATA.

a. **Description** (fig. 135). The chief structural member of the chassis, the frame, is composed of two parallel side rails kept in alinement by a series of properly spaced alligator, box and channel-type cross members. The pintle hook (fig. 136) is attached in the rear of the chassis frame for towing purposes. There are also four towing hooks—two at rear and two at front of chassis. The rear hooks are attached one to each outside face of side rail, close to rear end.

TM 9-818
160

FRAME, FENDERS AND RUNNING BOARDS

The front hooks are bolted to top of chassis side rail extension at front end, directly above bumper. The front bumper is a welded assembly consisting of a single pressed-steel channel reinforced with a flat plate, which serves also as a lower anchorage for the brush guard. The bumper is bolted to side rail extensions, both through the channel section and through the reinforcing plate. A hole in the center of bumper channel allows hand crank to be passed through to engage starting jaw on engine and also to provide a front bearing for hand crank. The right-hand end of channel is perforated to permit mounting the fender guide. The front fenders are of the conventional crown design, forming a unit with skirt extending to frame. The inner skirts, forming lower side sections of engine enclosure, are

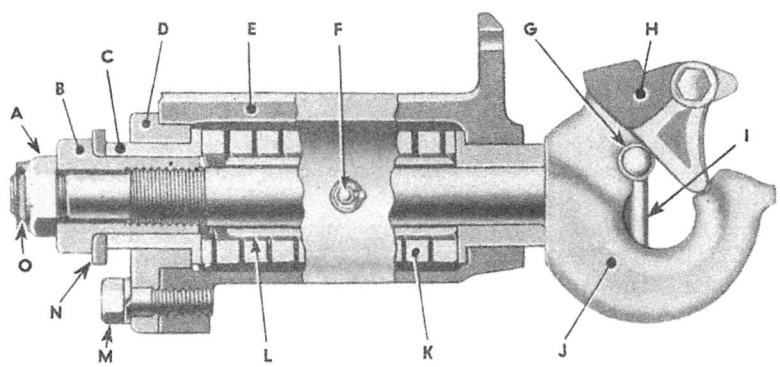

A	PINTLE HOOK NUT	H	LATCH
B	HOUSING BUSHING NUT	I	PIN HANDLE
C	HOUSING BUSHING	J	PINTLE HOOK
D	HOUSING CAP	K	PINTLE HOOK SPRING
E	PINTLE HOOK HOUSING	L	SPRING TUBE
F	LUBRICANT FITTING	M	HOUSING CAP SCREW
G	LATCH PIN	N	THRUST WASHER
		O	NUT COTTER PIN

RA PD 310655

Figure 136 — Pintle Hook

louvered to permit free air passage. Fenders are rigidly supported at front by channel brackets extending from chassis frame, and at rear by running boards to which they are fastened. Running boards are mounted at each side of cab, and are supported by channel-type step hangers which are strengthened by triangular shaped plates attached to side rails.

b. **Data.**
Wheelbase:
    Front axle to forward rear axle............................173 in.
    Forward rear axle to rearward rear axle...................55 in.

279

10-TON 6 x 4 TRUCK (MACK MODEL NR)

Frame width .... $33^{5}/_{16}$ in.
Number of cross members 7
Side rail material .... Chrome-manganese steel
Cross member material .... Plain carbon steel
Side rails, type .... Pressed channel
Side rails, size .... $10^{9}/_{16}$ x $3^{1}/_{4}$ x $^{9}/_{32}$ in.

161. FRONT BUMPER.

a. Removal.

(1) REMOVE TOWING HOOKS. Remove towing hooks as outlined in paragraph 162 a.

(2) REMOVE BRUSH GUARD. Remove the four bolts, lock washers, and nuts from the bumper brush guard mounting. Remove the two bolts, lock washers, and nuts from the brush guard braces. Remove the brush guard.

(3) REMOVE FENDER GUIDE. Remove nut from bottom of fender guide and remove the lock washers, pulling the fender guide out of the hole.

(4) REMOVE BUMPER. Remove the eight carriage bolts, lock washers, and nuts from the bumper mounting. Lift bumper off frame and remove spacer plates.

b. Installation.

(1) INSTALL BUMPER. Place spacer plates on side rail extension with the larger hole forward, and place the bumper on the plates, lining up the holes with drift pins. Install eight carriage bolts, lock washers, and nuts.

(2) INSTALL FENDER GUIDE. Place fender guide in hole on right-hand upper bumper flange, with nut and lock washer above and below flange, apply nut and tighten.

(3) INSTALL BRUSH GUARD. Mount brush guard on bumper. Insert four bolts in bumper mounting. Install lock washers and nuts. Place two bolts through brush guard braces with flat washers at brush guard, apply lock washers and nuts and tighten.

(4) INSTALL TOWING HOOKS. Install towing hooks as outlined in paragraph 162 b.

162. TOWING HOOKS.

a. Removal.

(1) REMOVE BOLTS. Remove the ¾-inch bolt, lock washer, and nut, using socket wrench to turn, and open-end wrench to hold. Remove the ⅝-inch bolt, lock washer, and nut, using socket wrench to turn, and open-end wrench to hold.

(2) REMOVE TOWING HOOK. Lift towing hook and remove spacer washer.

b. Installation.

(1) INSTALL TOWING HOOK. Place spacer washers so that the larger is at the front and place towing hook upon them.

TM 9-818
162-163

FRAME, FENDERS AND RUNNING BOARDS

(2) INSTALL BOLTS. Insert the 3/4-inch bolt through the forward hole and the 5/8-inch bolt through the rear hole. Apply lock washers and nuts and tighten.

163. PINTLE HOOK.

a. **Maintenance.** Inspect pintle hook latch pin and latch pin lock for wear. If there is any danger of the pin slipping out it should be replaced with the following:

(1) One hex head SAE steel bolt 3 3/4 inches long by 7/8 inch in diameter, two tapered washers, one 7/8-inch SAE castellated nut, and one cotter pin. NOTE: *If this bolt is not available it can be made out of round bar stock threaded on each end.* Install a nut on one end and weld solid to form the head of the bolt.

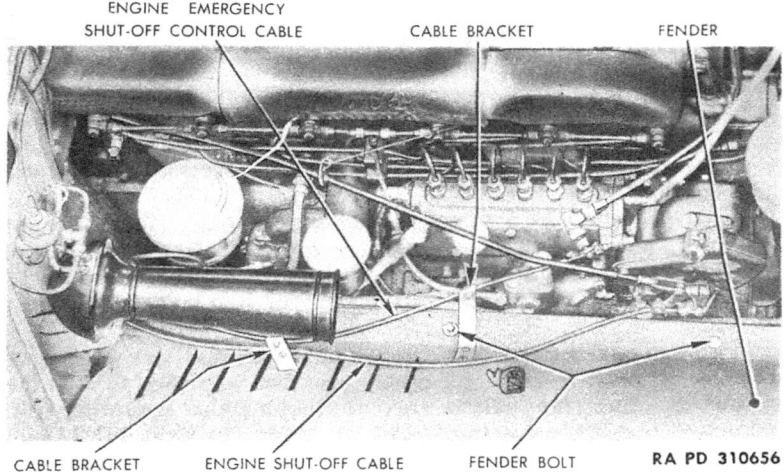

Figure 137—Fender Bolts and Cable Clamps

(2) When installing bolt and tightening nut, care should be taken to clamp the pintle latch. Lock with cotter pin.

(3) Discard old latch pin and latch pin lock screw. Plug latch pin lock screw hole with wooden plug.

b. **Removal.**

(1) REMOVE BOLTS. Remove 11 mounting bolts, nuts, and lock washers, and the three tapered washers on the bolts on the rear flange, using two wrenches.

(2) REMOVE PINTLE HOOK. Remove pintle hook.

c. **Installation.** Place pintle hook assembly on frame member and install the three forward bolts on each side. Apply lock washers

281

TM 9-818

10-TON 6 x 4 TRUCK (MACK MODEL NR)

and nuts, and tighten. Install three bolts through the flange at rear. Place tapered washers, with thin edge up, inside of the pintle hook cross member. Apply lock washers and nuts, and tighten. Install the remaining two bolts with lock washers and nuts, and tighten.

164. FRONT FENDERS.

a. Removal.

(1) REMOVE BOND STRAPS. Detach bond straps from front of left fender to radiator shell and strap from under fender to frame, removing the bolts, nuts and toothed lock washers.

(2) DETACH BLACKOUT LIGHT WIRE. Disconnect wire from blackout fender light by pulling out plug. Remove the clamp and screw with cross-recessed screwdriver and wrench.

(3) REMOVE HEADLIGHT (right only). Remove nut and lock washer, using socket wrench, remove headlight and disconnect wire by turning plug one-quarter turn and pulling out.

(4) UNFASTEN HEADLAMP BRACKET. Remove the two cap screws from headlight bracket, leaving it suspended by the wire or laid on the spring.

(5) REMOVE CLAMPS. Remove the engine clamps attaching emergency shut-off valve cable to the two cable brackets, using cross-recessed screwdriver and wrench. Remove the clamp for the engine shut-off cable from the rear bracket, using cross-recessed screwdriver and wrench.

(6) REMOVE FENDER BOLTS. Remove three bolts (four on right side), nuts and lock washers attaching fender to frame flange and remove the two bolts, nuts, and lock washers as well as spacer attaching fender to running board. Remove the four step bolts from fender bracket.

(7) REMOVE FENDER. Remove one bolt, nut, and lock washer at lower end of brush guard angle brace, using two wrenches. Loosen the one nut on upper rear radiator support cross member mounting bolt, and move the brace rearward. Lift up fender to clear brace, and remove.

(8) REMOVE SPONGE RUBBER. Open the split rivets attaching sponge rubber antisqueak to fender and remove sponge rubber.

(9) REMOVE HOOD CENTER BUMPER BRACKET. Use cross-recessed screwdriver and wrench to remove two screws, lock washers and nuts from hood center bumper bracket, and remove bracket from fender.

(10) REMOVE BLACKOUT LAMP. Remove nut and blackout lamp mounting, and remove lamp and wire with rubber washer.

(11) REMOVE CONTROL BRACKETS (right only). Remove two screws, nuts, and lock washers from each of two brackets, and remove control brackets.

b. Installation.

(1) INSTALL CABLE BRACKETS (right only). Place the longer cable bracket on right-hand fender, ahead of vent louvers, with short end toward outside, and attach with two screws, nuts, and lock washers. Place the shorter cable bracket between the fourth and

FRAME, FENDERS AND RUNNING BOARDS

fifth louvers, with the shorter side toward the inside, and attach each with two screws, nuts, and lock washers.

(2) INSTALL BLACKOUT LAMP. Placing wire and rubber washer in hole in fender, mount the blackout fender lamp on fender and attach with nut.

(3) INSTALL HOOD BUMPER BRACKET. Place hood bumper bracket on fender forward of louvers, with felt end upward and outward, and attach with two cross-recessed head screws, lock washers, and nuts, using cross-recessed screwdriver and wrench to tighten.

(4) ATTACH SPONGE RUBBER. Place sponge rubber antisqueak on fender where running board attaches, and fasten in place with split rivets.

(5) INSTALL FENDER. Lift fender and drop into place on side rail, with the brush guard angle brace through the hole in forward end of fender, and with antisqueak fabric in place along side rail.

(6) INSTALL FENDER BOLTS. Install three bolts on left and four on right side, attaching fender to frame flange, and install the two bolts and spacer attaching each fender to running board. Install lock washers and nuts. Install the four step bolts in each fender support bracket, and apply lock washers and nuts.

(7) INSTALL CONTROL CABLE CLAMPS. Install engine shut-off control cable in lower clamp on rear cable bracket and tighten, using cross-recessed screwdriver and wrench. Using the same tools, attach engine emergency shut-off valve control cable in top clamp on rear bracket, and in clamp on forward bracket.

(8) INSTALL HEADLIGHT BRACKETS AND LIGHT. With the upper end of each headlight bracket through hole in fender, attach bracket to front spring front bracket with two cap screws. Attach headlight to right-hand bracket with nut and lock washer. Place wire plug in socket on rear of light, and turn one-quarter turn to engage.

(9) INSTALL BLACKOUT PARKING LAMP WIRE. Connect blackout parking lamp wire by inserting wire in plug. Clamp wire to bracket with clamp, and tighten cross-recessed head screw with cross-recessed screwdriver and wrench.

(10) ATTACH BOND STRAPS. Attach bond strap to front of left fender from radiator shell and to frame at generator with bolts, nuts and toothed lock washers under heads and nuts.

165. RUNNING BOARDS.

a. Removal.

(1) REMOVE NUTS. Remove five nuts from the running board bolts, and use two wrenches to remove the two bolts, nuts, lock washers and felt spacers attaching running board to fender.

(2) REMOVE RUNNING BOARD. Move running board outward to free from bolts and remove.

b. Installation.

(1) Place running board on brackets and install the five nuts.

(2) Attach running board to fender with the two bolts, nuts, lock washers and felt spacers, using two wrenches.

TM 9-818
166

10-TON 6 x 4 TRUCK (MACK MODEL NR)

Section XXXII

BATTERY AND LIGHTING SYSTEM

	Paragraph
Description and tabulated data	166
Batteries	167
Battery cables	168
Lamp assemblies	169
Instrument panel switches	170
Headlight dimmer switch (on toeboard)	171
Stop light switch	172

166. DESCRIPTION AND TABULATED DATA.

a. Description.

(1) BATTERIES (fig. 138). The batteries act as reservoirs of electricity with a twofold function. They provide current for the lighting system and other electrical accessories, and also supply power to crank the engine. The 12-volt lighting system uses four 6-volt batteries connected in series parallel, with one positive terminal of each series pair grounded. They are located on the left-hand frame

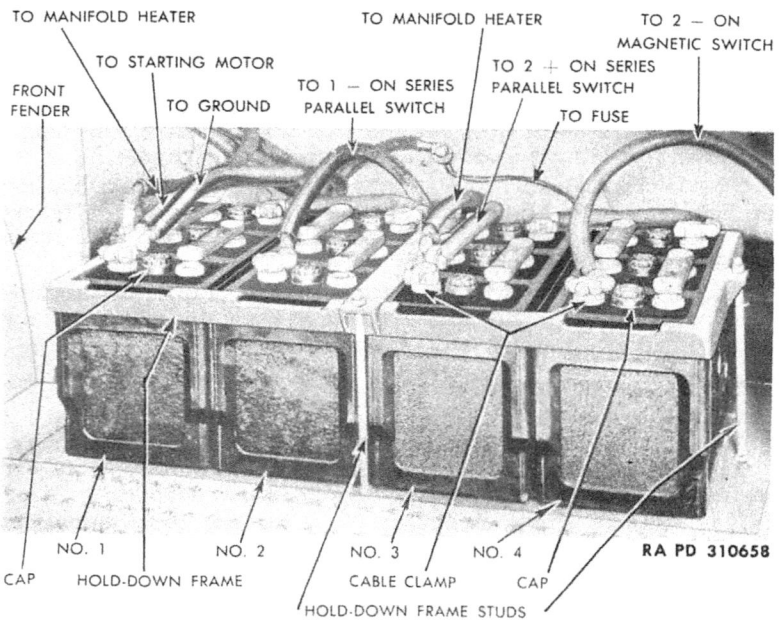

Figure 138—Batteries and Hold-down Frame

TM 9-818
166

BATTERY AND LIGHTING SYSTEM

side rail inside the running board apron door. The batteries have a built-in feature to prevent overfilling (fig. 139).

(2) LIGHTING. Lighting units are connected as shown by the diagram (fig. 140). They are controlled by switches within easy reach of the operator. The instrument panel lamp switch is a push-pull type switch located on the right-hand side of the instrument board. It is used to throw on the instrument board lights if the blackout switch is in first "ON" position. The main light and blackout switch is mounted to the right center of the instrument board. It controls the headlight, stop, and tail lamps and blackout equipment. It has one "OFF" and three "ON" positions. With the control knob all the way in, all lights are off. Pulling the knob to the first "ON" position throws on the blackout fender lights and the blackout tail-

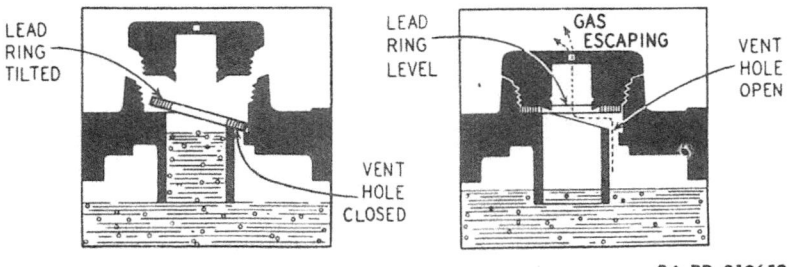

Figure 139—Sectional Views of Battery No-Over-Flo

light, and connects the circuit to the stop light switch. To light the blackout headlight, instrument panel lights, and tail and stop lights the switch must be pulled out to the second "ON" position, but this cannot be done until the switch-lock button is pressed. This lockout-control button is incorporated on this switch so that the running lights cannot be thrown on in error. Only by pressing this lockout plunger button can service lights be put on. Pulling the switch knob out to the third "ON" position lights the same lamps as position two. A blackout fender light is mounted on each fender and controlled by the blackout lighting switch. The lamps are equipped with 12-volt, 3-cp, single-contact bulbs, replaceable by removing the lens rim. Two combination blackout tail and stop lights are mounted, one on each side of the frame at the rear. They are controlled by the blackout lighting switch on the instrument board. All bulbs are 12-volt, 3-cp. There is also one sealed-beam, hooded, blackout headlight mounted on the right side. It is a 12-volt unit, serviced only by replacing the sealed unit. This unit contains two filaments. The upper filament provides dim light projected a short distance in advance of vehicle, whereas the lower filament projects a more intense beam to a greater distance. Its two beams are controlled by the headlight dimmer switch on the toeboard. The spherical mounting of the headlight permits the beam to be aimed

TM 9-818
166

10-TON 6 x 4 TRUCK (MACK MODEL NR)

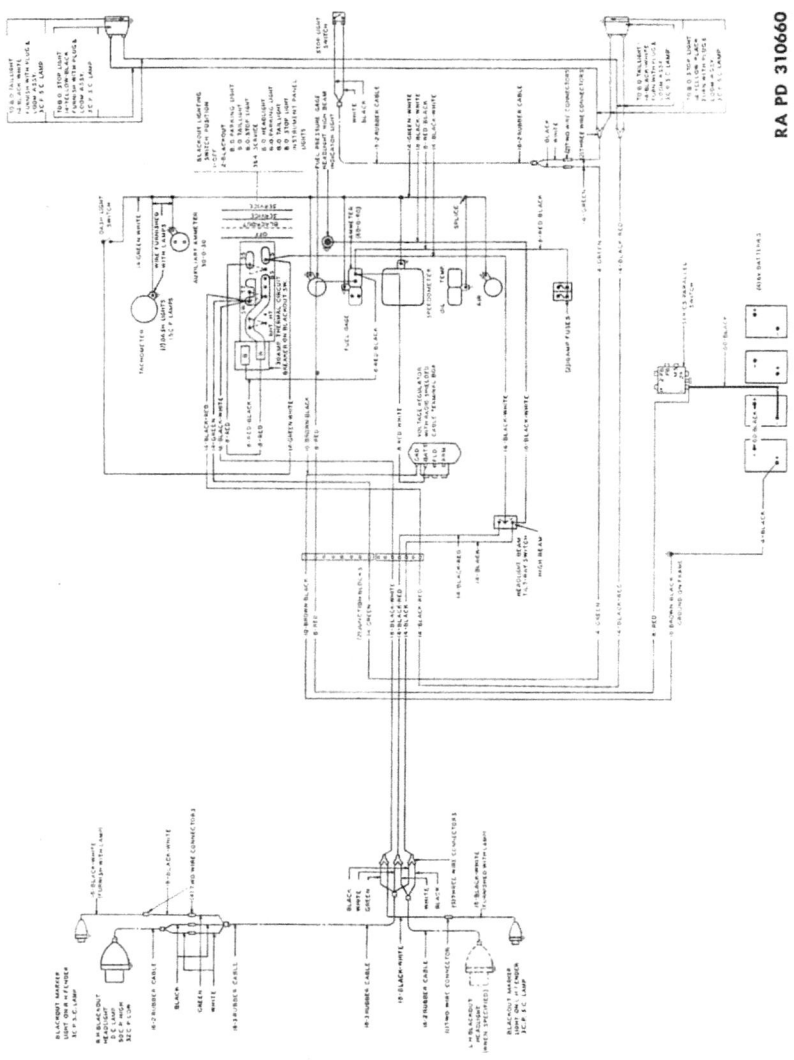

Figure 140—Lighting Circuit Diagram

BATTERY AND LIGHTING SYSTEM

where most desirable. It is important that the beam be aimed correctly. The headlight may be adjusted accurately and quickly with a commercial headlight test equipment.

b. Data.

(1) BATTERY.

Make	Exide
Type	XHM-25
Number per truck	4
Lighting and charging connection	series parallel
Starting connection	series
Plates per cell	25
Voltage each unit	6
Terminal grounded	Positive
Ampere hours at 20-hour rate	200
Specific Gravity:	
Fully charged	1.275-1.300
Recharge at	1.200

(2) LAMP BULBS.

Headlight R. H. sealed beam	Blackout shielded
Driving beam (upper)	50-cp
Passing beam (lower)	32-cp
Blackout parking (two)	3-cp Mazda No. 67
Blackout stop (two)	3-cp
Blackout tail (two)	3-cp
Headlight beam ind. (one)	1-cp Mazda No. 57
Dashlight beam ind. (five)	1-cp Mazda No. 57

167. BATTERIES.

a. Determine Specific Gravity with Hydrometer.

b. Temperature Effects.

(1) WARM WEATHER. Check the battery for heating in normal warm weather. If battery feels more than blood-warm to the touch (approximately 100°F), inspect for short circuits and excessive charging.

(2) HOT REGIONS. In tropical regions, danger of overheating is much greater than in cooler climates. The battery, when fully charged, should have a gravity reading of 1.225 under such conditions.

(3) COLD REGIONS.

(a) The efficiency of batteries decreases sharply with decreasing temperatures, and becomes practically nil at −40°F. Do not try to start the engine with the battery when it has been chilled to temperatures below −30°F until battery has been heated, unless a warm slave battery is available. See that the battery is always fully charged, with the hydrometer reading between 1.275 and 1.300. A fully charged battery will not freeze at temperatures likely to

10-TON 6 x 4 TRUCK (MACK MODEL NR)

be encountered even in arctic climates, but a fully discharged battery will freeze and rupture at $+5°F$.

(b) Do not add water to a battery when it has been exposed to subzero temperatures unless the battery is to be charged immediately. If water is added and the battery not put on charge, the layer of water will stay at the top and freeze before it has a chance to mix with the acid.

c. **Electrolyte Level.** Add pure water to the battery as often as necessary to keep the battery plates covered with electrolyte. The No-Over-Flo feature (see fig. 139) allows water to be added to the cells only until it has reached the proper level. Screwing the cap into position again opens the vent hole for ventilation.

d. **Removal** (fig. 138).

(1) Remove the two cap screws, lock washers and plain washers attaching cover to apron and remove the cover.

(2) Loosen cable clamps and remove all battery cables from battery terminals.

(3) Remove four lock nuts and nuts from battery hold-down frame studs. Remove hold-down frame.

(4) Lift batteries out of tray and remove.

e. **Installation.**

(1) Place the batteries in the tray with the first and third batteries with their positive poles toward the outside and the second and fourth with their negative poles toward the outside.

(2) Place hold-down frame over batteries with studs through four holes and install nuts and lock nuts.

(3) Follow the wiring diagram to connect the batteries, and tighten the cable clamps after making sure that all connections are clean.

(4) Place battery cover in running board opening, and apply the two cap screws with plain and lock washers. Tighten with socket wrench.

f. **Battery Disconnection.**

(1) When it is necessary to disconnect the batteries to perform work on the various components of the electrical system or in the case of other disassemblies, the negative connection on the second and fourth batteries must be removed.

(2) When batteries have been disconnected as in step (1) above, reconnect them.

168. BATTERY CABLES.

a. **Removal** (fig. 138).

(1) Remove cable ends from various batteries, switches and other electrical units, following procedure outlined for replacement of the various units to which cables are attached:

(a) Batteries (par. 167 c (1) (b)).

(b) Series parallel switch (par. 90 a (2)).

BATTERY AND LIGHTING SYSTEM

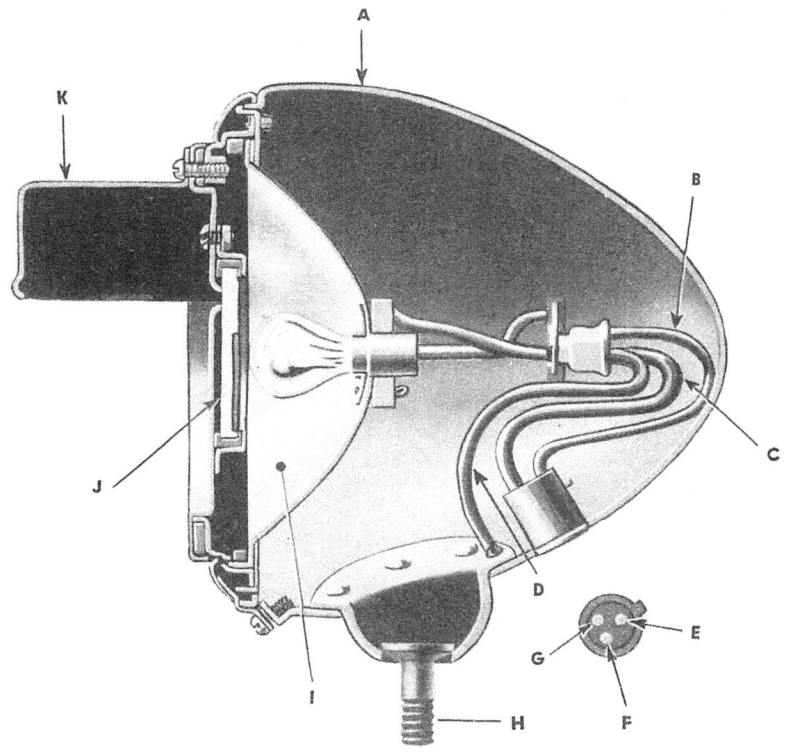

A	HOUSING ASSEMBLY	F	COMMON TERMINAL
B	LOWER BEAM WIRE	G	LOWER BEAM TERMINAL
C	HIGH BEAM WIRE	H	MOUNTING BOLT
D	GROUND WIRE	I	REFLECTOR
E	HIGH BEAM TERMINAL	J	LENS
		K	SHIELD

RA PD 310663

Figure 141—Sectional View of Headlight

(c) Magnetic starter switch (par. 89 a (2)).
(d) Heater relay magnetic switch (par. 54 a (2)).
(e) Manifold heaters (par. 54 a (2)).
(f) Cranking motor (par. 91 a (1)).

(2) If necessary to remove No. 00 cable from "1—" terminal on series parallel switch to heater relay magnetic switch, remove two clamps attaching cable to side rail.

(3) If necessary to remove No. 00 cable from magnetic switch to cranking motor or to heater relay magnetic switch, remove clamp attaching cable to side rail.

TM 9-818
168-169
10-TON 6 x 4 TRUCK (MACK MODEL NR)

b. **Installation.** To install cables, reverse removal procedure, making certain that cables are properly connected to their terminals and attached to frame where removed.

169. **LAMP ASSEMBLIES.**

a. **Headlights** (fig. 141). NOTE: *Sizes and types of bulbs are listed in paragraph* 166 a (2).

(1) REMOVAL. Remove nut and washer on headlight bracket beneath lamp. Disconnect the headlight wires by turning plug in back of lamp one-quarter turn, and remove the head lamp.

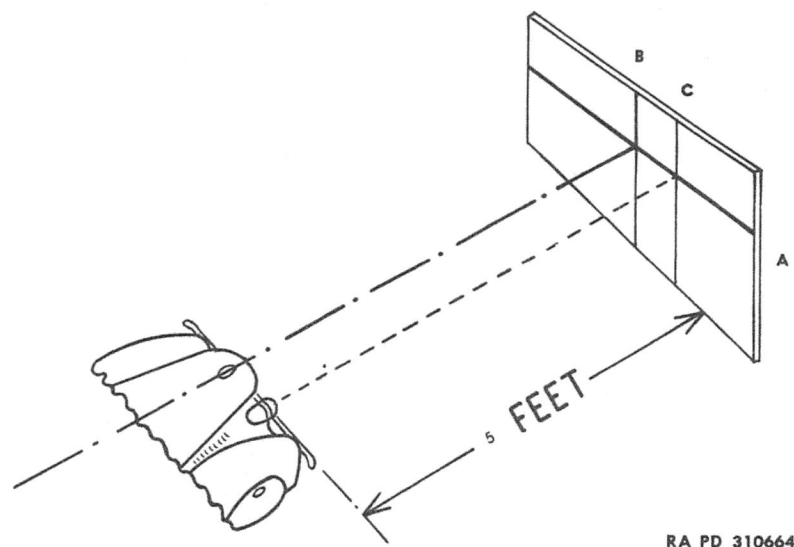

Figure 142—Aiming Headlight

(2) INSTALLATION. Place headlight on bracket and install nut and washer, using socket wrench. Insert the plug in under side of lamp, turning one-quarter turn to engage. See (4) below for headlight adjustment.

(3) HEADLIGHT ADJUSTMENT (figs. 142 and 143). The headlight may be adjusted accurately and quickly with a commercial headlight test equipment. If not available, proceed as follows: Place vehicle on level floor so that light is five feet from a vertical wall (or screen). The centerline of the vehicle should be perpendicular to the wall. Draw a horizontal line "A" on the wall at the same height from the floor as the slot in the headlight. Draw a vertical line "B" on the wall directly in line with the center of the vehicle and another vertical line "C" to the right of it the same distance the

290

BATTERY AND LIGHTING SYSTEM

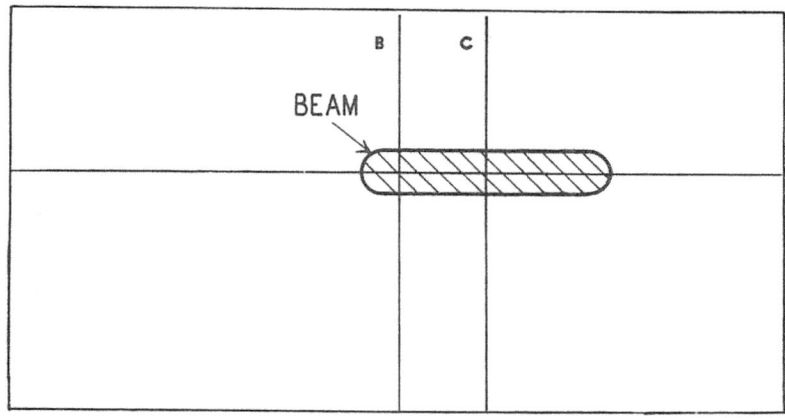

Figure 143—Headlight Beam of Chart

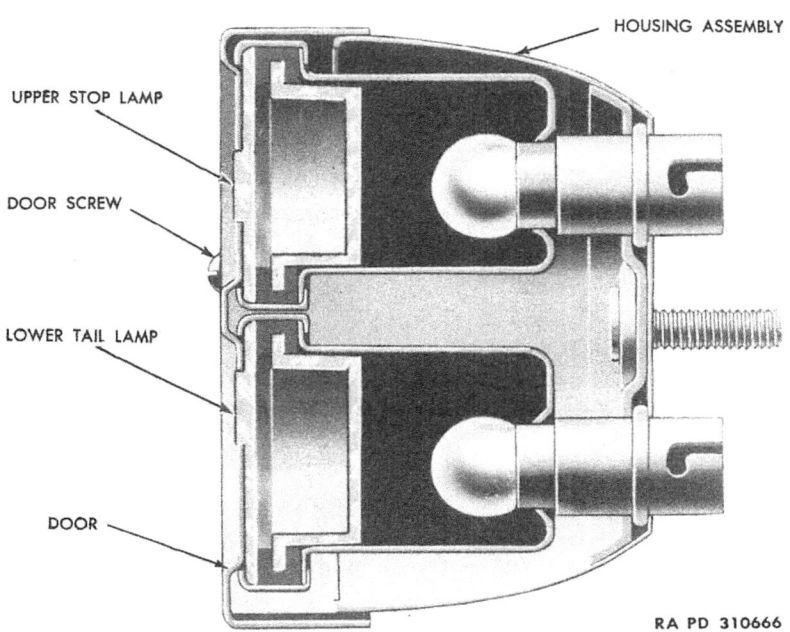

Figure 144—Tail and Stop Light

TM 9-818
169
10-TON 6 x 4 TRUCK (MACK MODEL NR)

light is to the right of the centerline of the vehicle. Turn on the high beam and aim the light so that the center of the beam path, both horizontally and vertically, coincides with the "A" and "C" lines. After tightening the lamp, check again to be sure it has not shifted in tightening.

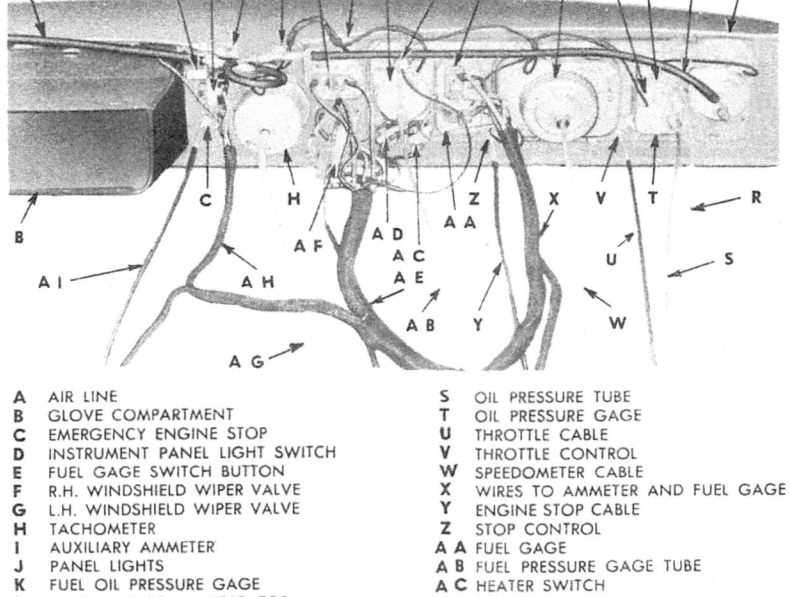

A	AIR LINE	S	OIL PRESSURE TUBE
B	GLOVE COMPARTMENT	T	OIL PRESSURE GAGE
C	EMERGENCY ENGINE STOP	U	THROTTLE CABLE
D	INSTRUMENT PANEL LIGHT SWITCH	V	THROTTLE CONTROL
E	FUEL GAGE SWITCH BUTTON	W	SPEEDOMETER CABLE
F	R.H. WINDSHIELD WIPER VALVE	X	WIRES TO AMMETER AND FUEL GAGE
G	L.H. WINDSHIELD WIPER VALVE	Y	ENGINE STOP CABLE
H	TACHOMETER	Z	STOP CONTROL
I	AUXILIARY AMMETER	A A	FUEL GAGE
J	PANEL LIGHTS	A B	FUEL PRESSURE GAGE TUBE
K	FUEL OIL PRESSURE GAGE	A C	HEATER SWITCH
L	HEADLAMP BEAM INDICATOR	A D	STARTER SWITCH
M	MAIN AMMETER	A E	WIRES TO BLACKOUT SWITCH AND AMMETER
N	SPEEDOMETER		
O	WATER TEMPERATURE GAGE	A F	BLACKOUT LIGHTING SWITCH
P	AIR PRESSURE GAGE	A G	TACHOMETER CABLE
Q	WINDSHIELD WIPER AIR LINE	A H	WIRES TO FUEL GAGE SWITCH
R	TEMPERATURE INDICATOR TUBE	A I	EMERGENCY STOP CABLE

RA PD 310667

Figure 145—Back of Instrument Panel

b. **Blackout Fender Lights.**

(1) REMOVAL. Disconnect wire at plug on fender support by pulling apart and remove the nut and washer, using a wrench and screwdriver. Remove the nut and washer under the fender, and remove the light.

(2) INSTALLATION. Place light on fender and install nut and washer from below it. Install the nut and washer, using wrench and screwdriver. Insert wire plug in receptacle on fender support.

292

BATTERY AND LIGHTING SYSTEM

c. **Tail and Stop Lights** (fig. 144).

(1) REMOVAL. Disconnect wire to light by pushing in plug on back of light and turning clockwise one-quarter turn to remove. Remove the two nuts, washers and ground wire on inside bolt on the lamp bracket. Remove light.

(2) INSTALLATION. Place light on bracket, and install the ground wire with two nuts and washers on the inside bolt. Insert plug in back of light and turn counterclockwise one-quarter turn to engage.

d. **Dashlights.**

(1) REMOVAL. All dashlight bulbs are accessible for bulb change by merely pulling the socket out of its mounting on the instrument panel from in back of the panel.

(2) INSTRUMENT PANEL LAMP BULB CHANGE. After removal of the socket, turn bulb one-quarter turn and remove from socket. Install new bulb and reinstall socket in panel by pushing into place.

170. INSTRUMENT PANEL SWITCHES (fig. 145).

a. **Blackout Switch.**

(1) REMOVAL. Disconnect battery following procedure outlined in paragraph 167. Remove wires from blackout switch, using socket wrench to remove the nuts on circuit breaker, and screwdriver to remove screws and lock washers on other terminals. Loosen the set screw in switch button and unscrew button. Loosen cap screw on left-hand side of switch, push the button down and remove cap by pulling off. Use socket wrench to remove nut and toothed lock washer from sleeve, and pull out switch.

(2) INSTALLATION. Installation of the switch is the reverse of the removal procedure. Wiring should be connected as follows: The No. 14 green wire with white tracer and No. 14 black wire with white tracer to "S" terminal, the No. 8 red wire to "SS" terminal, the No. 14 black, the No. 18 black wire with white tracer, No. 14 green wire and the No. 14 black wire with red tracer to the "SW" terminal; the No. 8 red wire with black tracer to the lower, and the No. 8 red wire to the upper circuit breaker terminals.

b. **Fuel Gage Switch.**

(1) REMOVAL. Remove the three wires from switch with cross-recessed screwdriver. Disconnect the pull-knob shaft from switch by prying out of hole with screwdriver. Remove the nut on instrument board and remove switch and plate.

(2) INSTALLATION. Installation is the reverse of removal procedure. Wiring should be connected as follows: The No. 14 black wire to the battery terminal, the No. 14 blue wire to "S" terminal, the No. 14 brown wire to the "H" terminal.

c. **Panel Light Switch.**

(1) REMOVAL. Remove two wires with screwdriver. Remove button, using small screwdriver to loosen set screw. Remove the nut and remove switch.

10-TON 6 x 4 TRUCK (MACK MODEL NR)

(2) INSTALLATION. Installation is the reverse of removal procedure. Wiring should be connected as follows: The No. 14 green wire with white tracer to the right-hand terminal and the No. 14 black wire from the panel lamps to the left-hand terminal.

171. HEADLIGHT DIMMER SWITCH (ON TOEBOARD).

a. Removal.

(1) REMOVE TOEBOARD. Follow procedure outlined in paragraph 159 a.

(2) REMOVE SWITCH. Remove two cross-recessed head screws attaching switch to floor, and pull switch through hole in reinforcing plate.

(3) DISCONNECT WIRING. With screwdriver, loosen screw terminals and remove wires.

b. Installation.

(1) ATTACH WIRING. Attach No. 14 black wire with white tracer to "BAT" terminal. Attach No. 18 black wire with white tracer and No. 14 black wire to one terminal and the No. 14 black wire with red tracer to the other terminal. Tighten with screwdriver.

(2) INSTALL SWITCH. Attach switch to floor with two cross-recessed head screws and tighten same.

(3) INSTALL TOEBOARD. Follow procedure outlined in paragraph 159 b.

172. STOP LIGHT SWITCH.

a. Removal.

(1) DISCONNECT WIRES. Remove two nuts from terminals and remove wires from terminals.

(2) REMOVE AIR LINE. Remove air line from switch.

(3) REMOVE SWITCH. Remove two bolts, nuts and lock washers attaching switch to cross member, and remove switch.

b. Installation.

(1) INSTALL SWITCH. Attach switch to cross member with two bolts, install lock washers and nuts, and tighten.

(2) INSTALL AIR LINE. Attach air line to switch, and tighten.

(3) CONNECT WIRES. Place white wire on left-hand and black on right-hand terminals. Apply washers and nuts, and tighten.

TM 9-818
173

Section XXXIII

INSTRUMENTS, CONTROLS, SWITCHES AND HORN

	Paragraph
Description and tabulated data	173
Instruments and buzzer	174
Engine controls	175
Windshield wiper control valves	176
Starter and manifold heater switches	177
Horn	178

173. DESCRIPTION AND TABULATED DATA.

a. Description.

(1) INSTRUMENTS AND BUZZER. See paragraph 4 c and figure 27.

(a) Air Pressure Gage. This gage indicates the pressure in the reservoirs and air lines.

(b) Water Temperature and Oil Pressure Gages. The temperature of the water in the cooling system is indicated by the upper gage, and the lubricating oil pressure by the lower gage.

(c) Speedometer. This instrument indicates the speed of the vehicle and records the total mileage traveled.

(d) Main Ammeter and Fuel Level Gage. The upper instrument is a zero-center ammeter which shows the charging current of the four batteries on the upper side of zero and their discharge current on the lower side. The lower instrument is a gage which shows the level of the fuel in one or the other of the two tanks, as determined by the position of the fuel gage switch button on the instrument panel.

(e) Fuel Pressure Gage. This gage indicates the pressure in the line between the fuel supply pump and the injection pump. During operation it should show between 3 and 18 pounds.

(f) Auxiliary Ammeter. This zero-center instrument shows the charging current of the second pair of batteries on the right of zero, and the discharge current on the left.

(g) Tachometer. Engine speed in revolutions per minute is indicated by the dial hand. The instrument also has a total revolution recorder which is geared to record also the total engine hours based on an assumed average speed of 1667 revolutions per minute.

(h) Low-pressure Buzzer. The electrical buzzer on the dash sheet inside of the cab is connected to the low-pressure indicator in the air brake system, and is actuated by the pressure falling below 60 pounds per square inch. It is connected across the ignition switch so that it is inoperative when the ignition is off.

(2) ENGINE CONTROLS.

(a) Throttle Control. This control is a pull-knob control on the instrument panel which is used when a setting for greater than idling speed is wanted.

10-TON 6 x 4 TRUCK (MACK MODEL NR)

(b) Engine Stop Control. This is a pull-knob control on the instrument panel for stopping the engine through the governor control.

(c) Emergency Engine Stop Control. This is a pull-knob control for stopping the engine by closing a valve in the fuel line.

(d) Accelerator. This control consists of a pedal at the right of the steering post which is connected by linkage to the governor. It is operated by the driver's right foot to control the speed and power of the engine, by varying the amount of fuel fed through the injection pump.

(3) WINDSHIELD WIPER CONTROL VALVES. Two valves mounted on the instrument panel provide for the independent control of the windshield wipers, including speed regulation.

(4) CRANKING MOTOR AND MANIFOLD-HEATER SWITCHES.

(a) Cranking Motor Switch. This is a push-button switch, mounted on the instrument panel, for actuating the magnetic series parallel switch which changes the battery connections from the 12-volt series-parallel combination to the 24-volt series combination for operating the cranking motor. This action closes the control circuit of the magnetic switch in the cranking motor circuit, thereby closing it and cranking the engine.

(b) Manifold-heater Switch. This is a push-button switch, mounted on the instrument panel, for actuating two manifold-heater magnetic relay switches.

(c) Manifold-heater Magnetic Relay Switches. There are two of these switches, mounted on the front of the dash sheet and connected in the circuits of the six intake manifold heaters. They are controlled by the manifold-heater switch (step d (2) above).

(5) HORN. The 12-volt vibrator-type horn is mounted on the front of the dash sheet. It is operated by a push-button in the center of the steering wheel.

b. **Data.**

Air pressure gage	Stewart-Warner, Model 103454
Water-temperature and oil-pressure gages	Stewart-Warner, Model 103453
Speedometer	Stewart-Warner, Model 596
Main ammeter and fuel-level gage	Stewart-Warner, Model 103452
Fuel pressure gage	Stewart-Warner, Model 105880
Auxiliary ammeter	Stewart-Warner, Model EX-22684-B
Tachometer	Stewart-Warner, Model 584 CP
Low-pressure buzzer	Electric Service Supply Co., Model 22984
Throttle control	Pull-knob and wire
Engine-stop control	Pull-knob and wire
Engine emergency stop control	Pull-knob and wire
Accelerator	Pedal and linkage
Windshield wiper control valves	Trico, Model 75470

INSTRUMENTS, CONTROLS, SWITCHES AND HORN

Starter switch .... Leese-Neville, Model 110-SS
Manifold-heater switch .... Leese-Neville, Model 133-SS
Manifold-heater magnetic relay switch .. Delco-Remy, Model 1422
Horn .... Delco-Remy, Model K26L1637

174. INSTRUMENTS AND BUZZER.

a. Except as outlined below, the installation procedure for the various instruments is the reverse of the removal procedure.

b. Air Pressure Gage.

(1) REMOVAL. Disconnect air line and remove adapter. Remove lamp by turning one-quarter turn and pulling out. Remove two nuts and lock washers with socket wrench. Remove gage.

c. Water-temperature and Oil-pressure Gages.

(1) REMOVAL. Disconnect oil pressure line and remove adapter. Drain enough water from radiator to clear the adapter hole in rear of water manifold, and remove the gage element from the adapter. Pull the cable through the dash sheet, removing the grommet. Remove the lamp from back of instrument by turning one-quarter turn and pulling out. Remove three nuts and lock washers with socket wrench. (It will be necessary to remove the air gage mounting stud nuts and loosen gage from mounting as outlined in step b, above, to remove the nut and lock washer on left-hand side.) Remove instrument.

d. Speedometer.

(1) REMOVAL. Disconnect speedometer cable from back of unit with pliers. Remove lamp by turning one-quarter turn and pulling out. Remove four nuts and lock washers with socket wrench and remove speedometer.

e. Main Ammeter and Fuel-level Gage.

(1) REMOVAL. Disconnect ammeter wires by removing two palnuts, nuts and lock washers. Disconnect the wires from the fuel-level gage. Remove lamp by turning one-quarter turn and pulling out. Remove the three nuts. (It will be necessary to remove the fuel-pressure gage from its mounting, removing only the two nuts and lock washers to remove the right-hand nut.)

(2) INSTALLATION. Installation is the reverse of removal. The wiring connections are as follows: The No. 8 plain red and the No. 10 plain black cables to top ammeter terminal, the No. 14 red wire to the right-hand and the No. 14 black wire to the left-hand fuel-level gage terminals. All these wires and cables are contained in one branch of the wiring harness.

f. Fuel-pressure Gage.

(1) REMOVAL. Remove gage pressure line at adapter, using two wrenches, one to turn and one to hold. Remove the adapter. Remove the lamp by turning one-quarter turn and pulling out. Remove the two nuts and lock washers with socket wrench and remove the gage.

10-TON 6 x 4 TRUCK (MACK MODEL NR)

g. **Auxiliary Ammeter.**

(1) REMOVAL. Disconnect wires by removing two terminal nuts and lock washers. Remove lamp by turning one-quarter turn and pulling out. Remove two nuts and lock washers and remove ammeter.

(2) INSTALLATION. Installation is the reverse of removal. Wiring connections are as follows: The No. 8 red cable with black tracer to left-hand terminal, and the No. 8 brown wire with black tracer to the right-hand terminal.

h. **Tachometer.**

(1) REMOVAL. Remove cable from back of tachometer. Remove lamp by pulling out. Remove two nuts and lock washers and remove tachometer.

i. **Low-pressure Buzzer.**

(1) REMOVAL.

(a) Remove screw from buzzer cover, and remove cover.

(b) Loosen terminal screws, and detach wires.

(c) Remove the two nuts, lock washers and screws, using socket wrench, and remove buzzer.

(2) INSTALLATION.

(a) Place buzzer on dash sheet inside cab and install two screws. Apply lock washers and nuts and tighten with socket wrench.

(b) Attach the two wires, one to each terminal, and tighten terminal screws.

(c) Place cover on buzzer, attach with screw and tighten.

175. ENGINE CONTROLS.

a. Installation procedure for the following units is the reverse of removal procedure.

b. **Throttle Control.**

(1) REMOVAL. Remove stop from end of hand-throttle wire. Pull the wire through the dash sheet. Remove the nut and lock washer on instrument panel, and pull out cable.

c. **Engine-stop Control.**

(1) REMOVAL. Disconnect wire from governor lever and clamp. Remove wire from fender clamp bracket by removing nut and lock washer. Pull cable through dash sheet, removing the self-tapping screw at the clamp on the firewall. Remove the nut and lock washer on instrument panel and remove cable.

d. **Engine Emergency Stop Control.**

(1) REMOVAL. Disconnect wire at valve on injection pump by removing the stop with screwdriver. Remove the two nuts and lock washers on the clamps on the right-hand fender, and remove clamps. Pull cable through dash sheet. Remove the nut on instrument panel and pull out control.

TM 9-818
175

INSTRUMENTS, CONTROLS, SWITCHES AND HORN

e. Accelerator Pedal and Rod.

(1) REMOVAL.

(a) Remove Pedal. Remove nut and lock washer from pedal ball joint stud. Remove cotter from pedal hinge pin, and remove pin. Remove pedal.

(b) Remove Pedal Rod. Detach from accelerator as in step *(a)* above. Remove cotter pin from cross shaft lever end of rod, and remove rod.

(2) INSTALLATION. Reverse removal procedure to install pedal and rod.

f. Accelerator Cross Shaft and Levers.

(1) REMOVAL.

(a) Remove Toeboard. Follow procedure outlined in paragraph 159 a.

(b) Remove Levers. Disconnect throttle control wire by removing stop screw with screwdriver and removing stop. Remove wire. Detach rod (e (1) *(b)* above). Remove cotter pin and clevis pin from clevis and detach from lever. Drive pin out of lever and remove lever. Remove two cotter pins from shaft center.

(c) Remove Shaft. Pull shaft out of mounting brackets to the left.

(d) Disassemble Shaft and Levers. Remove cotter pin at left-hand end of shaft and remove throttle control lever. Drive pin out of accelerator pedal rod lever and remove.

(2) INSTALLATION.

(a) Assemble Shaft and Levers. Place accelerator rod lever on shaft, insert pin and drive into position. Place hand throttle lever on shaft, and install washer and cotter pin at shaft end.

(b) Install Shaft. Insert shaft in left shaft bracket, install two washers, and insert in right bracket. Install two cotter pins in shaft to retain washers against inner sides of brackets.

(c) Install Lever. Place right-hand lever on shaft end in opposite position to that on left end of shaft, and drive pin into place through shaft. Attach clevis to lever and install clevis pin and cotter pin. Attach accelerator rod to left-hand lever and install cotter pin. Attach throttle hand control wire to throttle lever, and attach stop and screw.

(d) Install Toeboard. Follow procedure in paragraph 159 b.
NOTE: *Accelerator shaft brackets can be removed only after engine has been removed, excepting in emergency where it may be feasible to remove toeboard and cut a slot in dash sheet at each end of the cross shaft, bending the dash sheet up to clear shaft brackets.*

g. Accelerator Rod and Spring.

(1) REMOVAL.

(a) Remove Spring. Detach spring from lugs on rod and on exhaust manifold stud.

299

10-TON 6 x 4 TRUCK (MACK MODEL NR)

(b) Remove Rod. Remove cotter pins from clevis pins at both ends of rod, and remove pins. Remove rod.

(c) Remove Clevises. Back off clevis locking nuts. Remove clevises and nuts.

(2) INSTALLATION.

(a) Install Clevises. Place nuts on ends of rod and install clevises so that, when attached, the governor lever and the lever on right-hand end of cross shaft are parallel or nearly so.

(b) Install Rod. Install rod between levers with short bend to rear. Insert clevis pins and install cotter pins. Tighten clevis locking nuts.

(c) Install Spring. Attach spring to lug on exhaust manifold stud on No. 1 cylinder. Place spring rod end in spring end loop and attach rod end to lug on accelerator rod.

176. WINDSHIELD WIPER CONTROL VALVES.

a. **Removal.** Disconnect air lines by removing nuts. Remove each control button by removing two set screws and nut on instrument panel. Remove valve. Remove tee from valve, and remove straight fitting from adapter.

177. CRANKING MOTOR AND MANIFOLD-HEATER SWITCHES.

a. **Cranking Motor Switch.**

(1) REMOVAL. Remove wires, using screwdriver. Remove screw between terminals and remove clamp and switch.

(2) INSTALLATION. Installation is the reverse of removal. If the heater and cranking motor switches are removed at the same time, be sure that the green and yellow wires are attached to the cranking motor switch terminals and the black wires to the heater-switch terminals.

b. **Manifold-heater Switch.**

(1) REMOVAL. Follow procedure outlined in steps a (1) and (2) above.

c. **Manifold-heater Magnetic Relay Switches.**

(1) REMOVAL.

(a) Disconnect Batteries. Follow procedure outlined in paragraph 167 f (1).

(b) Disconnect Wires. Remove No. 00 cables by removing two nuts and lock washers from terminal on bottom of switch. Remove two nuts from terminals on side of switch and remove wires.

(c) Remove Switch. Remove two nuts and lock washers attaching switch to dash sheet, and remove switch.

(2) INSTALLATION.

(a) Install Switch (fig. 50). Attach switch bracket to dash with two bolts, tightening nuts with lock washers in place.

TM 9-818
177-178

INSTRUMENTS, CONTROLS, SWITCHES AND HORN

(b) Connect Wires. Attach No. 10 black wire (two on upper, one on lower switch) to left-hand side terminal. Attach No. 10 brown wire with black tracer (two on upper, one on lower switch) to right-hand side terminal with No 8-32 nuts and lock washers. Attach No. 00 cable with bent terminal to right-hand lower pole, and No. 00 cable from No. 1 manifold heater to left-hand lower pole on upper switch. Attach No. 00 cable from No. 6 manifold heater to left-hand lower pole, and remaining No. 00 cable to right-hand lower pole on lower switch. Install lock washers and nuts and tighten.

(c) Connect Batteries. Follow procedure outlined in paragraph 167 f *(1) (2)*.

178. HORN.
a. Removal.
(1) DISCONNECT WIRES. Remove the two screws and lock washers attaching wires to horn, and remove wires.

(2) REMOVE HORN. Remove the two nuts and lock washers, and remove horn from the mounting bracket.

b. Installation.
(1) INSTALL HORN. Place horn on mounting bracket and install the two nuts with lock washers.

(2) ATTACH WIRES. Place the yellow-and-black coded wire on the inner terminal and the yellow wire on the outer terminal, attaching them with the screws and lock washers.

10-TON 6 x 4 TRUCK (MACK MODEL NR)

Section XXXIV

FIRE EXTINGUISHER

	Paragraph
Description and tabulated data	179
Operation, care, test and refilling	180

179. DESCRIPTION AND TABULATED DATA.

a. **Description.** The fire extinguisher is mounted in a clamping bracket attached to the left side of cowl. It is a carbon tetrachloride, vaporizing extinguisher and is effective for extinguishing fires resulting from burning wood, fabric, oil, grease, gasoline, liquids, or from short circuits. It should also be used on burning electrical equipment such as the generator, cranking motor, wiring and other insulation.

b. **Data.**
```
Make ........................................ Pyrene
Model ....................................... C 21 T
Type ................................. Heavy vehicle
Classification ............................. B-2, C2
Number per truck ............................... One
Capacity ...................................... 1 qt
```

180. OPERATION, CARE, TEST AND REFILLING.

a. **Operation.** To operate extinguisher, turn handle to the left, pull it out, and proceed to pump with it. Best results are attained by directing the stream at the base of the flame. When used on burning liquids in a container, direct the stream against the inside of the container just above the burning liquid.

b. **Care.** Keep extinguisher clean. After use, either refill immediately (step d (2) below) or exchange for an extinguisher that is fully charged.

c. **Test.** Every month test the extinguisher by pumping some of the fluid into a clean, dry container. If the pump works satisfactorily, and the fluid is clean and free from foreign matter, pour the fluid back into the extinguisher. Fill to proper level (step d (2) below). If the fluid is dirty, exchange the extinguisher, or empty and refill it. If pump is inoperable, exchange the extinguisher.

d. **Refill Extinguisher.**

(1) EMPTY EXTINGUISHER. Pump all fluid out of extinguisher.

(2) REFILL EXTINGUISHER. Remove filler-hole plug and gasket. Pour approved fire extinguisher fluid (Specification O.F. 380 or 51 F 352 (1 gal container) or 51 F 353 (55 gal container) into extinguisher until level is approximately one-half inch below filler cap opening. Lock pump handle by turning to the right, and replace filler plug and gasket. Shake the extinguisher to remove any fluid remaining in the discharge tube, and install extinguisher in clamping bracket.

TM 9-818
181-182

Section XXXV

SHIPMENT AND TEMPORARY STORAGE

	Paragraph
General instructions	181
Preparation for temporary storage or domestic shipment	182
Loading and blocking for rail shipment	183

181. GENERAL INSTRUCTIONS.

a. Preparation for domestic shipment of the vehicle is the same as preparation for temporary storage or bivouac. Preparation for shipment by rail includes instructions for loading and unloading the vehicle, blocking necessary to secure the vehicle on freight cars, number of vehicles per freight car, clearance, weight, and other information necessary to properly prepare the vehicle for rail shipment. For more detailed information, and for preparation for indefinite storage, refer to AR 850-18.

182. PREPARATION FOR TEMPORARY STORAGE OR DOMESTIC SHIPMENT.

a. Vehicles to be prepared for temporary storage or domestic shipment are those ready for immediate service, but not used for less than thirty days. If vehicles are to be indefinitely stored after shipment by rail, they will be prepared for such storage at their destination.

b. If the vehicles are to be temporarily stored or bivouacked, take the following precautions:

(1) LUBRICATION. Lubricate the vehicle completely (par. 19).

(2) COOLING SYSTEM. If freezing temperature may normally be expected during the limited storage or shipment period, test the coolant with a hydrometer and add the proper quantity of antifreeze compound, to afford protection from freezing at the lowest temperature anticipated during the storage or shipping period. Completely inspect the cooling system for leaks.

(3) BATTERY. Check battery and terminals for corrosion and, if necessary, clean and thoroughly service battery (par. 167).

(4) TIRES. Clean, inspect, and properly inflate all tires. Replace with serviceable tires, all tires requiring retreading or repairing. Do not store vehicles on floors, cinders, or other surfaces which are soaked with oil or grease. Wash off immediately any oil, grease, gasoline, or kerosene which comes in contact with the tires under any circumstances.

(5) ROAD TEST. The preparation for limited storage will include a road test of at least five miles, after the battery, cooling system, lubrication, and tire services, to check on general condition of the vehicle. Correct any defects noted in the vehicle operation, before the vehicle is stored, or note on a tag attached to the steering wheel, stating the repairs needed, or describing the condition present. A written report of these items will then be made to the officer in charge.

TM 9-818
182

10-TON 6 x 4 TRUCK (MACK MODEL NR)

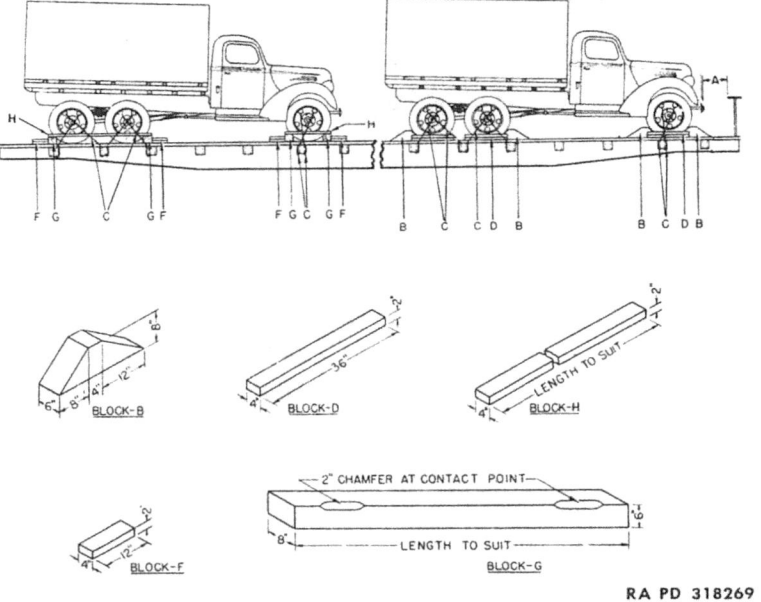

Figure 146—Blocking Requirements for Securing Truck to Railroad Car

TM 9-818
182-183

SHIPMENT AND TEMPORARY STORAGE

(6) FUEL IN TANKS. It is not necessary to remove the fuel from the tanks for shipment within the United States, nor to label the tanks under Interstate Commerce Commission Regulations. Leave fuel in the tanks except when storing in locations where fire ordinances or other local regulations require removal of all gasoline before storage.

(7) EXTERIOR OF VEHICLE. Remove rust appearing on any part of the vehicle exterior with sandpaper. Repaint painted surfaces whenever necessary to protect wood or metal from deterioration. Exposed polished metal surfaces which are susceptible to rust, such as winch cables, chains, and, in the case of track-laying vehicles, metal tracks, should be coated with a protective medium-grade lubricating oil. Close firmly all cab doors, windows, and windshields. Vehicles equipped with open-type cabs with collapsible tops will have the tops raised, all curtains in place, and the windshield closed. Make sure tarpaulins and window curtains are in place and firmly secured. Leave rubber mats, such as floor mats, where provided, in an unrolled position on the floor, and not rolled or curled up. Equipment, such as Pioneer and truck tools, tire chains, and fire extinguishers, will remain in place in the vehicle.

(8) INSPECTION. Make a systematic inspection just before shipment or temporary storage, to insure all above steps have been covered, and that the vehicle is ready for operation on call. Make a list of all missing or damaged items and attach it to the steering wheel. Refer to Before-operation Service (par. 14).

(9) ENGINE. To prepare the engine for storage, remove the air cleaner from the carburetor. Start the engine and set the throttle to run the engine at a fast idle. Pour one pint of medium grade, preservative lubricating oil, Ordnance Department Specification AXS-674, of the latest issue in effect, into the carburetor throat, being careful not to choke the engine. Turn off the ignition switch as quickly as possible after the oil has been poured into the carburetor. With the engine switch off, open the throttle wide, and turn the engine five complete revolutions by means of the cranking motor. If the engine cannot be turned by the cranking motor with the switch off, turn it by hand, or disconnect the high-tension lead and ground it before turning the engine by means of the cranking motor. Then reinstall the air cleaner.

(10) BRAKES. Release brakes and chock the wheels or tracks.

c. **Inspections in Limited Storage.** Vehicles in limited storage will be inspected weekly for condition of tires and battery. If water is added when freezing weather is anticipated, recharge the battery with a portable charger or remove the battery for charging. Do not attempt to charge the battery by running the engine.

183. LOADING AND BLOCKING FOR RAIL SHIPMENT.

a. **Preparation.** In addition to the preparation described in paragraph 182, when ordnance vehicles are prepared for domestic shipment, the following preparations and precautions will be taken:

TM 9-818
183

10-TON 6 x 4 TRUCK (MACK MODEL NR)

(1) EXTERIOR. Cover the body of the vehicle with a canvas cover supplied as an accessory.

(2) TIRES. Inflate pneumatic tires from 5 to 10 pounds above normal pressure.

(3) BATTERY. Disconnect the battery to prevent its discharge by vandalism or accident. This may be accomplished by disconnecting the positive lead, taping the end of the lead, and tying it back away from the battery.

(4) BRAKES. The brakes must be applied and the transmission placed in low gear, after the vehicle has been placed in position with a brake wheel clearance of at least 6 inches ("A", fig. 146). The vehicles will be located on the car in such a manner as to prevent the car from carrying an unbalanced load.

(5) All cars containing ordnance vehicles must be placarded "DO NOT HUMP."

(6) Ordnance vehicles may be shipped on flat cars, end-door box cars, side-door box cars, or drop-end gondola cars, whichever type car is the most convenient.

b. **Facilities for Loading.** Whenever possible, load and unload vehicles from open cars under their own power, using permanent end ramps and spanning platforms. Movement from one flat car to another along the length of the train is made possible by cross-over plates or spanning platforms. If no permanent end ramp is available, an improvised ramp can be made from railroad ties. Vehicles may be loaded in gondola cars without drop ends by using a crane. In case of shipment in side-door box cars, use a dolly-type jack to fit the vehicles into position with the car.

c. **Securing Vehicles.** In securing or blocking a vehicle, three motions (lengthwise, sidewise, and bouncing), must be prevented. There are two approved methods of blocking the vehicles on freight cars, as described below. When blocking dual wheels, all blocking will be located against the outside wheel of the dual.

(1) FIRST METHOD (fig. 146). Locate eight blocks "B," one to the front and to the rear of each front wheel and to the front of each forward rear wheel, and to the back of each rearward wheel. Nail the heel of each block to the car floor with five 40-penny nails, and toe-nail the portion of each block under the tire to the freight car floor with two 40-penny nails. Locate two cleats "D" against the outside face of each wheel. Nail the lower cleat "D" to the freight car floor with three 40-penny nails and the top cleat to the cleat below with three 40-penny nails. Pass four strands, two wrappings, of No. 8 gage, black annealed wire (C, figure 146) through the holes in the wheels and pass through the stake pockets. Tighten wires enough to remove slack. NOTE: *When a box car is used, this strapping must be applied in similar fashion and attached to the floor by the use of blocking or anchor plates. This strapping is not required when gondola cars are used.*

(2) SECOND METHOD (fig. 146). Place two blocks "G," one to the front and one to the rear of the front wheels. Place two blocks "G",

SHIPMENT AND TEMPORARY STORAGE

one to the front of the forward rear wheels and one to the back of the rearward rear wheels. NOTE: *These blocks "G" must be at least eight inches wider that the over-all width of the vehicle at the freight car floor.* Locate 16 cleats "F", 2 against blocks "G" to the front and rear of each blocked wheel. Nail lower cleats to freight car floor with five 40-penny nails, then nail top cleat "F" to lower cleat "F" with five 40-penny nails. Position four cleats "H", one over two cleats "G" and against the outside of each blocked wheel. Nail each end of cleat "H" to cleats "G" with three 40-penny nails. Pass four strands, two wrappings, of No. 8 gage, black annealed wire (c, fig. 146) through the holes in the wheels and pass through the stake pockets. Tighten wires enough to remove slack. NOTE: *When a box car is used, this strapping must be applied in similar fashion and attached to the floor by the use of blocking or anchor plates. This strapping is not required when gondola cars are used.*

d. Shipping Data.

Length, over-all	26 ft 10⅛ in.
Width, over-all	8 ft 7 in.
Height, over-all	10 ft 2¾ in.
Area of car floor occupied per vehicle	231 sq ft
Approximate volume occupied per vehicle	1,499 cu ft
Shipping weight per vehicle	20,750 lb

TM 9-818

10-TON 6 x 4 TRUCK (MACK MODEL NR)

REFERENCES

STANDARD NOMENCLATURE LISTS.

10-ton 6 x 4 truck (Mack)	SNL G-528
Cleaning, preserving and lubricating materials; recoil fluids, special oils, and miscellaneous related items	SNL K-1
Soldering, brazing and welding materials, gases, and related items	SNL K-2
Tool sets—motor transport	SNL N-19
Current Standard Nomenclature Lists are listed above.	
An up-to-date list of SNL's and other publications is maintained in the Index to Ordnance Publications	OFSB 1-1

EXPLANATORY PUBLICATIONS.

General.

List of publications for training	FM 21-6
Military motor vehicles	AR 850-15
Driver's manual	TM 10-460
Standard military motor vehicles	TM 9-2800

Related Technical Manuals.

Ordnance maintenance: Power plant for 10-ton 6 x 4 truck (Mack)	TM 9-1818A
Ordnance maintenance: Power train, chassis, and body for 10-ton 6 x 4 truck (Mack)	TM 9-1818B

Maintenance and Repair.

Basic maintenance manual	TM 38-250
Automotive electricity	TM 10-580
Chassis, body, and trailer units	TM 10-560
Automotive power transmission units	TM 10-585
Automotive brakes	TM 10-565
Automotive lubrication	TM 10-540
Sheet metal work, body, fender, and radiator repairs	TM 10-450
Electrical fundamentals	TM 1-455
Ordnance maintenance: Electrical equipment (Delco-Remy)	TM 9-1825A
Motor vehicle inspections and preventive maintenance services	TM 9-2810
Ordnance maintenance: Power brake systems (Bendix-Westinghouse)	TM 9-1827A

REFERENCES

Ordnance maintenance: Speedometers and tachometers (Stewart-Warner)	TM 9-1829A
Maintenance and care of pneumatic tires and rubber treads	TM 31-200
Cleaning, preserving, lubricating and welding materials and similar items issued by the ordnance department	TM 9-850
Cold weather lubrication and service of combat vehicles and automotive materiel	OFSB 6-11
Detailed lubrication instructions for ordnance materiel	OFSB 6-series

Protection of Materiel.

Camouflage	FM 5-20
Explosives and demolitions	FM 5-25
Defense against chemical attack	FM 21-40
Decontamination of armored force vehicles	FM 17-59
Chemical decontamination, materials and equipment	TM 3-220

Storage and Shipment.

Registration of motor vehicles	AR 850-10
Storage of motor vehicle equipment	AR 850-18
Ordnance storage and shipment chart, group G— Major items	OSSC-G
Rules governing the loading of mechanized and motorized army equipment, also, major caliber guns, for the United States Army and Navy, on open top equipment published by Operations and Maintenance Department of Association of American Railroads.	

TM 9-818

10-TON 6 x 4 TRUCK (MACK MODEL NR)

INDEX

A

	Page No.
Axles	
front	208
rear	209

B

	Page No.
Battery and lighting system	
batteries	
determine specific gravity with hydrometer	287
disconnection	288
electrolyte level	288
installation	288
removal	288
temperature effects	287
battery cables	
installation	290
removal	288
data, tabulated	287
description	284
headlight dimmer switch (on toeboard)	
installation	294
removal	294
instrument panel switches	
blackout switch	293
fuel gage switch	293
panel light switch	293
lamp assemblies	
blackout fender lights	292
dashlights	293
headlights	290
tail and stop lights	293
stop light switch	
installation	294
removal	294
Body and cab	
bow assemblies	268
cab seat cushions	271
cab top and curtains	
bow assembly replacement	270
cab top back replacement	269
curtain storage	269
door curtain replacement	269
rear side curtains replacement	269
top replacement	269
data, tabulated	268
description	267
floor and toeboards	276
tarpaulin	268

	Page No.
troop seats	268
windshield assembly	
installation	273
removal	271
windshield frame assembly	
installation	274
removal	274
windshield wiper assemblies	
installation	275
removal	275
Bogie (rear axle assembly)	
bogie assembly	
installation	215
removal	214
data, tabulated	214
description	214
rear spring seat assembly	
installation	218
removal	217
torque rod assemblies	
installation	216
removal	216
service	216
Brake system	
air compressor assembly	
installation	232
removal	230
replace compressor head	233
service	230
air pressure governor	
installation	233
removal	233
air reservoirs	
No. 1 air reservoir	239
No. 2 air reservoir	239
No. 3 air reservoir	240
air supply valve	
installation	238
removal	238
brake application valve	
installation	235
removal	235
data, tabulated	225
description	
parking brake	224
service brakes	224
flexible air hose, lines and connections	238

TM 9-818

INDEX

B—Cont'd	Page No.
Brake system—Cont'd	
front brake chambers	
installation	234
removal	234
front brake shoe assemblies	
adjustment	227
installation	228
removal	227
parking brake controls and linkage	
adjustment	242
replacement	242
parking brake disk	
installation	242
removal	242
parking brake shoe assemblies	
installation	241
removal	240
pressure indicator	
installation	238
removal	237
quick release valve	
installation	237
removal	237
rear brake chambers	
installation	234
removal	234
rear brake shoe assemblies	
adjustment	229
installation	230
removal	229
relay valve	
installation	236
removal	236
safety valve	
installation	237
removal	237
slack adjuster assembly	
installation	235
removal	235

C

	Page No.
Clutch	
adjustment	141
controls and linkage	
installation	144
removal	142
data, tabulated	140
description	140
installation	142
removal	141

	Page No.
Cooling system	
cooling system connections	185
data, tabulated	181
description	181
fan assembly	
installation	186
removal	186
fan belts	
adjustment	188
installation	188
removal	188
radiator assembly installation	184
radiator assembly removal	183
service cooling system	
antifreeze solutions	183
draining	182
filling	183
flushing	183
thermostat	
installation	190
removal	189
water manifold	
installation	188
removal	188
water pump assembly	
installation	187
removal	186

D

	Page No.
Data on vehicle	
capacities	25
performance	24
specifications	22
wheels and tires	25
Data, tabulated	
air compressor	226
air heater switches	195
air heaters	195
air reservoirs	226
application valve	226
battery	287
body and cab	268
bogie (rear axle assembly)	214
brake chambers	226
brake drums	225
clutch	140
controls	296
cooling system	181
cranking motor	195
energy cells	174
engine	114

TM 9-818
10-TON 6 x 4 TRUCK (MACK MODEL NR)

D—Cont'd

Data, tabulated—Cont'd
- exhaust system 179
- fire extinguisher 302
- frame, fenders, and running boards 279
- front axle 208
- front springs 251
- fuel filters
  - No. 1 filter 167
  - No. 2 filter 169
  - No. 3 filter 171
- fuel supply pumps 154
- fuel system 146
- fuel tanks 172
- generator 191
- governor 167
- horn 297
- injection nozzles 155
- injection pump 147
- instruments 296
- intake system 177
- lamp bulbs 287
- magnetic starting switch 195
- parking brake 226
- propeller shafts 204
- rear axles 219
- rear springs 251
- regulator 195
- series-parallel starting switch .. 195
- service brakes 225
- shipping 307
- shock absorbers 252
- steering 258
- switches 297
- Synchrovance 161
- tires 264
- trailer air connections 226
- transmission 200
- vehicle 22
- wheels 244

Description of vehicle
- body 15
- cab 14
- chassis
  - axles 7
  - brakes 12
  - brush guard, bumper, and towing connections 14
  - drive 7
  - electrical system 13
  - power plant 7
- spare tires 14
- suspension 7
- identification
  - chassis serial number 15
  - engine serial number 15
  - identification plate 16
  - publication data plate 16
- mobility 16

Driving controls
- controls
  - accelerator pedal 27
  - accelerator pedal stop 27
  - auxiliary transmission shifter lever 26
  - brake pedal 27
  - cab top latch 29
  - clutch pedal 26
  - cowl ventilator handle 29
  - door handle 29
  - engine stop control and engine emergency stop control 29
  - headlight beam control 29
  - parking brake lever 26
  - seat adjustment lever 29
  - starter switch and hand throttle control 26
  - steering 26
  - transmission shifter lever 26
  - windshield adjustment 29
  - windshield wipers 29
- general 26
- hand priming pump 33
- headlight beam indicator 33
- instrument panel
  - air pressure gage 31
  - auxiliary ammeter 32
  - blackout lighting switch 32
  - engine emergency stop control .. 31
  - engine stop control 31
  - fuel gage switch 30
  - fuel level gage 30
  - fuel pressure gage 30
  - gearshift diagram plate 31
  - hand throttle control 30
  - main ammeter 32
  - oil pressure gage 31
  - panel light switch 33
  - speedometer 31
  - starter switch 29
  - starting heater switch 29
  - tachometer 31

312

INDEX

D—Cont'd

	Page No.
Driving controls—Cont'd	
instrument panel—Cont'd	
water temperature gage	31
windshield wiper control valves	32
instrument panel (miscellaneous)	33
compartment latch button	33
Diesel instruction plate	33
gearshift diagram plate	33
glove and map compartment	33

E

	Page No.
Engine	
adjust valves	
adjustment	121
breather pipe removal	121
breather tubes installation	121
valve covers installation	121
valve covers removal	121
crankcase breather	
installation	126
removal	126
service breather assembly	126
cylinder head and gasket replacement	
installation	118
removal	117
data, tabulated	114
description	114
engine tune-up	115
exhaust manifold	
installation	124
removal	124
installation	133
intake manifold	
installation	123
removal	123
manifold air heaters	
installation	125
removal	125
oil filter	
cartridge replacement	126
installation	126
removal	126
oil pan assembly	
oil pan gasket replacement	125
oil pan sump replacement	125
push rods	
installation	122
removal	122

TM 9-818

	Page No.
removal	127
valve rocker arm assembly	
installation	122
removal	121
Exhaust system	
data, tabulated	179
description	178
installation	179
removal	179

F

	Page No.
Fire extinguisher	
care	302
data, tabulated	302
description	302
operation	302
refilling	302
test	302
Frame, fenders, and running boards	
data, tabulated	279
description	278
front bumper	
installation	280
removal	280
front fenders	
installation	282
removal	282
pintle hook	
installation	281
maintenance	281
removal	281
running boards	
installation	283
removal	283
towing hooks	
installation	280
removal	280
Front axle	
data, tabulated	208
description	208
front axle assembly	
installation	210
removal	209
steering knuckle assemblies	
installation	211
removal	211
steering lever	212
wheel toe-in adjustment	208

TM 9-818

10-TON 6 x 4 TRUCK (MACK MODEL NR)

F—Cont'd	Page No.
Fuel system	
bleed fuel system	158
data, tabulated	146
description	146
energy cells	
cleaning	174
data, tabulated	174
description	173
installation	176
removal	174
fuel filters	
No. 1 filter	167
No. 2 filter	167
No. 3 filter	169
fuel supply pump	
data, tabulated	154
description	153
installation	154
removal	154
fuel tanks	
data, tabulated	172
description	171
filter bag removal	172
installation	173
removal	172
governor	
data, tabulated	167
description	167
high-pressure fuel lines	
installation	155
removal	154
injection nozzles	
cleaning	156
data, tabulated	155
description	155
installation	157
removal	156
injection pump	
adjustment (timing of pump to engine)	149
data, tabulated	147
description	147
installation	152
removal	147
timing of pump to engine procedure	150
Synchrovance	
data, tabulated	161
description	160
installation	162
removal	161

G	Page No.
Governor	167

I	
Inspection and preventive maintenance service	
after-operation service and weekly service procedures	52
at-halt service procedures	51
before-operation service procedures	47
during-operation service procedures	50
purpose	46
Instruments, controls, switches and horn	
cranking motor and manifold heater switches	
cranking motor switch	300
manifold heater magnetic relay switches	300
manifold heater switch	300
data, tabulated	296
description	295
engine controls	
accelerator cross shaft and levers	299
accelerator pedal and rod	299
accelerator rod and spring	299
engine emergency stop control	298
engine stop control	298
throttle control	298
horn	
installation	301
removal	301
instruments and buzzer	
air pressure gage	297
auxiliary ammeter	298
fuel pressure gage	297
low-pressure buzzer	298
main ammeter and fuel-level gage	297
speedometer	297
tachometer	298
water temperature and oil pressure gage	297
windshield wiper control valves	
removal	300

INDEX

I—Cont'd

	Page No.
Intake system	
data, tabulated	177
description	177
installation	178
removal	178
Introduction to manual	
scope of manual	5

L

Lubrication	
introduction	56
Lubrication Guide	
general	56
lubrication notes	56
supplies	56

M

Modification records	
FSMWO and major unit assembly replacement record	
description	73
early modifications	73
instructions for use	73

N

New vehicle run-in test	
correction of deficiencies	74
purpose	74
run-in test procedures	
preliminary service	74
run-in test	77

O

Operation of vehicle	
driving vehicle	36
road speed chart	36
starting engine	34
stopping engine	39
stopping vehicle	39
Operation under unusual conditions	
in flood	45
in high temperature	41
in low temperature	
antifreeze	42
antifreeze table	42
effects of low temperatures on metals	44
fuel	41
initial movement of vehicle	43

TM 9-818

	Page No.
lubrication	42
starting instructions	43
in mud	44
in sandy or desert terrain	
air cleaners	44
general hints on desert operation	44
oil filter	44
reference	44
tires	44
in snow and ice	
safety hints	45
Organization tools and equipment	
special tool set	93
standard tool sets	93

P

Propeller shafts	
center bearing assembly	
installation	207
removal	207
data, tabulated	204
description	204
propeller shaft assemblies	
front propeller shaft	205
inter-axle propeller shaft	207
rear propeller shaft	206

R

Rear axles	
axle shafts	
installation	223
removal	223
data, tabulated	219
description	219
forward rear axle assembly	
installation	221
removal	219
rearward rear axle assembly	
installation	222
removal	222
References	
explanatory publications	
general	308
maintenance and repair	308
protection of materiel	309
related technical manuals	308
storage and shipment	309
standard nomenclature lists	308

TM 9-818
10-TON 6 x 4 TRUCK (MACK MODEL NR)

S

	Page No.
Second echelon preventive maintenance services	
first echelon participation	79
frequency	79
general procedures	79
road test chart	81
chassis, body, and attachments	87
engine and accessories	83
tools and equipment	92
Shipment and temporary storage	
general instructions	303
inspections in limited storage	305
loading and blocking for rail shipment	
facilities for loading	306
preparation	305
securing vehicles	306
shipping data	307
preparation for temporary storage or domestic shipment	303
Springs and shock absorbers	
data, tabulated	251
description	251
front spring assemblies	
installation	252
removal	252
rubber shock insulators	253
rear spring assemblies	
installation	255
removal	254
shock absorber assemblies	
installation	254
removal	254
Starting and generating system	
cranking motor	
installation	199
removal	199
data, tabulated	191
description	191
generator assembly	
installation	196
removal	196
magnetic starter switch	
installation	198
removal	196
regulator	
installation	197
removal	196
series-parallel switch	
installation	198
removal	198

	Page No.
Steering	
data, tabulated	258
description	258
drag link assembly	
installation	262
removal	262
steering gear assembly	
removal	260
steering (Pitman) arm	
installation	261
removal	261
steering wheel	
installation	259
removal	258

T

	Page No.
Tires	
data, tabulated	264
description	263
tire and rim assembly	
installation	265
removal	265
tire casings and tubes	
installation	266
removal	265
Tools and equipment stowage on vehicle	
equipment	70
spare parts	71
tools	69
Towing vehicle	
towing disabled vehicle	40
towing to start vehicle	40
Transmission	
data, tabulated	200
description	200
transmission assembly	
installation	202
removal	201
Trouble shooting	
battery and lighting system	112
brake system	108
clutch	
diagnosis	98
trouble shooting chart	98
cooling	103
engine	
diagnosis	97
trouble shooting chart	94
exhaust	102
frame	112

INDEX

T—Cont'd	Page No.	W	Page No.
Trouble shooting—Cont'd		Wheels	
front axle	107	data, tabulated	244
fuel		description	244
diagnosis	102	front brake drums	247
trouble shooting chart	99	front wheel and hub assemblies	
general	94	installation	246
propeller shaft	106	removal	244
rear axle	108	front wheel bearing	246
springs and shock absorbers	111	front wheel grease retainers	246
starting and generating systems		rear brake drums	250
diagnosis	105	rear wheel and hub assemblies	
trouble shooting chart	103	installation	248
steering gear	111	removal	248
transmission	105	rear wheel bearings	249
wheels	110	rear wheel grease retainers	249

NOTES

NOTES

NOTES

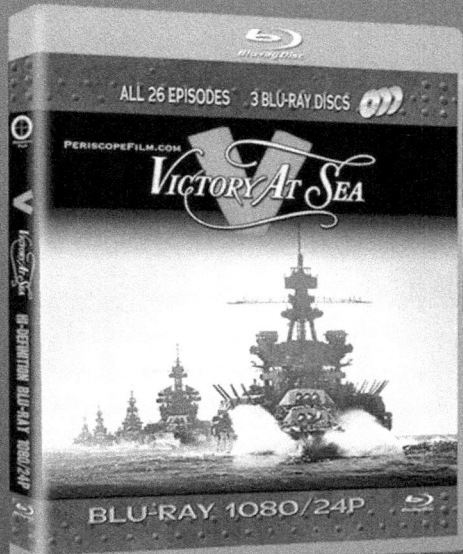

©2013 Periscope Film LLC
All Rights Reserved
ISBN#978-1-940453-21-7
www.PeriscopeFilm.com

www.ingramcontent.com/pod-product-compliance
Lightning Source LLC
Chambersburg PA
CBHW060414170426
43199CB00013B/2132